ADVANCED TOPICS IN POLYMER CHEMISTRY AND MATERIALS SCIENCE

ADVANCED TOPICS IN POLYMER CHEMISTRY AND MATERIALS SCIENCE

Editors

Tamara Tatrishvili, PhD

Marc Jean Médard Abadie, DSc

AAP | APPLE ACADEMIC PRESS

First edition published 2025

Apple Academic Press Inc.
1265 Goldenrod Circle, NE,
Palm Bay, FL 32905 USA

760 Laurentian Drive, Unit 19,
Burlington, ON L7N 0A4, Canada

CRC Press
2385 NW Executive Center Drive,
Suite 320, Boca Raton FL 33431

4 Park Square, Milton Park,
Abingdon, Oxon, OX14 4RN UK

Library and Archives Canada Cataloguing in Publication

Title: Advanced topics in polymer chemistry and materials science / Tamara Tatrishvili, PhD, Marc Jean Médard Abadie, DSc, editors.
Names: Tatrishvili, Tamara, editor | Abadie, Marc J. M., editor
Description: First edition. | This volume presents some of the proeedings of the Eight International Caucasian Symposium on Polymers and Advanced Materials. | Includes bibliographical references and index.
Identifiers: Canadiana (print) 20250137267 | Canadiana (ebook) 20250137275 | ISBN 9781774919385 (hardcover) | ISBN 9781774919392 (softcover) | ISBN 9781003594871 (PDF)
Subjects: LCSH: Polymers. | LCSH: Polymerization. | LCSH: Nanostructures. | LCSH: Polymeric composites.
Classification: LCC TA455.P58 A38 2025 | DDC 620.1/92—dc23

Library of Congress Cataloging-in-Publication Data

CIP data on file with US Library of Congress

ISBN: 978-1-77491-938-5 (hbk)
ISBN: 978-1-77491-939-2 (pbk)
ISBN: 978-1-00359-487-1 (ebk)

About the Editors

Tamara Tatrishvili, PhD

Assistant Professor, Department of Chemistry, Tbilisi State University; Director, TSU Faculty of Exact and Natural Sciences, Institute of Macromolecular Chemistry and Polymeric Materials; Director of the Institute of Macromolecular Chemistry and Polymeric Materials; TSU Main Specialist, TSU Faculty of Exact and Natural Sciences, Georgia

Tamara Tatrishvili, PhD, is an Assistant Professor and Main Specialist at the Office of Academic Process Management (Faculty of Exact and Natural Sciences) at Ivane Javakhishvili Tbilisi State University as well as Director of the Institute of Macromolecular Chemistry and Polymeric Materials at TSU, Tbilisi, Georgia. She is also a DAAD alumna and a member of the Georgian Chemical Society. Her research interests included polymer chemistry, polymeric materials, and chemistry of silicon-organic compounds. She is the executive editor of the *Journal of the Georgian Chemical Society.* Dr. Tatrishvili is the author of more than 200 scientific publications, 15 books, and several monographs.

Marc Jean Médard Abadie, DSc

Professor Emeritus, ICGM, University of Montpellier, CNRS, ENSCM, Montpellier, France

Marc J. M. Abadie, DSc, is an Emeritus Professor at the University of Montpellier, France. He was head of the Laboratory of Polymer Science and Advanced Organic Materials – LEMP/MAO. He is the "Michael Fam" Visiting Professor at the School of Materials Sciences and Engineering, Nanyang Technological University (NTU), Singapore. His present activity concerns high-performance polymers for PEMFCs, composites, nanocomposites, UV/EB coatings, and biomaterials. He has published 11 books and holds 11 patents. He has advised nearly 95 MS and 52 PhD students, with whom he has published over 402 papers. He has more than 40 years of experience in polymer science, with 10 years in industry (IBM, USA – MOD, UK &

SNPA/Total, France). He created, in the 1980s, the International Symposium on Polyimides and High-Temperature Polymers, a.k.a. STEPI, which takes place every three years in Montpellier, France.

Contents

Contributors

Marc Jean Médard Abadie
Department School of Materials Science and Engineering, Nanyang Technological University, Singapore; ICGM Institute Charles Gerhardt, Montpellier Université de Montpellier, Département Chimie des Matériaux, Nanostructures, UMR 5253 CNRS-UM ENSCM, Pôle Chimie Balard Recherche, Campus CNRS, Route de Mende, Montpellier, France

L. Sh. Abbasova
Institute of Polymer Materials, Ministry of Science and Education, Sumgayit, Azerbaijan

N. Z. Ahmedli
Azerbaijan Architecture and Construction University, Azerbaijan

Rômulo Simões Angélica
Universidade Federal do Pará (UFPA), Belem, Brazil

Tamar Archuadze
Ilia Vekua Sukhumi Institute of Physics and Technology, Georgia

Manana Areshidze
Institute of Cybernetics, Georgian Technical University, Tbilisi, Georgia

Ilgar Askerov
Institute of Polymer Materials, Ministry of Science and Education, Sumgait, Azerbaijan

Oktay Askerov
Institute of Polymer Materials, Ministry of Science and Education, Sumgait, Azerbaijan

Zurab Aslamazishvili
Ferdinand Tavadze Institute of Metallurgy and Materials Science, Tbilisi, Republic of Georgia

Marina Avaliani
R. Agladze Institute of Inorganic Chemistry and Electrochemistry, I. Javakhishvili Tbilisi State University, Georgia

A. Baishibekov
The Institute of Metallurgy and Ore Beneficiation, Almaty, Kazakhstan

Tamaz Batsikadze
Ferdinand Tavadze Institute of Metallurgy and Materials Science, Tbilisi, Republic of Georgia

George K. Belousov
Research Institute for Physical Chemical Problems, Belarusian State University, Minsk, Belarus

O. S. Berezhnytska
V.I. Vernadsky Institute of General and Inorganic Chemistry, NAS of Ukraine Akad., Kyiv, Ukraine

Vaja Chagelishvili
R. Agladze Institute of Inorganic Chemistry and Electrochemistry, I. Javakhishvili Tbilisi State University, Georgia

Andro Chanishvili
Institute of Cybernetics, Georgian Technical University, Tbilisi, Georgia

Roin Chedia
Semiconducting and Powder Composite Materials Laboratory, Ferdinand Tavadze Metallurgy and Materials Science Institute, Tbilisi, Georgia; Petre Melikishvili Institute of Physical and Organic Chemistry, Ivane Javakhishvili Tbilisi State University, Tbilisi, Georgia

Manuchar Chiqovani
Department of the Faculty of Exact and Natural Sciences, Akaki Tsereteli State University, Georgia

Ia Chitrekashvili
Petre Melikishvili Institute of Physical and Organic Chemistry, Ivane Javakhishvili Tbilisi State University, Tbilisi, Georgia

Levan Chkhartishvili
Department of Engineering Physics, Georgian Technical University, Tbilisi, Georgia; Semiconducting and Powder Composite Materials Laboratory, Ferdinand Tavadze Metallurgy and Materials Science Institute, Tbilisi, Georgia

Ketevan Chubinidze
Institute of Cybernetics of Georgian Technical University, Tbilisi, Georgia; Institute of Macromolecular Chemistry and Polymeric Materials, Ivane Javakhishvili Tbilisi State University, Tbilisi, Georgia

Olga Chudakova
G. Tsulukidze Mining Institute, Tbilisi, Georgia

Nino Darakhvelidze
Ilia Vekua Sukhumi Institute of Physics and Technology, Georgia

Margalita Darchiashvili
Ferdinand Tavadze Institute of Metallurgy and Materials Science, Tbilisi, Republic of Georgia

Maryam Dehghani
Universidade Federal do Pará (UFPA), Belem, Brazil

Lali Devadze
Institute of Cybernetics, Georgian Technical University, Tbilisi, Georgia

Nora Dokhturishvili
Petre Melikishvili Institute of Physical and Organic Chemistry, Ivane Javakhishvili Tbilisi State University, Tbilisi, Georgia

Darejan Dughashvili
Institute of Hydrometeorology, Georgian Technical University, Tbilisi, Georgia; Raphael Agladze Institute of Inorganic Chemistry and Electrochemistry, Ivane Javakhishvili Tbilisi State University, Tbilisi, Georgia

Tamar Dundua
Faculty of Agricultural Sciences and Biosystems Engineering, Georgian Technical University, Tbilisi, Georgia

Dali Dzanashvili
TSU Rafael Agladze Institute of Inorganic Chemistry and Electrochemistry, Tbilisi, Georgia

Natela Dzebisashvili
Institute of Hydrometeorology, Georgian Technical University, Tbilisi, Georgia; Raphael Agladze Institute of Inorganic Chemistry and Electrochemistry, Ivane Javakhishvili Tbilisi State University, Tbilisi, Georgia

Abdollah Esmaeili
Universidade Federal do Pará (UFPA), Belem, Brazil

D. E. Fischer
The Institute of Metallurgy and Ore Beneficiation, Almaty, Kazakhstan

Maia Gabrichidze
University Geomedi LLC, Tbilisi, Georgia

Vakhtang Gabunia
Petre Melikishvili Institute of Physical and Organic Chemistry, Ivane Javakhishvili Tbilisi State University, Tbilisi, Georgia

Eter Gavashelidze
Petre Melikishvili Institute of Physical and Organic Chemistry, Ivane Javakhishvili Tbilisi State University, Tbilisi, Georgia

Nazi Gelashvili
Petre Melikishvili Institute of Physical and Organic Chemistry, Ivane Javakhishvili Tbilisi State University, Tbilisi, Georgia

Tsintskaladze Giorgi
Petre Melikishvili Institute of Physical and Organic Chemistry, Ivane Javakhishvili Tbilisi State University, Tbilisi, Georgia

Bagrat Godibadze
Ferdinand Tavadze Institute of Metallurgy and Materials Science, Tbilisi, Republic of Georgia

Juozas Vidas Grazulevicius
Department of Polymer Chemistry and Technology, Kaunas University of Technology, Kaunas, Lithuania

Camilo Andres Guerrero
Universidade Federal do Pará (UFPA), Belem, Brazil

Lali Gurchumelia
Ferdinand Tavadze Institute of Metallurgy and Materials Science, Tbilisi, Georgia; TSU Rafael Agladze Institute of Inorganic Chemistry and Electrochemistry, Tbilisi, Georgia

Marina Gurgenishvili
Petre Melikishvili Institute of Physical and Organic Chemistry, Ivane Javakhishvili Tbilisi State University, Tbilisi, Georgia

Salome Gvazava
Ferdinand Tavadze Institute of Metallurgy and Materials Science, Tbilisi, Georgia

A. B. Huseynov
Institute of Geotechnological Problems of Oil, Gas, and Chemistry, Azerbaijan State University of Oil and Industry, Ministry of Science and Education, Republic of Azerbaijan; Nagiyev Institute of Catalysis and Inorganic Chemistry, Ministry of Science and Education, Republic of Azerbaijan

Dj. N. Huseynov
Azerbaijan Architecture and Construction University, Azerbaijan

Yaroslav Ivanechko
Department of Freshwater Hydrochemistry, Institute of Hydrobiology of the National Academy of Sciences, Kyiv, Ukraine

Natia Jalagonia
Ilia Vekua Sukhumi Institute of Physics and Technology, Georgia

Irma Jinikashvili
Department of Medical Chemistry, Tbilisi State Medical University, Tbilisi, Georgia

T. K. Jumadilov
A.B. Bekturov Institute of Chemical Sciences, Almaty, Kazakhstan

Leila Kalatozishvili
Ilia Vekua Sukhumi Institute of Physics and Technology, Georgia

Shalva Kekutia
Vladimer Chavchanidze Institute of Cybernetics, Georgian Technical University, Tbilisi, Georgia

Gyulgun Khanbabayeva
Institute of Polymer Materials, Ministry of Science and Education, Sumgait, Azerbaijan

Evgeny Khichua
A. Tvalchrelidze Caucasian Institute of Mineral Resources, I. Javakhishvili Tbilisi State University, Tbilisi, Georgia

Sergei V. Kostjuk
Research Institute for Physical Chemical Problems, Belarusian State University, Minsk, Belarus; Department of Chemistry, Belarusian State University, Minsk, Belarus

Tinatin Kuchukhidze
Ilia Vekua Sukhumi Institute of Physics and Technology, Georgia

Ia Kurashvili
Ilia Vekua Sukhumi Institute of Physics and Technology, Georgia

James Lee Wai Leong
Department School of Materials Science and Engineering, Nanyang Technological University, Singapore

Petro Linnik
Department of Freshwater Hydrochemistry, Institute of Hydrobiology of the National Academy of Sciences, Kyiv, Ukraine

Rostyslav Linnik
Faculty of Chemistry, Taras Shevchenko Kyiv National University, Kyiv, Ukraine

Eprikashvili Luba
Petre Melikishvili Institute of Physical and Organic Chemistry, Ivane Javakhishvili Tbilisi State University, Tbilisi, Georgia

M. Ya. Magerramova
Nagiyev Institute of Catalysis and Inorganic Chemistry, Azerbaijan

Dzagania Maia
Petre Melikishvili Institute of Physical and Organic Chemistry, Ivane Javakhishvili Tbilisi State University, Tbilisi, Georgia

Kopaleishvili Maka
Ivane Javakhishvili Tbilisi State University, Petre Melikishvili Institute of Physical and Organic Chemistry, Laboratory of Petrochemistry, Georgia

Shio Makatsaria
Department of Engineering Physics, Georgian Technical University, Tbilisi, Georgia; Deltamed Georgia, Official Representatives of Siemens Healthcare Diagnostics in Georgia, Tbilisi, Georgia

Giorgi Makharadze
Ivane Javakhishvili Tbilisi State University, Tbilisi, Georgia

Tamar Makharadze
R. Agladze Institute of Inorganic Chemistry and Electrochemistry, Ivane Javakhishvili Tbilisi State University, Tbilisi, Georgia; Institute of Cybernetics of Georgian Technical University, Tbilisi, Georgia; Institute of Macromolecular Chemistry and Polymeric Materials, Ivane Javakhishvili Tbilisi State University, Tbilisi, Georgia

Aynura Mamedova
Institute of Polymer Materials, Ministry of Science and Education, Sumgait, Azerbaijan

B. A. Mammadov
Institute of Polymer Materials, Ministry of Science and Education, Sumgayit, Azerbaijan

A. F. Mammadova
Institute of Polymer Materials, Ministry of Science and Education, Sumgayit, Azerbaijan

Zautashvili Marine
Petre Melikishvili Institute of Physical and Organic Chemistry, Ivane Javakhishvili Tbilisi State University, Tbilisi, Georgia

Jano Markhulia
Vladimer Chavchanidze Institute of Cybernetics, Georgian Technical University, Tbilisi, Georgia

Zaqaria Melashvili
Ferdinand Tavadze Institute of Metallurgy and Materials Science, Tbilisi, Republic of Georgia

Sophio Mikaberidze
Ilia Vekua Sukhumi Institute of Physics and Technology, Georgia

Archil Mikeladze
Ferdinand Tavadze Institute of Metallurgy and Materials Science, Tbilisi, Georgia

Vladimer Mikelashvili
Vladimer Chavchanidze Institute of Cybernetics, Georgian Technical University, Tbilisi, Georgia

Lela Mirtskhulava
Department of Computer Science, Ivane Javakhishvili Tbilisi State University, Tbilisi, Georgia

Teimuraz Namicheishvili
Ferdinand Tavadze Institute of Metallurgy and Materials Science, Tbilisi, Republic of Georgia

Pirtskhalava Nino
Petre Melikishvili Institute of Physical and Organic Chemistry, Ivane Javakhishvili Tbilisi State University, Tbilisi, Georgia

Afonso Cesar Rodrigues Nogueira
Universidade Federal do Pará (UFPA), Belem, Brazil

Dilbar Nurullayeva
Institute of Polymer Materials, Ministry of Science and Education, Sumgait, Azerbaijan

Zurab Pachulia
Sokhumi State University, Faculty of Natural Sciences, Mathematics, Technologies, and Pharmacy, Tbilisi, Georgia

Givi Papava
Petre Melikishvili Institute of Physical and Organic Chemistry, Ivane Javakhishvili Tbilisi State University, Tbilisi, Georgia

Giorgi Parunashvili
Ferdinand Tavadze Institute of Metallurgy and Materials Science, Tbilisi, Republic of Georgia

Akaki B. Peikrishvili
Ferdinand Tavadze Institute of Metallurgy and Materials Science, Tbilisi, Republic of Georgia

Gia Petriashvili
Institute of Cybernetics of Georgian Technical University, Tbilisi, Georgia; Institute of Macromolecular Chemistry and Polymeric Materials, Ivane Javakhishvili Tbilisi State University, Tbilisi, Georgia

Nana Pirtskheliani
Sokhumi State University, Faculty of Natural Sciences, Mathematics, Technologies, and Pharmacy, Tbilisi, Georgia; Institute of Macromolecular Chemistry and Polymeric Materials, Ivane Javakhishvili Tbilisi State University, Tbilisi, Georgia

A. A. Quvalov
Azerbaijan Architecture and Construction University, Azerbaijan

O. O. Rohovtsov
V.I. Vernadsky Institute of General and Inorganic Chemistry, NAS of Ukraine Akad., Kyiv, Ukraine

Laura Estefanía Garzón Rojas
Universidade Federal do Pará (UFPA), Belem, Brazil

Kakha Rukhaia
Doctor of Chemistry, Ministry of Environment Protection and Agriculture, Georgia

N. V. Rusakova
A.V. Bogatsky Physico-Chemical Institute NASU, Odessa, Ukraine

Maia Rusia
Department of General, Inorganic, and Metalorganic Chemistry, Faculty of Exact and Natural Sciences, Iv. Javakhishvili Tbilisi State University, Georgia

N. I. Salmanova
Institute of Geotechnological Problems of Oil, Gas, and Chemistry, Azerbaijan State University of Oil and Industry, Ministry of Science and Education, Republic of Azerbaijan

I. A. Savchenko
National Taras Shevchenko University, Kyiv, Ukraine

Nino Sepashvili
Institute of Cybernetics, Georgian Technical University, Tbilisi, Georgia

Lana Shamanauri
R. Dvali Institute of Machine Mechanics, Tbilisi, Georgia

Elena Shapakidze
A. Tvalchrelidze Caucasian Institute of Mineral Resources, I. Javakhishvili Tbilisi State University, Tbilisi, Georgia

S. S. Smola
A.V. Bogatsky Physico-Chemical Institute NASU, Odessa, Ukraine

Zurab Tabukashvili
Petre Melikishvili Institute of Physical and Organic Chemistry, Ivane Javakhishvili Tbilisi State University, Tbilisi, Georgia

Shaţakishvili Tamar
Ivane Javakhishvili Tbilisi State University, Petre Melikishvili Institute of Physical and Organic Chemistry, Laboratory of Petrochemistry, Georgia

Tamara Tatrishvili
Institute of Macromolecular Chemistry and Polymeric Materials, Ivane Javakhishvili Tbilisi State University, Tbilisi, Georgia; Ivane Javakhishvili Tbilisi State University, Tbilisi, Georgia

Svetlana Tavzarashvili
Institute of Cybernetics, Georgian Technical University, Tbilisi, Georgia

Kordzakhia Teimuraz
Petre Melikishvili Institute of Physical and Organic Chemistry, Ivane Javakhishvili Tbilisi State University, Tbilisi, Georgia

Gabunia Tinatin
Ivane Javakhishvili Tbilisi State University, Petre Melikishvili Institute of Physical and Organic Chemistry, Laboratory of Petrochemistry, Georgia

Sharashenidze Tinatin
Petre Melikishvili Institute of Physical and Organic Chemistry, Ivane Javakhishvili Tbilisi State University, Tbilisi, Georgia

Salome Tkemaladze
TSU Rafael Agladze Institute of Inorganic Chemistry and Electrochemistry, Tbilisi, Georgia

Gulnara Todradze
A. Tvalchrelidze Caucasian Institute of Mineral Resources, I. Javakhishvili Tbilisi State University, Tbilisi, Georgia

O. K. Trunova
National Technical University, Ukraine "Igor Sikorsky Kyiv Polytechnic Institute," 37, Prosp. Peremohy, Kyiv, Ukraine

Otar Tsagareishvili
Semiconducting and Powder Composite Materials Laboratory, Ferdinand Tavadze Metallurgy and Materials Science Institute, Tbilisi, Georgia

Murman Tsarakhov
TSU Rafael Agladze Institute of Inorganic Chemistry and Electrochemistry, Tbilisi, Georgia

Aliaksei A. Vaitusionak
Research Institute for Physical Chemical Problems, Belarusian State University, Minsk, Belarus

Gabunia Vakhtang
Petre Melikishvili Institute of Physical and Organic Chemistry, Ivane Javakhishvili Tbilisi State University, Tbilisi, Georgia

Molodinashvili Zaza
Ivane Javakhishvili Tbilisi State University, Petre Melikishvili Institute of Physical and Organic Chemistry, Laboratory of Petrochemistry, Georgia

E. B. Zeynalov
Institute of Geotechnological Problems of Oil, Gas, and Chemistry, Azerbaijan State University of Oil and Industry, Ministry of Science and Education, Republic of Azerbaijan; Nagiyev Institute of Catalysis and Inorganic Chemistry, Ministry of Science and Education, Republic of Azerbaijan

S. G. Zeynalova
Azerbaijan Medical University, Sumgayit, Azerbaijan

Vladyslav Zhezherya
Department of Freshwater Hydrochemistry, Institute of Hydrobiology of the National Academy of Sciences, Kyiv, Ukraine

Tsisana Zurabishvili
Institute of Cybernetics, Georgian Technical University, Tbilisi, Georgia

Abbreviations

ABS	Acrylonitrile-Butadiene-Styrene
AC	Activated Carbon
AIBN	Azobisisobutyronitrile
AIE	Aggregation-Induced Emission
AMP	Antimicrobial Polymer Materials
ASTM	American Society for Testing and Materials
BOD	Biochemical Oxygen Demand
BNCT	Boron Neutron Capture Therapy
CEmax	Maximum Current
CM	Carboxymethyl/Composition Materials
COD	Chemical Oxygen Demand
CO	Carbon Monoxide
CPAESAMM	Copolymer of Salicylic Acid Allyl Ether with Methyl Methacrylate
CVD	Chemical Vapor Deposition
CTC	Charge-Transfer Complex
DIC	Dainippon Ink & Chemicals
DEAE	Diethylaminoethyl
DIC	Dainippon Ink & Chemicals
DFT	Density Functional Theory
DMP	2-(Dodecylthiocarbonothioylthio)-2-methylpropionic acid
DMA	Dynamic Mechanical Analyzer
DOM	Dissolved Organic Matter
DRS	Diffuse Reflection Spectroscopy
DSC	Differential Scanning Calorimetry
DPC	Differential Photocalorimetry
EDX	Energy Dispersive X-ray
EAS	Electronic Absorption Spectroscopy
EI-MS	Electron Impact Mass Spectrometry
EL	Electroluminescent
EQE	External Quantum Efficiency
EQEmax	External Quantum Efficiencies
FCC	Fluid Catalytic Cracking
FA	Fulvic Acids
FML	Fiber Metal Laminate
FTIR	Fourier Transform InfraRed
GO	Graphene Oxide

HA	Humic Acids
HC	Hydrocarbons
HML	Hybrid Metal Laminate
HMW	High Molecular Weight
IMM	Institute of Mining Mechanics
IQE	Internal Quantum Efficiency
IR	Infrared
ISC	Intersystem Crossing
ITO	Indium-Tin-Oxide
KMnO4	Potassium Permanganate
LC	Liquid Crystal
LMW	Low Molecular Weight
Ln	Luminescence
MEK	Methyl-Ethyl Ketone
MDSC	Modulated Differential Scanning Calorimetry
MMFF	Merck Molecular Force Field
MPA	Meat-Peptone Agar
MT-NN	Multi-Task Deep Neural Networks
MSW	Municipal Solid Waste
MWCNT	Multiwalled Carbon Nanotubes
NOM	Natural Organic Matter
NOx	Nitrogen Oxides
OLEDs	Organic Light-Emitting Diodes
PCV	Positive Crankcase Ventilation
PEDOT	Poly(2,3-dihydrothieno-1,4-dioxin)-Poly(styrene sulfonate)
PEmax	Maximum Power
PEPA	Polyethylene Polyamine
PL	Photoluminescence
PLQYs	Photoluminescence Quantum Yields
PMMA	Polymethyl Methacrylate
PS	Polystyrene
PVK	Polyvinylcarbazole
PVA	Polyvinyl Alcohol
REM	Rare Earth Metals
RISC	Reverse Intersystem Crossing
rGO	Reduced Graphene Oxide
SEM	Scanning Electron Microscope
SHS	Self-Propagating High-Temperature Synthesis
SMILES	Simplified Molecular-Input Line-Entry System
SP	Spiropyrans
STA	Simultaneous Thermal Analysis
TADF	Thermally Activated Delayed Fluorescence
TGA	Thermogravimetric Analyzer

THF	Tetrahydrofuran
TSCT	Through-Space Charge-Transfer
UAER	Universal Automatic Experimental Robot
VSM	Vibrating Sample Magnetometer
WGS	Water Gas Shift
XRD	X-Ray Diffraction

Preface

Advanced polymers, also known as engineering polymers or high-performance polymers, are synthetic polymers with exceptional mechanical, thermal, chemical, and electrical properties (high heat and thermal stability, frost resistance, hydrophobicity, high dielectric indices, etc.).

The branch of science discussed in this book has a particular applied objective: the creation of new polymeric materials with improved properties. The Eighth International Caucasian Symposium on Polymers and Advanced Materials (ICSP&AM8), which took place at Ivane Javakhishvili Tbilisi State University, encourages scientists working in polymer chemistry and advanced materials to present their investigations dedicated to problems and discoveries in the different fields of polymer chemistry.

This meeting, gathering many participants, has provided a good platform for academic and industrial scientists to discuss recent advances in the area of polymers and advanced materials. The symposium was conducted with the support of Ivane Javakhishvili Tbilisi State University and the Shota Rustaveli National Foundation of Georgia (Project № MG-ISE-23-172).

This volume presents some of the proceedings of the Eight International Caucasian Symposium on Polymers and Advanced Materials. The symposium was special because it was dedicated to the memory and 75th birthday anniversary of its establisher and former organizing committee chairman, Professor Omar Mukbaniani.

This book, *Advanced Topics in Polymer Chemistry and Material Science*, highlights new concepts and achievements in polymer chemistry and chemical engineering. The book presents selected chapters based on the materials from symposium abstracts that discuss different fields of polymer chemistry. The chapters provide a survey of the important categories of polymers, including commodity thermoplastics and fibers, elastomers, thermosets, engineering, and specialty polymers.

Preface

[illegible]

PART I
Kinetics of Thermosets

CHAPTER 1

Cured Glycidyl Ethers with Condensed Carbon Cycles

IA CHITREKASHVILI, NORA DOKHTURISHVILI,
ETER GAVASHELIDZE, MARINA GURGENISHVILI, NAZI GELASHVILI,
GIVI PAPAVA, and ZURAB TABUKASHVILI

Petre Melikishvili Institute of Physical and Organic Chemistry, Ivane Javakhishvili Tbilisi State University, Tbilisi, Georgia

ABSTRACT

The unique properties of epoxy polymers have led to their wide application in various fields of modern technology. The influence of the structure of bisphenols on the properties of epoxy polymers is known. The chemical structure of the hardener also affects the properties of epoxy polymers. As hardeners, we used acid and amine hardeners of various chemical structures and the diol component – glycidyl ethers of polycyclic bisphenols synthesized by us.

Since the thermal stability of polymers, in addition to the chemical structure of bisphenols, also depends on the structure of the hardener used, amine and acid hardeners of various chemical structures are used to improve the thermal characteristics of polymers, including thermal resistance and heat resistance. The influence of the chemical structure of these hardeners on the properties of epoxy polymers is investigated.

It is shown that the following amine and acid hardeners provide high heat resistance: pyromellite and methyltetrahydrophthalic anhydrides, 4,4'-diaminodiphenylsulfone, benzidine, 4,4'-diaminodiphenyloxide, and other aromatic diamines. Polymers obtained by curing these components of diglycidyl ethers of bisphenols with norbornane-type substituents are

Advanced Topics in Polymer Chemistry and Materials Science: Tamara Tatrishvili and Marc Jean Médard Abadie (Eds)

characterized by high temperature resistance, deforming in the temperature range of 220–245°C. In all cases, polymers obtained on the basis of these hardeners decrease in weight by 10% in the temperature range of 340–400°C.

1.1 INTRODUCTION

The term epoxy resin refers to both the uncured and the cured forms of the resin. Industrial interest and use of these resins reside in both the valuable properties of the cured resin, which include good adhesion to many substrates, relatively high toughness (particularly when rubber modified), good environmental resistance, high electrical resistivity, low shrinkage, etc., and the ease with which the curing reaction may be adapted to suit the fabricating process.

The combination of good processing characteristics and useful properties has led to the use of epoxy resins in many applications, including adhesives, in electronics for encapsulation, potting, and printed circuit boards, and in the aerospace industries as matrices for composites. Epoxy resins—because of their reactivity that enables them to bond well to fibers and their toughness—are the thermoset resins that, combined with glass, carbon, or aramid fibers, produce composite materials with the best properties of most thermosets [1].

The epoxy group is characterized by its reactivity towards both nucleophilic and electrophilic species, and it is thus receptive to a wide range of reagents or curing agents. Such curing agents are of two types: they may be either catalysts or hardeners [2].

Elucidation of the mechanism of the curing reaction of epoxy resin is still of great interest for both fundamental and applied chemistry, but the mechanism of curing has not yet been fully disclosed [3–5]. Epoxy polymers obtained on the basis of various bisphenols have been described in references [6–13]. The purpose of this work was to study the effect of the chemical structure of the hardener on the properties of epoxy polymers and the synthesis of polymers with improved physical, mechanical, and chemical properties.

It was found that the presence of cyclic groups in the dipole component, both aromatic and alicyclic, by their nature, largely determines the properties of the structured polymer, contributing to the growth of thermal parameters of polymers.

Since the thermal stability of polymers, in addition to the chemical structure of bisphenols, also depends on the structure of the hardener used, amine and acid hardeners of various chemical structures have been used to improve the thermal properties of polymers. The influence of the chemical structure of these hardeners on the properties of epoxy polymers was studied.

1.2 EXPERIMENTAL METHODS AND MATERIALS

1.2.1 DETERMINATION OF EPOXY GROUPS

In two conical flasks (250 ml) with lapped stoppers, oligomer attachments (0.200g) were placed and dissolved in a neutral solvent. In the resulting solutions, 25 ml of 0.2 N HCl solution in absolute acetone was added and left to stand for 2 hours at room temperature, after which the excess hydrogen chloride was filtered out with 0.1 N caustic potassium solution in absolute ethyl alcohol in the presence of phenolphthalein. In parallel, a blank sample was prepared.

The content of epoxy groups is calculated by the following formula:

$$ЭZ = \frac{(V_1 - V_2).0,0043.100}{g}$$

where V_1 is the amount of 0.1 N solution of caustic potassium consumed for titration of a blank sample, ml; V_2 – the amount of 0.1 N solution of caustic potassium consumed for titration of a sample with a sample of the substance, ml; g – the sample of the substance, g; 0.0043 – the number of epoxy groups corresponding to 1 ml of 0.1 N solution of caustic potassium.

1.2.2 CURING OF DIGLYCIDYL ETHERS IN THE PRESENCE OF HARDENERS

Depending on the type of hardener, the structuring process of glycidyl ethers was carried out under appropriate temperature conditions. A suspension of glycidyl ether, heated to its softening temperature, was placed in a polymerization beaker, and a liquid hardener was added to it (the crystalline hardener was previously melted, thoroughly mixed until a homogeneous mixture was formed, and placed in a thermal cabinet). The amount of hardener required for curing is calculated using the following formula. [10].

$$A = \frac{M}{43}K$$

where A – is the amount of hardener per 100 g of glycidyl ether; 43 – the molecular weight of the epoxy group; M – the molecular weight of the hardener; K – is the percent content of epoxy groups in glycidyl ether.

In the presence of anhydride, structuring was carried out at a temperature of: 120°C – 2 h, 160°C – 3 h, 180°C – 3 h, and 200°C – 10 h.

A reduction in the duration of the anhydride curing process at 200°C was achieved by introducing 0.1% by weight of triethanolamine into the composition.

1.3 RESULTS AND DISCUSSION

On the basis of diglycidyl ethers, by curing them, we have obtained epoxy polymers. Both anhydride and amine-type hardeners were used as hardeners.

During curing, an equimolecular amount of the initial components was taken. To accelerate the curing process with anhydrides, triethanolamine was used in an amount of 0.1 wt. part per 100 mass parts of glycidyl ether.

The study of the thermal and heat resistance of the obtained polymers was carried out by methods of thermomechanical and thermogravimetric analysis. The results obtained are shown in Table 1.1 and in Figures 1.1a,b and 1.2a,b.

Thermomechanical and thermogravimetric analysis of polymers showed that epoxy polymers based on glycidyl ethers of polycyclic bisphenols of the norbornane type are characterized by high thermal and heat resistance. The change in the volume of the cord grouping does not significantly affect the heat resistance of the cured polymers.

TABLE 1.1 Characteristics of Di Glycidyl Ethers Obtained on the Basis of Polycyclic Bisphenols and Epichlorohydrin[(x)]

No.	Structure of Glycidyl Ether	Softening Temperature According to Ubbelode, °C	Total Chlorine Content,%	Content of Epoxy Groups,%		Molecular Weight		
				Calculated for Diglycidyl Ether	Found	Calculated for Di glycidyl Ether	Found	
							by End Groups	Determined by the Ebullioscopic Method
1.	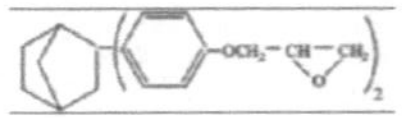	55	0.20	21.9	17.0	392	505	495
2.	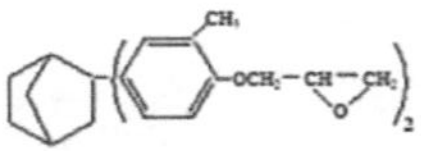	75	0.11	20.5	18.5	420	491	490
3.	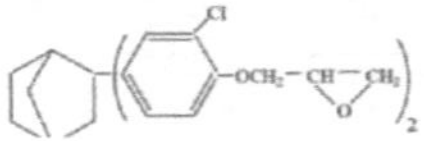	45	14.20	18.7	18.1	461	475	420
4.	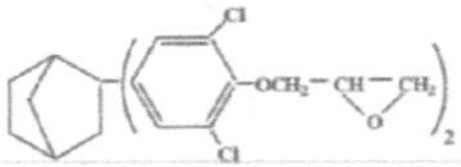	45	22.00	16.2	15.8	530	544	590
5.	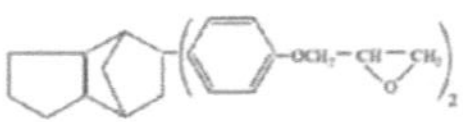	60	0.35	19.9	16.1	432	537	557
6.	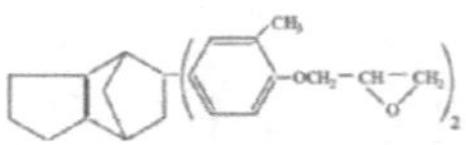	65	0.50	18.7	17.4	460	494	480
7.	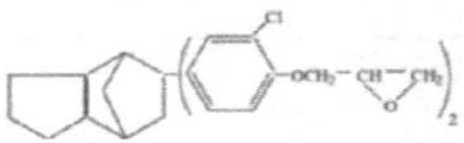	50	14.17	17.2	14.9	501	577	600
8.	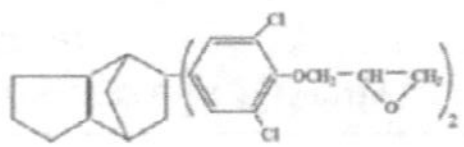	45	24.91	15.1	14.0	570	614	590

TABLE 1.1 *(Continued)*

No.	Structure of Glycidyl Ether	Softening Temperature According to Ubbelode, °C	Total Chlorine Content,%	Content of Epoxy Groups,%		Molecular Weight		
				Calculated for Diglycidyl Ether	Found	Calculated for Di glycidyl Ether	Found	
							by End Groups	Determined by the Ebullioscopic Method
9.	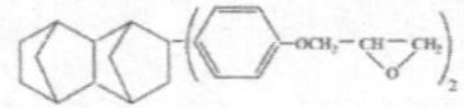	83	0.41	18.8	14.0	458	614	598
10.	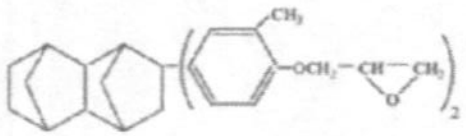	75	0.51	17.7	15.3	486	562	530
11.	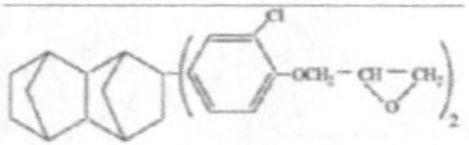	70	13.47	16.3	13.3	527	646	620
12.	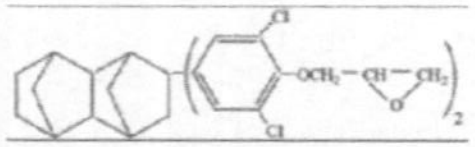	55	23.82	11.2	11,2	596	768	740

(x) Ratio of bisphenol: epichlorohydrin: NaOH =1:10:4, mol. concentration of aqueous NaOH solution – 50%, reaction temperature 95°C, duration 60 min.

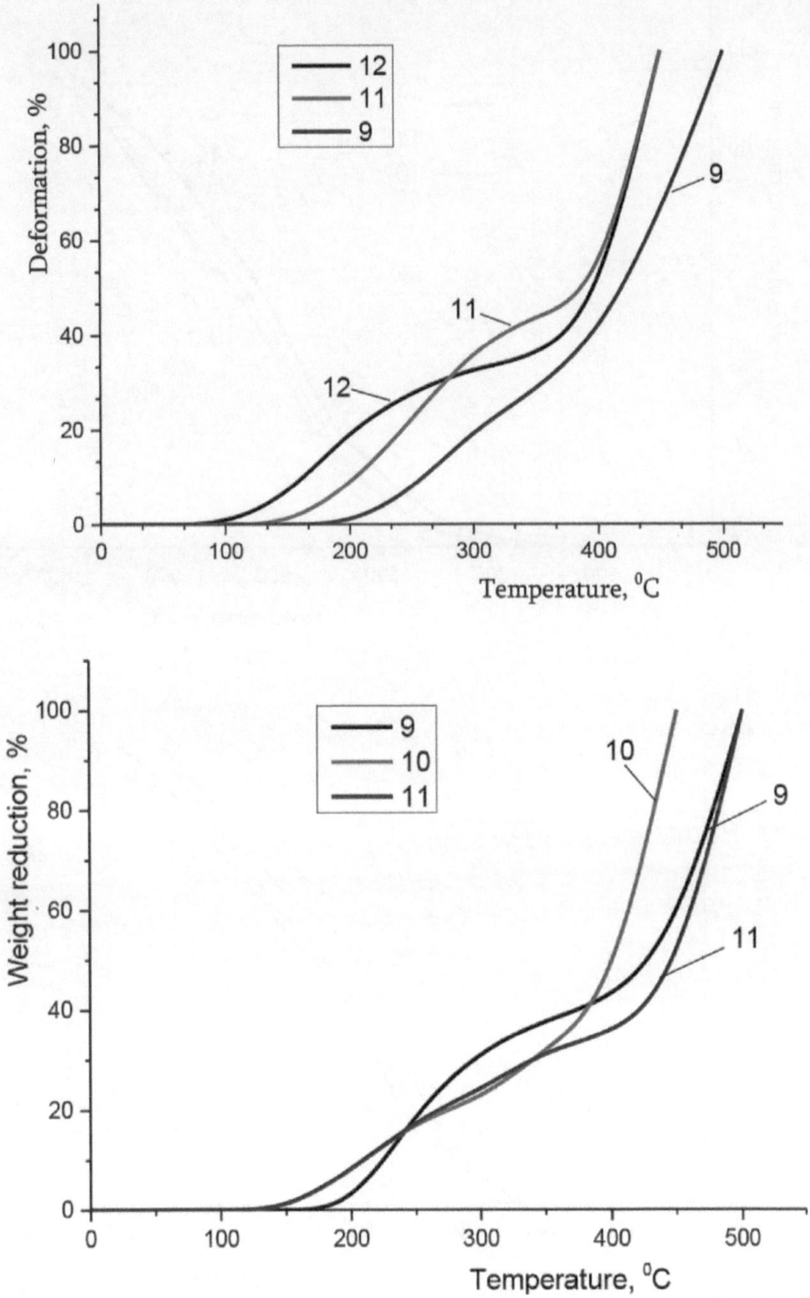

FIGURE 1.1 Thermomechanical curves of epoxy polymers cured with (a) methyltetrahydrophthalic anhydride; (b) 4,4-diaminodiphenylsulfone. (The numbers of the curves correspond to the ordinal numbers of the polymers in Table 1.1).

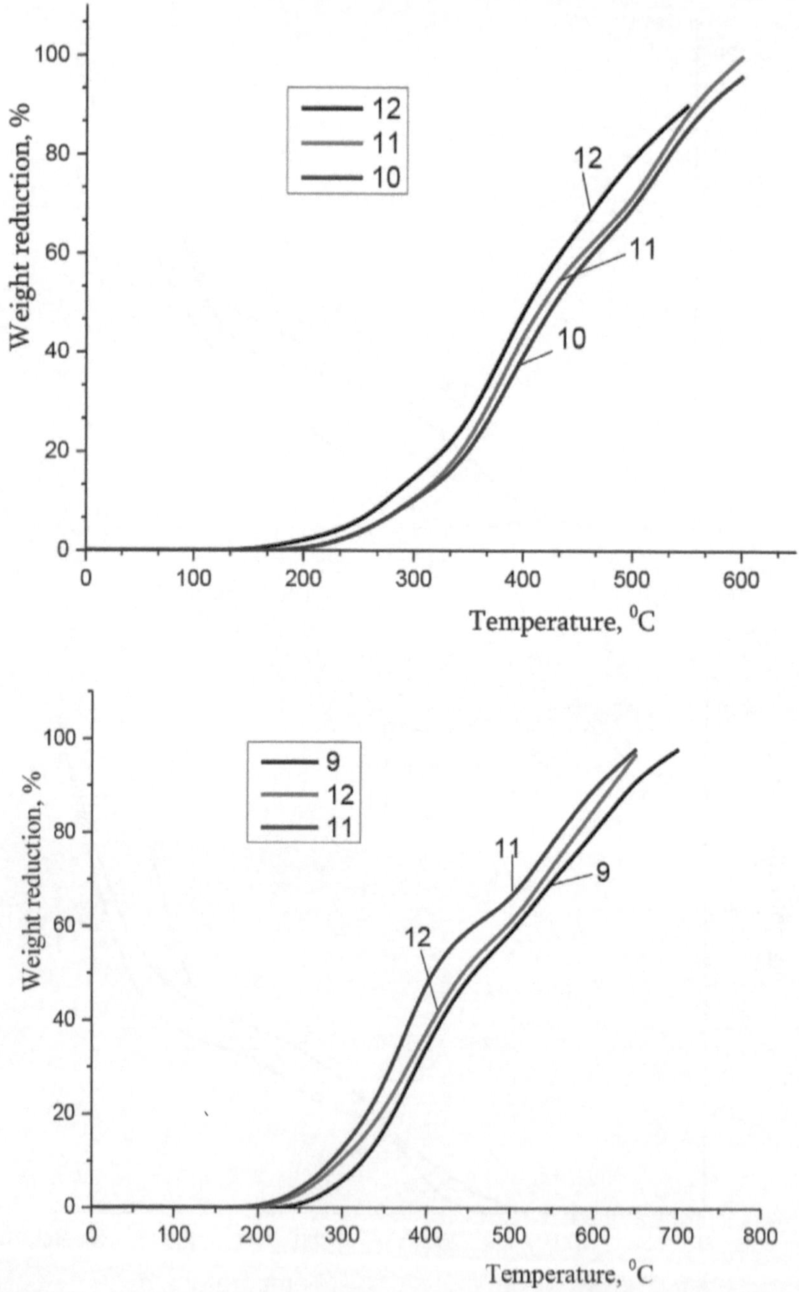

FIGURE 1.2 Thermogravimetric curves of epoxy polymers cured with (a) methyltetrahydrophthalic anhydride; and (b) 4,4-diaminodiphenylsulfone. (The numbers of the curves correspond to the ordinal numbers of the polymers in Table 1.1).

The influence of both the nature and structure of the hardener on the properties of cured polymers was studied. For this purpose, hardeners of various structures and chemical natures were used to cure the diglycidyl ether of 4,4′-(hexahydro-4,7-methylenindane-5-ylidene) diphenol. The obtained results are shown in Figures 1.3 and 1.4.

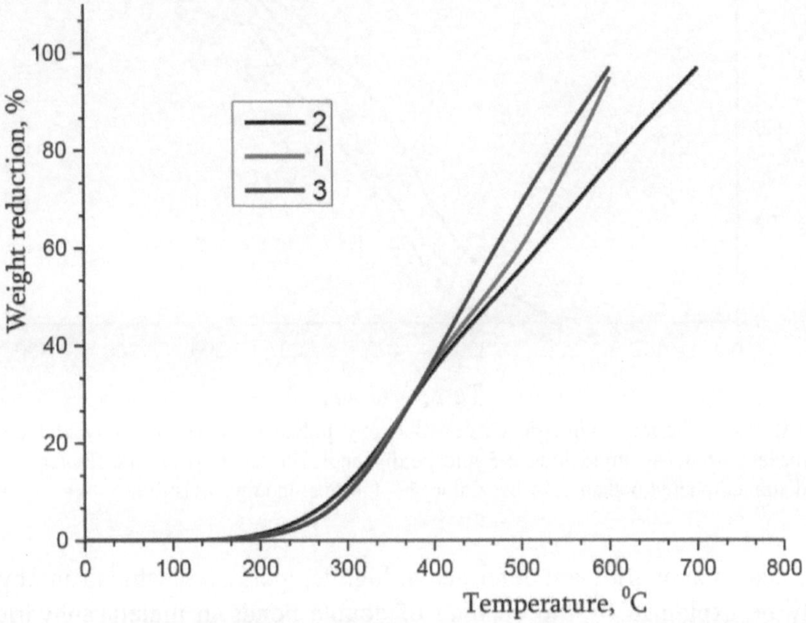

FIGURE 1.3 Thermogravimetric curves of epoxy polymers based on glycidyl ether 4,4′-hexahydro-4,7-methylenindane-5-ylidene)diphenol. Hardeners: 1 – methyltetrahydrophthalic anhydride, 2 – maleic anhydride, 3 – pyromellite anhydride.

As can be seen from the data in the table and figures, epoxy polymers are characterized by fairly high heat resistance. The only exceptions are polymers cured with 4,4′-diamino-3,3′-dimethoxydiphenylmethane and 4,4′-diaminodiphenylmethane. Their heat resistance is equal to 155°C. This is obviously due to the structure of the above-mentioned diamines, namely, the presence of methoxy and methylene groups in their molecules.

The following hardeners give particularly high results in heat resistance: 4,4′-diaminodiphenylsulfone, pyromellite, methyltetrahydrophthalic, and malein anhydrides. Polymers obtained by curing glycidyl ethers with these components begin to deform in the temperature range of 220–310°C. It should be noted that polymers formed during the curing of glycidyl ethers

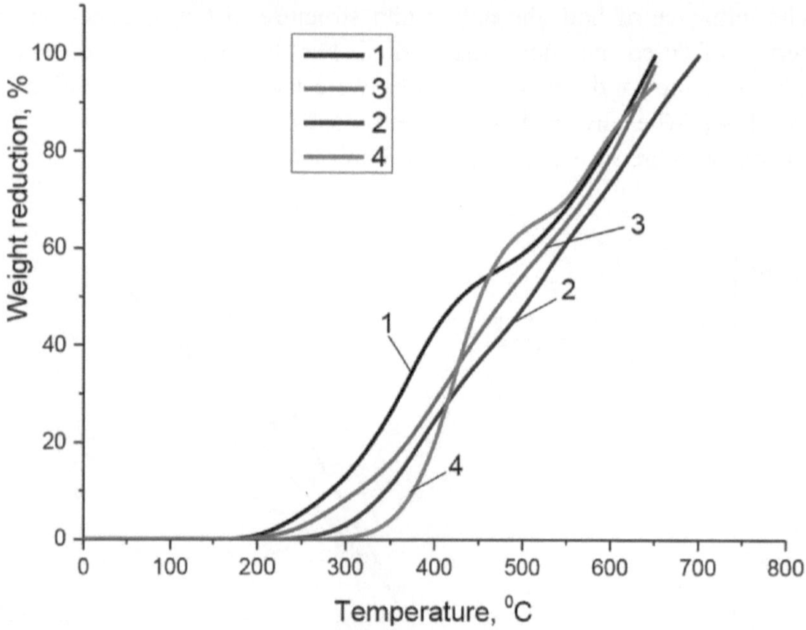

FIGURE 1.4 Thermogravimetric curves of epoxy polymers based on diglycidyl ether 4,4′-(hexahydro-4.7-methylenindane-5-ylidene)diphenol. Hardener: 1- aniline fluorene, 2 – 4,4′-diaminodiphenylmethane, 3 – benzidine, 4 – 4,4′-diaminodiphenylsulfone.

with malein anhydride are deformed at high temperatures, which can obviously be explained by the opening of double bonds in malein anhydride under the conditions of curing glycidyl ether and the formation of additional chemical bonds.

Of particular interest was the use of aliphatic silicon-containing diamine as a hardener. It could be assumed that the presence of siloxane links in the structured polymer would increase its heat resistance. To this end, we cured the diglycidyl ether of 4,4′-(hexahydro-4,7-methylenindane-5-ylidene) diphenol with α,ω-bis-(γ-aminopropylsiloxy) methylphenyl siloxane.

A mixture of Si-containing diamine and hexamethylenediamine was used to identify the contribution to increasing the heat resistance of siloxane units. Experimental data show that a polymer cured with hexamethylenediamine alone decreases in mass by 10% at 280°C, and by 50% at 355°C. These indicators, when silicon-containing diamine is used as a hardener, are 320°C and 480°C, respectively. An increase in the weight fraction of the siloxane component in the mixture of hardeners leads to an increase in the thermal stability of polymers (*see* Table 1.2 and Figure 1.5).

TABLE 1.2 Properties of Structured Polymers Obtained by Curing the Glycidyl Ether 4,4′-(Hexahydro-4,7-Methylenindane-5-Ylidene)Diphenol (I) with a Mixture of Hexamethylenediamine (III) and A,Ω-Bis-(Γ-Aminopropyldimethylsiloxy) Methylphenyl Siloxane (II)

S. No.	The Ratio of the Starting Materials, Weight %			10% Deformation, °C	Mass Reduction Temperature,°C		
	I	II	III		by 10%	by 20%	by 50%
1	1	0	100	310	280	320	355
2	1	10	90	280	290	330	400
3	1	20	80	270	295	335	400
4	1	30	70	270	300	340	410
5	1	100	0	255	320	375	480

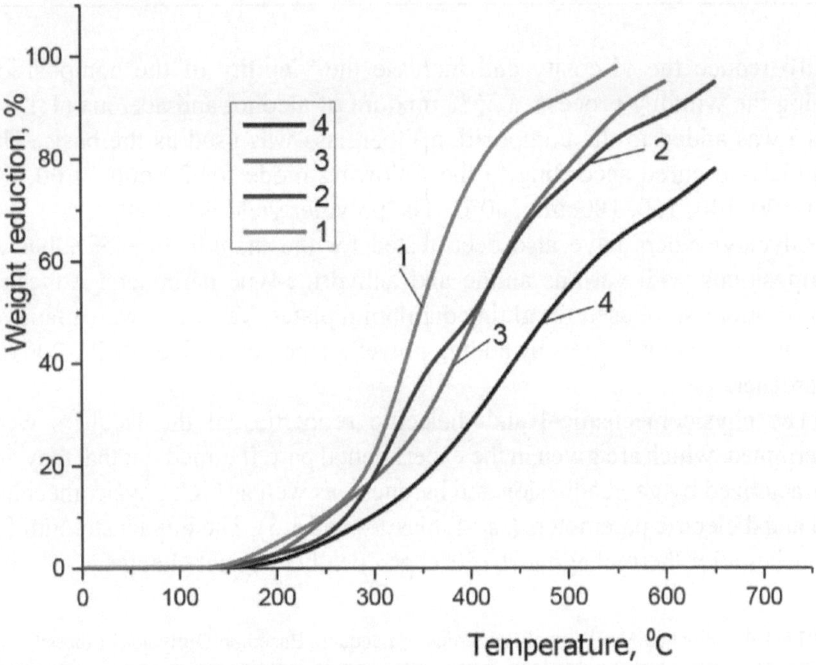

FIGURE 1.5 Thermogravimetric curves of epoxy polymers based on glycidyl ether 4,4′-(hexahydro-4,7-methylenindane-5-ylidene)diphenyl, silicon-containing diamine, hexamethylenediamine and their mixtures.

The diglycidyl ethers synthesized by us have been tested as binders for fiberglass. Curing was carried out with 3,3′-dimethoxy and 4,4′-diaminodiphenylmethane. From the test results given in Table 1.3, it can be seen that fiberglass is characterized by good strength indicators.

TABLE 1.3 Properties of Fiberglass Based on Diglycidyl Ethers Approved by 3,3′-Dimethoxy-4,4′-Diaminodiphenylmethane

S. No.	Structure of Diglycidyl Ether	The Amount of Hardener (Weight % of the Initial Diglycidyl Ether)	Tensile Strength, kgf /cm^2	Elongation at Break Tensile,°C	Modulus of Elasticity, kgf /cm^2
1	2	3	4	5	6
1	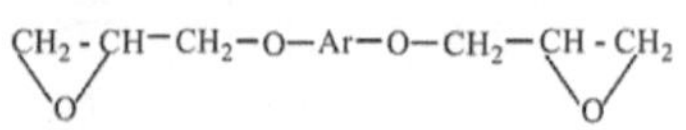	24	170	7	$2.6 \cdot 10^5$
2	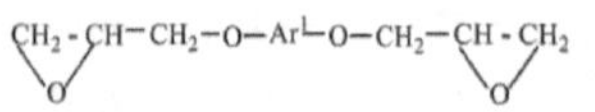	20	160	5	$3.1 \cdot 10^5$
3	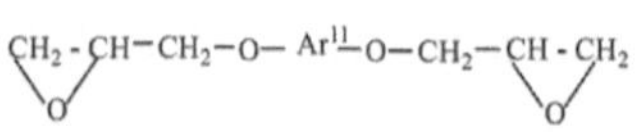	20	150	6	$2.6 \cdot 10^5$

To reduce the viscosity and increase the viability of the composition during the winding process, a 25% mixture of alcohol and acetone (1:1 wt. part.) was added to the composition Fiberglass was used as the basis. The samples are cured according to the following mode for 2 hours at 60, 80, 100, 120, 140, 160, 180 and 200°C. The polymer yield is 95–98%.

Glycidyl ethers have also been tested for the manufacture of adhesive compositions with various amine and anhydride-type hardeners. Adhesive compositions were used for gluing duralumin plates. Varnishes were made on the basis of glycidyl ethers by adding polyethylene polyamine (PEPA-200) as a hardener.

The physicomechanical and dielectric properties of the lacquers were determined, which are given in the experimental part. It turned out that they are characterized by good adhesion and hardness, as well as high physicomechanical and dielectric parameters (*see* Tables 1.4 and 1.5). The impact strength for varnishes after thermal aging at 150°C practically does not change.

TABLE 1.4 Physico-Mechanical Properties of Lacquers Based on Diglycidyl Ethers[x)]

S. No.	Composition of the Composition, Mol.	Impact Strength, kg/cm.	Bending Strength, mm	Hardness According to the Pendulum Device	Adhesion Determined by the Lattice Method in Points	Impact Strength After Thermal Aging at 150°C, kgf/cm.		
						After 5 Hours	After 10 Hours	After 17 Hours
1	I: PEPA-200 = 1:0.60	40	10	0.8	1	20	20	20
2	II: PEPA-200 = 1:0.65	50	8	0.9	2	45	40	35
3	III: PEPA-200 = 1:0.7	40	10	0.8	2	20	20	20

TABLE 1.5 Dielectric Properties of Lacquers Based on Diglycidyl Ethers[x)]

S. No.	Composition of the Composition, Mol.	Dielectric Constant	The Tangent of the Dielectric Loss Angle	Specific Volumetric Electrical Resistance at 25°C, Ω·cm	Electrical Strength, kV/mm
1	I: PEPA-200 = 1:0.60	1.65	0.111	$2.65 \cdot 10^{16}$	74
2	II: PEPA-200 = 1:0.65	2.81	0.019	$3.95 \cdot 10^{16}$	40
3	III: PEPA-200 = 1:0.7	3.09	0.018	$3.25 \cdot 10^{16}$	40

In Tables 1.4 and 1.5:

x) PEPA -200 – polyethylene polyamine, mol. wt. 200

I=
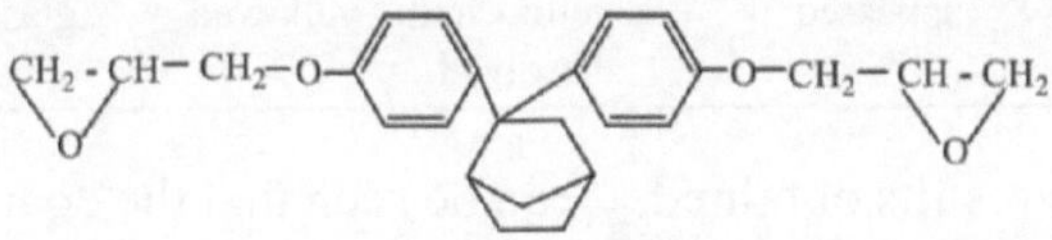

II=
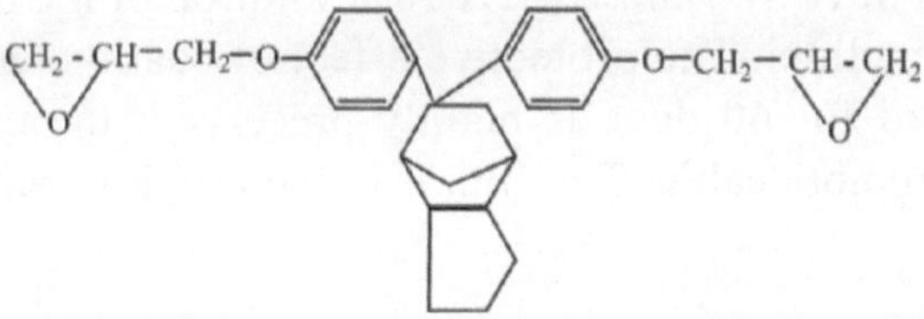

III=
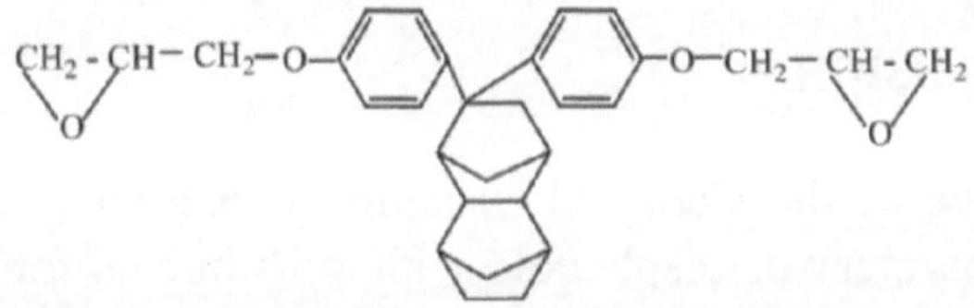

The water and chemical resistance, as well as UV resistance, of the coatings were studied. Chemical resistance to aggressive media was examined using a polymer obtained from 4,4'-(hexahydro-4,7-methylenindane-5-ylidene) diphenol and 4,4'-diaminodiphenylsulfone. The lacquer composition was prepared in a mixture of solvents (butyl acetate: xylene, volume ratio 7:3). Varnishes were applied to glass plates by dipping. To achieve the desired thickness (200–300 μm), a 3–4 layer coating was obtained.

The degree of curing of the lacquer compositions, determined by the content of the insoluble product in the film, was assessed through extraction with acetone in the Soxhlet apparatus for 12 hours. The chemical resistance of the coatings was tested at room temperature. The results of the coating tests are shown in Table 1.6.

TABLE 1.6 Chemical Resistance of Epoxy Polymer Coating Based on Diglycidyl Ether 4,4′-(hexahydro-4.7-methylenindane-5-ylidene)-Diphenol and 4,4′-Diaminodiphenylsulfone

30% Solution H2SO4	30% Solution HCl	10% Solution HNO_3	12% Acetic Acid Solution	Distilled Water	0.5% Solution $KMnO_4$
		After 30 days			
The film has not changed	Spots on the film	The film is clouded	The film has not changed	The film has not changed	Spots on the film
		After 60 days			
The film has not changed color, moved away from the glass	The film turned yellow; cracks appeared	The film turned yellow and dissolved	The film is slightly yellowed	moved away from the glass	the film dissolved

From the results obtained, it can be seen that the coatings are resistant to the action of H_2SO_4 and HCl. A 10% solution of nitric acid is a highly aggressive medium. The protective effect of coatings in a 12% solution of acetic acid for 60 days is mainly preserved; the effect of distilled water is more noticeable. The film is especially low-resistant in $KMnO_4$ solution.

1.4 CONCLUSION

The influence of the chemical structure of hardeners on the properties of epoxy polymers of bisphenols with substituents for norbornane type has been studied. It has been shown that the properties of cured epoxy polymers, especially heat resistance, are affected not only by the structure of bisphenol but also by the chemical structure of the hardener. The synthesis of polymers with increased thermal parameters was carried out. It was found that 4,4′-diaminodiphenylsulphone, benzidine, 4,4′-diaminodiphenyloxide, and other aromatic diamines give high results in thermal and heat resistance. Polymers obtained by curing diglycidyl ethers of polycyclic bisphenols with these hardeners are deformed in the temperature range of 220–245°C. Among acid hardeners, acid anhydrides, such as pyromellitic and methyltetrahydrophthalic, are particularly effective. Polymers obtained using these compounds as hardeners deform in the temperature range of 220–245°C and decrease in mass by 10% in the temperature range of 340–400°C.

KEYWORDS

- **bisphenol**
- **epichlorohydrin**
- **epoxy polymer**
- **ether**
- **hardener**
- **pyromellitic**

REFERENCES

1. Gibson, G. (2017). *Brydson Plastics Materials* (8th ed., pp. 773–797).
2. Hodd, K. (1989). Comprehensive polymer science and supplements (Vol. 5, pp. 667–669).
3. Ooi, S. K., Cook, W. D., Simon, G. P., & Such, C. H. (2000). DSC studies of the curing mechanisms and kinetics of DGEBA using imidazole curing agents. *Polymer, 41*(9), 3639–3649. https://doi.org/10.1016/S0032-3861(99)00600-X.
4. Liu, W. B., Qiu, Q. H., Wang, J., Huo, Z. C., & Sun, H. (2008). Curing kinetics and properties of epoxy resin–fluorenyl diamine systems. *Polymer, 49*(19), 4399–4405. https://doi.org/10.1016/j.polymer.2008.08.004.
5. Pan, G. Y., Du, Z. J., Zhang, C., Li, C. J., Yang, X. P., & Li, H. Q. (2007). Synthesis, characterization, and properties of novel novolac epoxy resin containing naphthalene moiety. *Polymer, 48*(12), 3686–3693. https://doi.org/10.1016/j.polymer.2007.04.032.
6. Mukbaniani, O., Markarashvili, E., Iskakova, M., & Mindiashvili, G. (2008). Synthesis and investigation of properties of diepoxydiorganosiloxane and composite materials on their basis. *Oxidation Communications, 31*(1), 116.
7. Kozlov, G. V., Bejev, A. A., & Zaikov, G. E. (2002). The physical reasons of homogeneous and nonhomogeneous reactions of haloid-containing epoxy polymers curing. *Oxidation Communications, 25*(4), 529.
8. Paken, A. M. (1962). Epoxy compounds and epoxy resins. State Chemical Publishing House.
9. Mukbaniani, O. V., Markarashvili, E. G., Iskakova, M. I., Aneli, J. N., & Gventsadze, L. D. (2008). Modification reactions of novolac resins by epoxy-containing polyphenylsilsesquioxanes. *Oxidation Communications, 31*(2), 300–310.
10. Papava, G., Dokhturishvili, N., Gurgenishvili, M., Chitrekashvili, I., & Chubinishvili, Z. (2020). Epoxy polymers based on diglycidyl ethers with cyclic groups. *World Science, 7*(59), 100. https://doi.org/10.31435/rsglobal_ws/30092020/7208.
11. Papava, G., Gurgenishvili, M., Chitrekashvili, I., Chubinishvili, Z., & Tabukashvili, Z. (2018). Synthesis of thermoreactive oligomers on the basis of policyclic bisphenols. *World Science, 38*(10), 42–47. https://doi.org/10.31435/rsglobal_ws/31102018/6178.

12. Gavashelidze, E., Maisuradze, N., Dokhturishvili, N., Papava, G., Gelashvili, N., Molodinashvili, Z., Gurgenishvili, M., & Chitrekashvili, I. (2012). Polyurethanes on the basis of card-type polycyclic bisphenols different diisocyanates. *Bulletin of the Georgian National Academy of Sciences, 6*(1), 113–116.
13. Gelashvili, N., Papava, G., Gurgenishvili, M., Chitrekashvili, I., Molodinashvili, Z., & Tabukashvili, Z. (2009). Phenol-formaldehyde type polymers on the basis of bisphenols. *Bulletin of the Georgian National Academy of Sciences, 3*(2), 84.

PART II
Synthesis

CHAPTER 2

Synthesis and Hydrosilylation of 2-Methyl(Ethyl)-1-Propargyl Pyrrole

OKTAY ASKEROV, AYNURA MAMEDOVA, DILBAR NURULLAYEVA, ILGAR ASKEROV, AND GYULGUN KHANBABAYEVA

Institute of Polymer Materials, Ministry of Science and Education, Sumgait, Azerbaijan

ABSTRACT

It has been developed the method of the synthesis of unsaturated organosilicon pyrroles by the interaction of trialkyl(aryl)hydridsilanes with 2-methyl(ethyl)-1-propargyl pyrrole in the presence of the catalyst – platin hydrochloric acid – in a medium of absolute benzene (yield 75–80%), and their properties have been studied. It has been shown that the hydrosilylation reaction proceeds on a triple bond according to Farmer's rule, resulting in the formation of unsaturated organosilicon pyrroles while preserving the pyrrole ring. These compounds are highly reactive and can interact with both nucleophilic and electrophilic reagents. In this context, the synthesis of functionally substituted 2-methyl(ethyl)-1-propargyl pyrroles, along with their silicon derivatives, and the investigation of their antimicrobial activity, represents an important area of scientific and practical research. The composition and structure of the synthesized unsaturated organosilicon pyrroles have been confirmed by IR and PMR spectroscopy data.

2.1 INTRODUCTION

It is known from the analysis of native and foreign literature that pyrrole and its derivatives occur in nature in a free state, and also as fragments

Advanced Topics in Polymer Chemistry and Materials Science: Tamara Tatrishvili and Marc Jean Médard Abadie (Eds)

in the composition of complex natural compounds. However, they are present in the body when performing the most complex physiological functions (Filippov, 1985; Lucet, 1998; Smith, 1987; Tsuda, 1999; Uchida, 1990).

In connection with the above-mentioned information and in continuation of investigations in the field of synthesis of organosilicon 2-methyl(ethyl)-1-propargyl pyrrole-containing compounds, the study of the addition reaction of trialkyl (aryl) hydridsilanes with 2-methyl(ethyl)-1-propargyl pyrrole in the presence of a catalyst – 0.1 N platin hydrochloric acid or rhodium dicarbonyl acetylacetonate – and, thereby, revealing the influence of the nature of substituents in the multiple bond of the propargyl radical on the yield and structure of the reaction products is of scientific interest (Flett, 2005; Nedolya, 2002; Yuryev, 1982).

2.2 EXPERIMENTAL METHODS AND MATERIALS

5.95 g (0.05 mol) of 2-methyl(ethyl)-1-propargyl pyrrole, 30 ml of absolute benzene, and 5.104 g (0.05 mol) of methyldiethylsilane were loaded into a three-necked flask equipped with a reflux condenser, a thermometer, and a drop funnel, and boiled for 26 hours. After distillation of the low-boiling components, a compound (III) was isolated from the residue by vacuum. Yield – 16.36 g (74%).

B.p. 107–108°C (1 mm merc. c.), n_D^{20} 1.4965, d_4^{20} 0.8966, MR_D (found) 76.98, MR_D (calculated) 71.90. Brutto-formula: $C_{13}H_{23}NSi$. Found,%: C 69.90; H 11.01; N 5.95; Si 12.95. Calculated,%: C 70.00; H 11.00; N 6.00; Si 13.00.

Similarly, the compounds (IV-VIII) have been obtained, and the physical-chemical constants of which are given in Table 2.1.

Similarly, the compound (IX) has been obtained from 6.65 g (0.05 mol) of 2-ethyl-1-propargyl pyrrole and 5.95 g (0.05 mol) of methyldiethylsilane. Yield – 18.33 g (78%).

B.p. 113–114°C (1 mm merc. c.), n_D^{20} 1.4930, d_4^{20} 0.8900, MR_D (calculated) 76.61, MR_D (found) 76.56. Found,%: C 71.74; H 10.28; N 5.96; Si 11.96. Calculated,%: C 71.76; H 10.29; N 5.98; Si 11.99.

Similarly, there have been obtained the compounds (X-XIV), the physical-chemical constants of which are also given in Table 2.1.

TABLE 2.1 Physical-Chemical Constants of Unsaturated Organosilicon Pyrroles

No. of Compounds	Yield, %	B.p., °C (mm merc. c.)	n_D^{20}	d_4^{20}	MRD Calculated	MRD Found	Brutto Formula
III	77	107–108 (1)	1.4965	0.8966	71.98	71.90	$C_{13}H_{23}NSi$
IV	75	121–122 (1)	1.4935	0.8960	81.48	81.40	$C_{15}H_{27}Nsi$
V	74	125 (1)	1.4905	0.8900	90.58	90.50	$C_{17}H_{31}Nsi$
VI	80	98–99 (1)	1.4925	0.9100	97.50	97.40	$C_{19}H_{35}Nsi$
VII	82	145–146 (1)	1.5510	1.0200	102.60	102.50	$C_{13}H_{23}Nsi$
VIII	78	103–104 (1)	1.5030	0.9251	70.09	70.05	$C_{13}H_{21}Nsi$
IX	79	113–114 (1)	1.4930	0.8900	76.61	76.56	$C_{14}H_{25}Nsi$
X	81	115–116 (1)	1.4820	0.8801	85.11	85.00	$C_{16}H_{29}Nsi$
XI	83	118–119 (1)	1.4925	0.8902	95.21	95.20	$C_{18}H_{35}Nsi$
XII	85	96–97 (1)	1.4900	0.8811	104.63	104.60	$C_{20}H_{37}Nsi$
XIII	78	137–138 (1)	1.5640	1.2400	107.01	107.00	$C_{22}H_{25}Nsi$
XIV	84	102–103 (1)	1.5025	0.9200	74.72	74.60	$C_{14}H_{22}Nsi$

2.3 RESULTS AND DISCUSSION

We have developed the method for the synthesis of pyrrole-containing organosilicon compounds by the interaction of 2-methyl(ethyl)-1-propargyl pyrrole with trialkyl(aryl)silanes and 0.1 N platin hydrochloric acid as a catalyst. As a result, one of the two π-bonds (C≡C) is opened in the propargyl radical, and trialkyl(aryl)hydridsilanes are attached to it according to the following scheme:

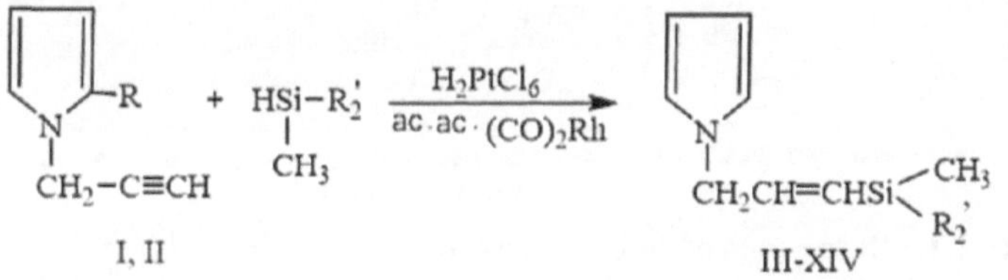

R=CH$_3$ (I); R$_2$' =C$_2$H$_5$ (III); C$_3$H$_7$ (IV); C$_4$H$_9$ (V); C$_5$H$_{11}$ (VI); C$_6$H$_5$ (VII); (CH$_2$)$_4$ (VIII);

R=C$_2$H$_5$ (II); R$_2$'=C$_2$H$_5$(IX); C$_3$H$_7$(X); C$_4$H$_9$ (XI); C$_5$H$_{11}$ (XII); C$_6$H$_5$ (XIII); (CH$_2$)$_4$ (XIV

An attempt to hydrosilylated the second π-bond was unsuccessful, although from the available literature data, one could expect the proceeding of the hydrosilylation reaction also on the pyrrole ring. The presence of a pyrrole ring in the spectrum of the compound (III) is confirmed by the availability of an absorption band in the field of 3400–3440 cm^{-1} belonging to the vibration of the pyrrole ring, without affecting the pyrrole cycle in this case (Gordon, 1976; Kazitsina, 1971; Silverstein, 1977).

The structure of the obtained compounds (III-XIV) has been studied by modern spectral methods. The study of spectra of the compound (III) (Figure 2.1) showed that there are the absorption bands in the field of 1251–1231 cm^{-1} referring to the valence vibrations of C-N bond, and the absorption band in the field of 1620 cm^{-1} refers to the valence vibration of Si-C- alkyl bond. The deformation (1358, 1421, 1458 cm^{-1}) and valence (2873, 2911, 2951 cm^{-1}) vibrations are referred to C-H bond in CH_2-grouping (Gordon, 1976; Kazitsina, 1971; Silverstein, 1977).

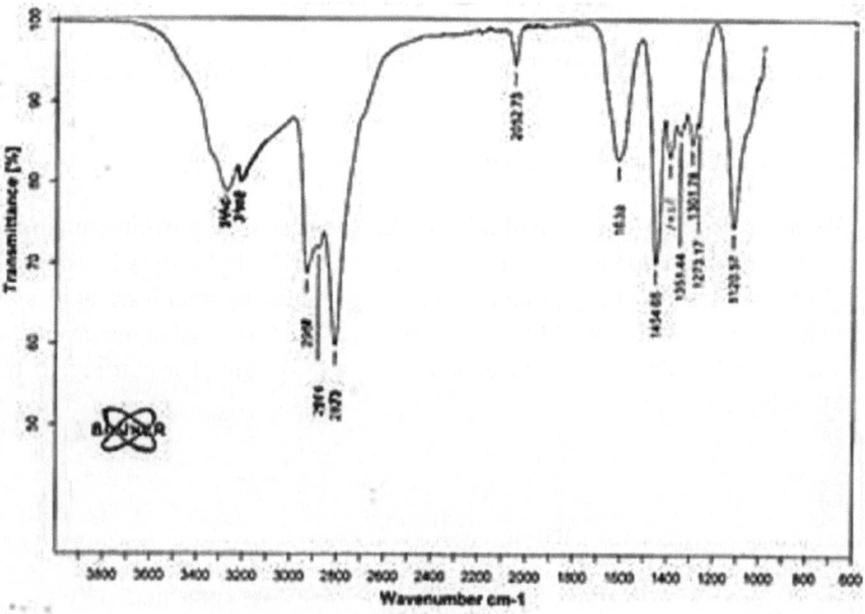

FIGURE 2.1 IR spectrum of the compound III.

Thus, under the conditions accepted by us, the hydrosilylation reaction of 2-methyl(ethyl)-1-propargyl pyrrole with trialkyl(aryl)silanes catalyzed by rhodium acetylacetonate dicarbonyl or 0.1 N platin hydrochloric acid proceeds exclusively on the C-C bond of the propargyl radical, resulting in the formation of unsaturated organosilicon pyrroles, with trialkyl(aryl) silanes fixed at the peripheral carbon atom.

2.3.1 DETERMINATION OF ANTIMICROBIAL ACTIVITY OF UNSATURATED ORGANOSILICON PYRROLES (I-IV)

The antimicrobial activity of unsaturated organosilicon pyrroles of the following structure has been studied by serial dilutions in a liquid nutrient medium in relation to gram-positive (*Staphylococcus aureus*), gram-negative (*Escherichia coli, Pseudomonas aeruginosa*) bacteria and yeast-like fungi (*Candida albicans*).

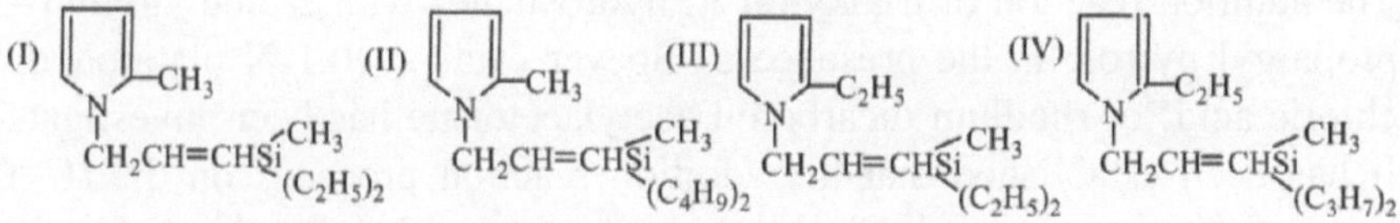

In experiments with bacteria, the meat-peptone agar was used, and Saburo medium was taken for fungi. The microbial load in all tests was $5 \cdot 10^6$ microbial bodies in 1 ml. The seeds were carried out after 10, 20, 30, 40, 60 min. The incubation duration for bacteria was 18–24 h at 38°C, and for fungi – 1–2 days at 28°C. The minimum suppressive concentration (MSC) for compounds (I-IV) has been determined. The indices of the antimicrobial activity of pyrroles (I-IV) are given in the Table 2.2.

TABLE 2.2 Antimicrobial Activity of Unsaturated Organosilicon Pyrroles

Test Cultures	Exposure of Seeds, Min.	Minimum Suppressive Concentration of Unsaturated Tested Organosilicon Pyrroles, Mcg/Ml			
		I	II	III	IV
Staphylococcus aureus	10	124	250	124	250
	20				
Escherichia coli	10	124	250	124	250
	20				
Pseudomonas aeruginosa	10	124	500		
	20			60.0	
	30		62.5		
Candida albicans	10			60.0	
	30		125		
	40	62.5			
	60			60.0	

As can be seen from Table 2.2, the compounds I and III at MSC 126 mcg/ml destroy *Staphylococcus aureus* and *Escherichia coli*, while the compounds

II and IV at the same concentration do not destroy these cultures. Compound II destroys *Candida albicans* at MSC 124 mcg/ml, and compounds I, III, and IV do so at 62 mcg/ml. Thus, the electron-donor substituents in the radicals of the initial pyrroles decrease, while electron-acceptor substituents, on the contrary, increase the antimicrobial activity against *Candida albicans*.

2.4 CONCLUSION

The addition reaction of trialkyl(aryl) hydrosilanes with 2-methyl(ethyl)-1-propargyl pyrrole in the presence of Speyer catalyst (0.1 N platinohydrochloric acid) or rhodium dicarbonyl acetylacetonate has been investigated. It has been established that the addition reaction proceeds on the (C≡C) bond of the propargyl radical. It has also been revealed that the yield of the addition reaction using rhodium dicarbonyl acetylacetonate as a catalyst is higher than that using the Speyer (platinum) catalyst. It has been revealed that the unsaturated organosilicon 2-methyl(ethyl)-1-propargyl pyrroles show significant antimicrobial activity in relation to *Escherichia coli* and *Pseudomonas aeruginosa.*

KEYWORDS

- **acetylene bonds**
- **antimicrobial activity**
- ***Candida albicans***
- **hydrosilylation**
- **platinohydrochloric acid**
- **trialkyl (aryl) hydridsilanes**
- **unsaturated organosilicon pyrroles**

REFERENCES

1. Filippov, V. V. (1985). *Functions and synthesis of biotin in a living organism* (Rep. ed. L. N. Andreev). Moscow: Nauka.
2. Lucet, D., Le Gal, T., & Mioskowski, C. (1998). The chemistry of vicinal diamines. *Angewandte Chemie International Edition, 37*(19), 2580–2627.

3. Tsuda, M., Uemoto, H., & Kobayashi, J. (1999). Slagenins A-C, novel bromopyrrole alkaloids from marine sponge Agelas nakamurai. *Tetrahedron Letters, 40*(30), 5709–5712.
4. Smith, R. C., & Reaves, J. C. (1987). Antioxidant properties of 2-imidazolones and 2-imidazolthiones. *Biochemical Pharmacology, 36*(9), 1457–1460.
5. Uchida, K., & Kawakishi, S. (1990). Site-specific oxidation of angiotensin I by copper(II) and L-ascorbate: Conversion of histidine residues to 2-imidazolones. *Archives of Biochemistry and Biophysics, 283*(1), 20–26.
6. Nedolya, N. A., Brandsma, L., & Shlyakhtina, N. I. (2002). An elegant synthesis of 1-(2-vinyloxyethyl)- and 1-(2-hydroxyethyl)pyrrole-3-carbaldehydes. *Russian Journal of Organic Chemistry, 38*(7), 1070–1072.
7. Yuryev, V. P., & Salimgareeva, I. M. (1982). *Hydrosilylation reaction of olefins* (Metal complex catalysis in hydrosilylation processes, stereochemistry, and reaction mechanism). Moscow: Nauka.
8. Flett, D. S. (2005). Solvent extraction in hydrometallurgy: The role of organophosphorus extractants. *Journal of Organometallic Chemistry, 690*(11), 2426–2438.
9. Silverstein, G., Bassler, T., & Morrill, C. (1977). *Spectrometric identification of organic compounds* (Translated from English by N. A. Donskoy & B. N. Tarasevich; Ed. A. A. Maltseva). Moscow: Mir.
10. Kazitsina, P. A., & Kupletskaya, N. B. (1971). *Use of UV, IR, NMR Spectroscopy in Organic Chemistry*. Vysshaya Shkola Publ. House.
11. Gordon, A., & Ford, R. (1976). *Chemist's satellite* (Translated from English by E. L. Rosenberg & S. I. Kopnel). Moscow: Mir.

CHAPTER 3

Petroleum Sorbents Obtained by Hybridization of Organic Synthetic and Inorganic Materials

MOLODINASHVILI ZAZA, KOPALEISHVILI MAKA, GABUNIA TINATIN, and SHATAKISHVILI TAMAR

Ivane Javakhishvili Tbilisi State University, Petre Melikishvili Institute of Physical and Organic Chemistry, Laboratory of Petrochemistry, Georgia

ABSTRACT

For the obtaining of porous materials, natural sorbents – perlite, zeolite, and diatomite – were chosen. Thermal and chemical modification, followed by hydrophobization of the sorbents, was carried out. The optimal conditions were determined; amido aldehyde oligomers were synthesized with the aim of their usage as a matrix in the composition of porous materials characterized by high sorption activity and the ability to float on the surface of water together with the absorbed compounds. Their removal from the water surface is possible mechanically. Hybrid porous polymers are characterized by selectivity, fire resistance, and a wide range of variability of properties depending on the structure and ratio of initial substances.

3.1 INTRODUCTION

The emergence of the oil industry gave a great push to the development of various industrial sectors. Over 2.5 billion tons of crude oil are produced annually in the world. However, the oil industry has negative consequences, namely the pollution of the natural environment by oil and

Advanced Topics in Polymer Chemistry and Materials Science : Tamara Tatrishvili and Marc Jean Médard Abadie (Eds)

products of its processing. The intensification of oil extraction, evaporation of hydrocarbons during oil refining, and loss during production and transportation – all of this makes a significant contribution to the worsening of the ecological situation in the world.

As a result of pollution, large areas become unsuitable for agricultural use. The oil entering the soil divides into fractions. At that point, light fractions of oil gradually evaporate into the atmosphere, and some are mechanically carried out by water into various water reservoirs. Part of the oil undergoes chemical and biological oxidation. A source of pollution is also the wastewaters from petrochemical, oil, and gas industries, as well as surface runoffs from the adjacent territories. Accidents that occur at oil-producing, oil-pumping, and oil-refining enterprises become a source of catastrophic contamination. As a result, a large amount of carcinogenic compounds enters the atmosphere, causing many animals and plants to die [1, 2]. In this regard, there is a need to protect the environment from the negative components of oil. Disposal of oil wastes is one of the priorities for 2023. Research in this direction will help in solving problems like the rehabilitation of areas exposed to the negative impact of enterprises with oil-containing wastes and the prevention of the accumulation of pollution in the future.

Development of modern and effective methods for the removal of oil and oil products from the surface of water and soil is currently very relevant. Therefore, every year, more and more methods that can be used to clean the environment from the harmful effects of oil products are being created. At the moment, such methods as mechanical (settling, filtration, and centrifugal removal of water pollutants), chemical (chemical ozonation), biological methods, as well as methods of physical and chemical orientation (coagulation and flotation) are known.

Numerous studies have shown that the most environmentally friendly, efficient, and cost-effective treatment is using sorbents. At the same time, sorbents must have a number of specific indicators: hydrophobicity, the possibility of regeneration, significant adsorption capacity, floatability, and chemical and thermal stability. They should also be environmentally friendly and have a low cost [3].

Sorbents, according to the method of exposure, can be divided into inorganic, organic (natural), organic mineral, and synthetic. Inorganic sorbents most often include clays and diatomites, as they are inexpensive and can be produced in large volumes. Sand is usually used for the sorption of small area spills. However, according to environmental considerations, the use of sorbents of this type is inefficient since their oil capacity is very low (from 70 to 150%). Moreover, they are unable to retain light oil fractions such as kerosene, gasoline, and diesel fuels. It is impossible and pointless to use them for cleaning water, as they sink together with oil products. Another reason for the inefficiency of inorganic sorbents is their disposal, which can be carried out by washing them with water containing surfactants and extractants or by burning them out. However, progress does not stand still, and getting rid of these shortcomings is on its way [1, 5].

Hence, the development of new, highly effective sorbents with high sorption ability in respect of petroleum and petroleum products, which do not sink in water together with the absorbed compounds, and are available due to the simplicity of production technology and low cost, is one of the urgent tasks of the study.

Porous sorbents are also effectively used as heat and sound insulation materials. According to literary data, both organic (fibrous synthetic polymers: polyesters and carbocyclic compounds) and mineral (inorganic) raw materials—such as talc, calcites, basalt fiber, etc.,—are used in polymeric compositions. We think it is prospective to use zeolites, diatomite, and perlite in polymeric compositions. As polymeric matrices in compositions, phenol-aldehyde oligomers, polyurethanes, polyethylene, polystyrene, polypropylene, polyamides, polyacrylates, and other classes of polymers are used. We consider amido-aldehyde oligomers as polymers to be used for the intended purpose because of their availability and low cost [4–7].

3.2 EXPERIMENTAL METHODS AND MATERIALS

Studies for the selection of natural mineral zeolites were conducted. Natural sorbents have an advantage in comparison with synthetic sorbents: cost-effectiveness and availability. However, in terms of effectiveness, purity, and homogeneity, natural sorbents fall behind synthetic sorbents. From this point of view, mineral sorbents—sorbents of the silicon series (diatomites), bentonite clays, and natural zeolites resistant to acids (clinoptilolite, perlite, etc.)—are of great interest. When choosing natural sorbents, their pore sizes for complete sorption of substances, as well as their thermal stability (to ensure that the deformation of the pores does not occur), must be taken into account. Three main objects were distinguished: natural zeolite, diatomite, and perlite [8–10]. On the initial stage, the natural sorbent—zeolite—was chosen, which was resistant to aggressive media (acids) and was thermally stable as well, a significant factor from the point of view of exploitation parameters.

The optimal conditions for the thermal modification of the natural sorbent were determined—heating at 240–700°C for 6 hours.

Research to determine the optimal conditions for the chemical modification of natural zeolite was carried out. Work to obtain the H-form of natural zeolite (clinoptilolite) was conducted. The natural zeolite was treated with different concentrations of hydrochloric acid (0.05N; 0.1N; 0.2N; 0.3N) at 25°C. Chemical modification was performed to increase the size and activity of the zeolite pores. Optimal conditions (temperature, time, concentration of solutions, and ratio of the components) for the chemical modification of the zeolite were determined. Several methods of chemical modification were identified, particularly using acids, ammonium chloride (NH_4Cl), potassium chloride (Kcl), and other compounds.

Experiments in thermal modification of the diatomite were carried out. Thermal modification of diatomite can be carried out by:

1. Heating at 900–1000°C.
2. Treatment with NaCl and heating at 1000°C.
3. Treatment with SiF_4 and heating at 1000°C.

To determine the optimal temperature for thermal modification, the diatomite was heated for 3 hours at 250, 300, 400, 450, 500, 550, 600, 700, 900, and 1000°C. During heating up to 250°C, mainly the release of absorbed water takes place. An increase in temperature leads to the destruction of organic components and the widening of pores. At 700°C, the size of the pores reaches its maximum, and the capacity of the modified diatomite pores increases. It was determined that the introduction of thermally (heating at 700°C) and chemically (using silicon organic compounds) modified perlite into the polymeric composition sharply improves the properties of the porous polymer – its density, ability to float, sorption properties, etc. Natural sorbents – zeolite, diatomite, and perlite – can be used as remedies for cleaning water polluted with petroleum. Since they are hydrophilic and sink in water, it is necessary to carry out their hydrophobization. The processes of hydrophobization of diatomite, perlite, and zeolite were studied. In particular, hydrophobization of both natural unmodified and thermally modified (at 400°C) diatomite was carried out at 250°C for 6 hours in a silicon medium.

The experiment showed that thermally modified diatomite after hydrophobization:

1. Does not get wet in water and does not sink.
2. Effectively absorbs oil from the surface of polluted water.

Diatomite is more prospective sorbent for purification of water-oil emulsions. Heated at 450–700°C diatomite acts more effectively in case of low concentrations of oil ($1.0 \cdot 10^3$ mg/L – $6 \cdot 10^3$ mg/L). If the concentration of oil is high the heating at 1000–1200°C diatomite is used.

To obtain polymeric compositions, the complex carbamide-formaldehyde (a), melamine-formaldehyde (b), and carbamide-melamine-formaldehyde (c) oligomers were synthesized, which are prospective materials for use as a matrix in compositions. The synthesis of oligomers was carried out mainly in aqueous solutions, based on the amide component and formaldehyde:

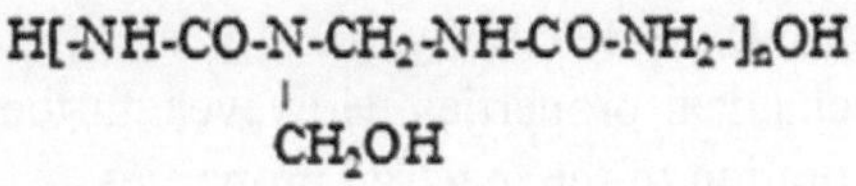
(a)

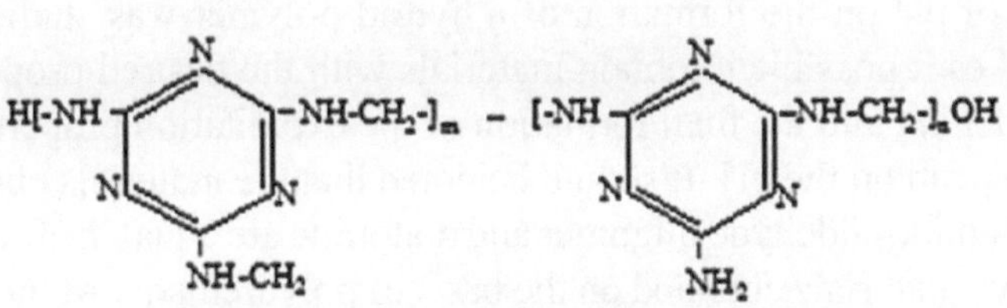
(b)

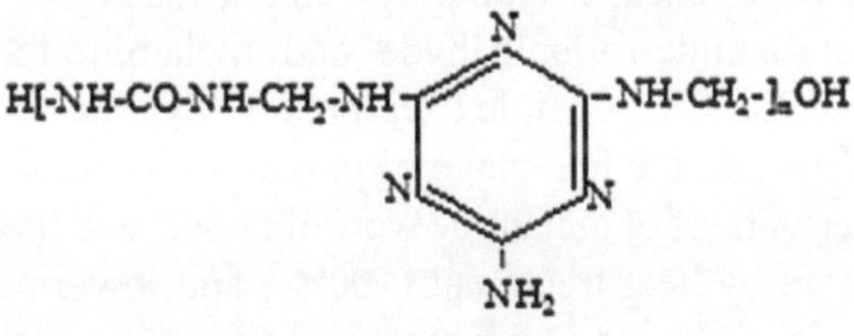
(c)

The method provides for the interaction of the amide component (carbamide, melamine) and formaldehyde in a water solution under defined reaction media (pH = 7.5–5) and temperature (30–100°C) conditions. Correction of the reaction media is carried out using a 10% NaOH solution and formic acid. The ratio of the resulting components (amide component: formaldehyde) is 1:1.7, respectively. A water solution of formaldehyde was used [11–14].

Amide oligomers were obtained by polycondensation in an alloy with a ratio of the obtained components (carbamide: aldehyde) of 1:1.1, respectively, at 100°C, with a reaction duration of 1 hour.

The composites were prepared from diatomite (both natural and modified) and amido-aldehyde oligomer (with different percentages) according to the technology developed by us: a foaming agent, resorcin, and water mixture were placed into the utensil. After thorough mixing, 2.5 ml of the mixture were foamed by an air stream to obtain stable foam. A certain amount of amido-aldehyde oligomer was added to the foam, with a pre-introduced specific amount of natural sorbent. Foaming was continued by an air stream until full effervescence and the formation of stable foam.

It was determined that the properties of the obtained porous hybrid polymer were affected by the ratio of amide oligomer to diatomite and the reaction media. The results allowed us to conclude that the optimal mass ratio of diatomite to amido-aldehyde oligomer for obtaining an effective material is 0.3:0.7, respectively. Increasing the diatomite amount up to 50% (in relation to the oligomer) improves heat stability, density, and fire resistance, as well as physical and mechanical properties. However, further increases in the ratio result in a deterioration of the material properties.

The effect of pH on the formation of a hybrid polymer was studied. Its correction makes it possible to obtain materials with the desired properties. The rapid hardening and the final formation of the exploitation properties of the material depend on the pH. It should be noted that the materials obtained on the basis of amido-aldehyde oligomer and diatomite are equal in efficiency to the expensive materials obtained on the basis of polyurethane. At the same time, they are 10–12 times cheaper than polyurethane materials.

Together with carbamide-formaldehyde and melamine-formaldehyde homogeneous polymers, the complex carbamide-hyaline-formaldehyde polymers were obtained. At the first stage of polymer formation in aqueous solution, methylol derivatives of melamine were obtained, and then they were converted to oligomers by heat treatment (100°C) and lowering the pH to 5–6. By interaction of the obtained oligomers and a foaming agent (foaming agent, 60% H_3PO_4, resorcin, and H_2O), a foam-polymer was obtained, from which polymers with a porous structure were prepared.

3.3 RESULTS AND DISCUSSION

To study the sorption properties of the polymer, water artificially polluted with a petroleum product (benzene) was taken. Porous polymers (sorbents) were placed into the resulting emulsion for a defined time (1, 3, 5, 7, 10 days). The sorbents were hybrid materials derived from carbamide-formaldehyde oligomers and modified diatomite. Afterward, the extraction of porous samples (sorbents) from the aqueous medium was performed.

Using spectral analysis, the concentration of the petroleum product remaining in the aqueous emulsion was determined.

The study showed that porous composites are characterized by much higher sorption properties than natural and modified diatomite. Thermally modified, hydrophobized diatomite almost completely absorbs the petroleum product if its concentration does not exceed $6 \cdot 10^3$ mg/L, and at higher concentrations – $14 \cdot 10^3$ mg/L, the amount of petroleum product absorbed by diatomite decreases. Under these conditions, high sorption properties are shown by hybrid composites obtained on the basis of the amido-aldehyde oligomer and containing 30% of natural or modified diatomite. Studies of the properties of sorbents were carried out using spectral and chromatographic methods.

Samples of porous polymer with different percentages of diatomite – 20%, 25%, 30%, and 40% – were obtained. The best option was the 25% content of diatomite. The resulting polymer has high strength and increased porosity. The obtaining of the polymer occurred in a shorter time than in other cases.

The influence of various factors (reaction medium, oligomer concentration, air flow intensity, mixing frequency, amount of natural sorbent, etc.) on the formation of a hybrid porous polymer was studied. It was established that the optimal conditions for polymer formation were the following: pH of the reaction mixture < 7, oligomer concentration – 55%, intensive air flow, maximum mixing frequency, temperature 20 ± 5°C, and content of natural sorbent – 25%.

The obtained samples of mixed polymer containing different percentages of diatomite were tested for sorption capacity, the results of which are shown in the Table 3.1. It has been established that for the improvement of the sorption properties of a polymer, both thermal and chemical modification of diatomite are of great importance.

The study showed that if the concentration of petroleum products in the polluted water is $6 \cdot 10^3$ mg/L, then a sample of mixed porous polymer containing 25% thermally and chemically modified diatomite completely absorbs petroleum products from the surface of the water in five days. In case of polymer sample containing only thermally (1000°C) modified diatomite the sorption capacity is well manifested at a low concentration of contamination with petroleum products $1 \cdot 10^3$ mg/L^{-6}. Work continues. Table 3.1 summarizes the results of sorption by hybrid porous sorbent from the surface of water polluted with benzene.

TABLE 3.1 The Results of Sorption by Hybrid Porous Sorbent from the Surface of Water Polluted with Benzene (Light Fraction of Petroleum)

S. No.	Samples of Porous Hybrid Materials	Sample Weight, G		Amount of Absorbed Benzene2x	
		Before Sorption	After Sorption	g	%
1	Carbamide-formaldehyde polymer	3.24	5.42	2.18	67
2	Hybrid porous polymer (carbamide-formaldehyde oligomer – 70%, natural diatomite – 30%)	3.53	6.32	2.79	79
3	Hybrid porous polymer (carbamide-formaldehyde oligomer – 70%, modified diatomite – 30%)[3x]	4.28	7.85	3.57	83
4	Hybrid porous polymer (carbamide-formaldehyde oligomer – 60%, modified diatomite – 40%)	5.15	8.67	3.53	68

[x] – Size of the sample 3x2x1 cm^3, ρ=0,2–0,7g/cm^3.

[2x] – Time of sample contact with benzene – 5 hours (in water-benzene emulsion); amount of benzene – 5ml.

(d=0.879g/cm^3, $t_{boiling}$= 800C). Water – 100 ml.

[3x] – Diatomite is thermally modified (900°C) and hydrophobized with polymethylphenylsiloxane.

3.4 CONCLUSION

Natural sorbents—zeolite, diatomite, and perlite—were selected and thermally and chemically activated to enhance their properties. Optimal conditions for thermal modification were established: diatomite was heated at 1000°C for 3 hours, zeolite at 240–700°C for 6 hours, and perlite at 400–600°C for 3 hours. Chemical modifications included treating zeolite with 0.1N ammonium chloride, diatomite with SiF_4, and perlite with silicon organic compounds. Hydrophobization processes for diatomite, perlite, and zeolite were also investigated, with hydrophobization of unmodified and thermally (400°C) modified diatomite conducted at 250°C for 6 hours in a silicon environment. Amide oligomers were synthesized using a 1:1.7 ratio of the amide component to formaldehyde, with a formaldehyde water solution. Additionally, amide oligomers were obtained through polycondensation in a melt at a 1:1.1 molar ratio of carbamide to aldehyde at 100°C, with a reaction duration of 1 hour. The optimal mass ratio of diatomite to amido-aldehyde oligomer was found to be 0.3:0.7. Various factors influencing the formation of hybrid porous polymers were examined, and samples of mixed polymer with different percentages of diatomite were tested for sorption ability. The results indicated that porous composites exhibited significantly higher sorption properties than either the polymer or natural and modified diatomite alone.

KEYWORDS

- **hydrophobization**
- **modification**
- **natural sorbents**
- **oil products**
- **porous material**
- **synthesis**

REFERENCES

1. Peters, A. (2006). Oil spills and the environment. *Ecology, 4*, 11.
2. Gridin, O. M. (2000). About oil sorbents. *Metallurgist, 10*, 25–31.
3. Zheltov, M. (2012). Sorbents in the fight against oil spills. *Aquaterm, 1*(65), 98–101.
4. Toroptseva, A. M., Belogorodskaya, K. V., & Bondarenko, V. M. (1972). *Laboratory Workshop in Chemistry and Technology of High-Molecular Compounds*. Leningrad: Chimia.
5. Gladun, V. D., Andreeva, N. N., Akatieva, L. V., & Dragina, O. G. (2000). Inorganic adsorbents from industrial wastes for treatment of industrial wastewaters. *Ecology and Industry of Russia, 5*, 17–20.
6. Abbasov, V. M. (2002). Ecological problems of the Caspian Sea associated with the extraction and processing of oil. In *Theses of the V Baku International Conference* (pp. 3–6).
7. Dvadnenko, M. V., Privalova, N. M., Lyavina, E. B., Protsai, A. A., & Dinchenko, Y. V. (2009). The use of sorption technology for the treatment of oil-containing wastewaters. *Fundamental Researches, 55*, 45–46.
8. Melkozerov, V. M., Vasilyev, S. I., & Lapushova, L. A. (2016). Composition for obtaining of sorbent based on carbamide-formaldehyde resin. *Patent 2587440 of the Russian Federation*. Bull. No. 17.
9. Melkozerov, V. M., Vasilyev, S. I., & Ortman, A. S. (2016). Polymer composition for plastic foam. *Patent 2593160 of the Russian Federation*. Bull. No. 21.
10. Lapushova, L. A., & Vasiliev, S. I. (2015). Results of the study of the structure of the polymer sorbents "Uni-Polimer-M" for liquidation of anthropogenic spillage of oil and oil products. *Protection of the Environment in the Oil and Gas Complex, 6*, 17–21.

11. Vasilyev, S. I., Lapushova, L. A., Melkozerov, V. M., Matveykina, Y. V., & Gorbunova, L. N. (2016). Results of the study of hydrophilia of polymer sorbents from the "Unipolymer" series. *Systems. Methods. Technologies, 1*(29), 135–139.
12. Melkozerov, V. M., Vasilyev, S. I., Velp, A. Y., Krylyshkin, R. N., & Maryanchik, D. I. (2011). Operational properties of polymeric sorbents. *Journal of Siberian Federal University: Engineering & Technologies, 4*, 369–379.
13. Levchenko, A. G., Vitkovsky, M. I., Kurkin, V. A., & Fedotova, A. S. (2013). Recultivation of agricultural soils using the "Unipolymer-M" sorbent. *Protection of the Environment in the Oil and Gas Complex, 10*, 42–54.
14. Bazunova, M. V., Idrisov, I. F., Bazunov, A. A., & Akhmetkhanov, R. M. (2017). Sorption-active hybrid polymer nanocomposites based on polyethylene and inorganic components. *Scientific Almanac, 2, 3*(28), 422–426.

CHAPTER 4

Obtaining Antibacterial Composition Material Based on the Copolymer of Salicylic Acid Allyl Ether with Methyl Methacrylate and ABS and Studying Properties

L. SH. ABBASOVA,[1] A. F. MAMMADOVA,[1] B. A. MAMMADOV,[1] and S. G. ZEYNALOVA[2]

[1]*Institute of Polymer Materials, Ministry of Science and Education, Sumgayit, Azerbaijan*

[2]*Azerbaijan Medical University, Sumgayit, Azerbaijan*

ABSTRACT

ABS (acrylonitrile-butadiene-styrene copolymer) based antibacterial composition materials (CM) were obtained, and their properties were studied using the copolymer of allyl ester of salicylic acid with methyl methacrylate as an antibacterial additive. It was determined that the addition of 4, 12, and 20 wt% salicylic acid allyl ester copolymer with methyl methacrylate to the composition of the ABS composite material practically does not affect its physical-mechanical properties, but the thermal resistance properties increase with the amount of the additive. The antimicrobial properties of the obtained composition materials were studied. *Staphylococcus aureus* (*S. aureus*) from Gram-positive microorganisms, *E. coli* from Gram-negative bacteria, blue-green pus bacillus from pigment-forming bacteria (*P. aeruginosa*), and *C. albicans* from the Candida genus were taken as test cultures for their antifungal and antibacterial properties. MPA (meat-peptone agar) was used

Advanced Topics in Polymer Chemistry and Materials Science: Tamara Tatrishvili and Marc Jean Médard Abadie (Eds)

for the cultivation of bacteria, and Sabouraud nutrient medium was used for the cultivation of fungi. The experiments showed that no change occurred in any of the obtained compositions, either visually or microscopically. It was determined that the obtained polymer composite materials have both bactericidal and fungicidal properties.

4.1 INTRODUCTION

Polymer materials used in natural climatic conditions are exposed to the influence of various external factors, including microbiological aggression. In order to protect polymer materials and products made from them from microbiological aggression, the development of technology for the preparation of antimicrobial polymer materials (AMP) is being carried out in several directions, one of which is the inclusion of natural and synthetic antimicrobial additives in their composition during the processing of polymer materials [1].

A number of methods have been proposed for obtaining antimicrobial composite materials of widely used polymers such as polyethylene, polystyrene, and acrylonitrile-butadiene-styrene co-polymers. However, the most widely used metallic silver powder and its salts are distinguished by their high antimicrobial properties and non-toxicity. Currently, extensive research is being conducted in the direction of obtaining new antimicrobial polymer materials from silver and silver-based preparations, and there is extensive information in the literature in this field [2–4]. One of the common missing aspects in almost all of the known works is the technological difficulties encountered in mixing silver nanoparticles with any amount of polymer in the preparation of polymer compositions, and the other is the decrease in efficiency due to the washing away of nanoparticles from the surface during the operation of such polymers [5–7].

When the oligomers obtained from the polycondensation of guanidine hydrochloride salts with diamines are added to polymers such as polyethylene, polypropylene, and polystyrene, composites with high antibacterial properties are obtained [8]. It should be noted that as a result of the dissolution of antibacterial oligomeric additives included in the composition of the products, their antimicrobial properties weaken or completely disappear after a certain period of time. Another promising direction in obtaining antimicrobial materials based on antibacterial additives is the use of water-insoluble antimicrobial polymers for this purpose. Since these additives have a longer shelf life in the product, the quality and exploitation life of the antimicrobial product is longer. However, the poisonous and toxic nature of these polymers leads to a decrease in their application areas.

When acrylic and methacrylic esters of salicylic acid are used as antimicrobial additives, these additives are gradually hydrolyzed under the influence of water vapor in the environment. A layer of salicylic acid is formed on the surface of the polymer, and the surface of the product becomes sterile for a certain period of time. However, in this case, the formed salicylic acid is gradually washed away from the surface of the product, which has a negative effect on its antibacterial properties. At the same time, it should be noted that in order to use homo- and co-polymers obtained based on methacryloylsalicylic acid as an antimicrobial additive, the synthesis technology of methacryloyl salicylate has been improved, the technology for obtaining homo- and co-polymers has been developed, and PE-based bactericidal and fungicidal composition materials have been obtained from these co-polymers. [9–11].

The presented article is devoted to the study of the properties of ABS-based antibacterial composition materials using the co-polymer formed by the allyl ester of salicylic acid with methyl methacrylate as an antibacterial and antifungal additive.

4.2 EXPERIMENTAL METHODS AND MATERIALS

4.2.1 *ADSORBENT*

Initial substances: a copolymer of salicylic acid allyl ether and methylmethactylate: T_m = 159°C, T_b = 211°C. It is partially soluble in alcohol, acetone, diethyl ether in all proportions, water and carbon disulfide. This copolymer was synthesized according to the following scheme [12]:

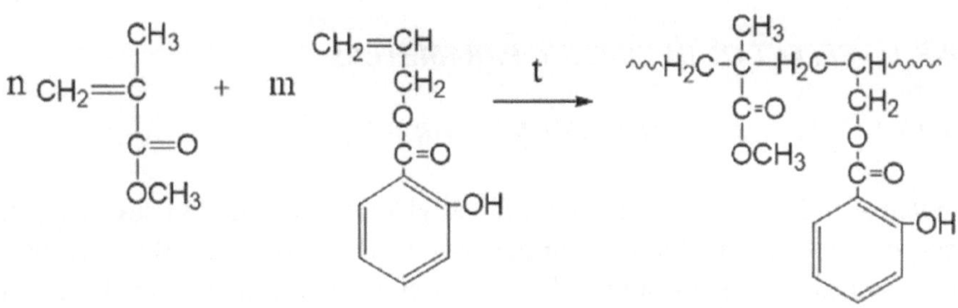

4.2.2 ADSORBATE

4.2.2.1 PREPARATION OF COMPOSITE MATERIALS AND DETERMINATION OF THEIR PHYSICAL-MECHANICAL PROPERTIES

The acrylonitrile-butadiene-styrene (ABS) composite material used contains 70% ABS co-polymer and 30% calcite mineral ($CaCO_3$), with a viscosity-flow temperature of 180°C. The initial ABS is soluble in ketone, complex ether, acetone, methyl ethyl ketone, 1,2-dichloroethane, and ethyl acetate. To prepare the antibacterial composite materials, ABS and a copolymer of salicylic acid allyl ether with methyl methacrylate (CPAESAMM) are mixed and passed through an extruder at 170°C. Subsequently, the mixture is pressed into a standard sheet form under a pressure of 15 MPa. The composition of the prepared materials is as follows: CM1 – ABS composite; CM2 – ABS composite + 4 wt% CPAESAMM; CM3 – ABS composite + 12 wt% CPAESAMM; and CM4 – ABS composite + 20 wt% CPAESAMM. Ultimate tensile stress (σ, MPa) and elongation at break (ε, %) were determined according to GOST 11262–80. The melt flow index of the composites was determined using the ASTM D1238–10 standard in a CEAST MF50 capillary rheometer (INSTRON, Italy) at 200°C under a load of 2.160 kg. Thermal analysis of the samples was conducted on a NETZSCH STA 449F3 device (Germany) using DTA and TGA methods, with results processed by NETZSCH Proteus software. The analysis was performed in a nitrogen environment from room temperature to 600°C. The antimicrobial properties of the composition samples were tested by the serial dilution method [13–16] at the Department of Medical Microbiology and Immunology, Azerbaijan Medical University.

4.3 RESULTS AND DISCUSSION

4.3.1 EFFECT OF OPERATING PARAMETERS

4.3.1.1 EFFECT OF AMOUNT OF ADSORBENT

When 4, 12, 20 wt% salicylic acid allyl ether and methyl methacrylate co-polymer are added to the composition of ABS-based CM, their physical-mechanical properties practically do not change or change very little. The results obtained from the studies are presented in Figure 4.1.

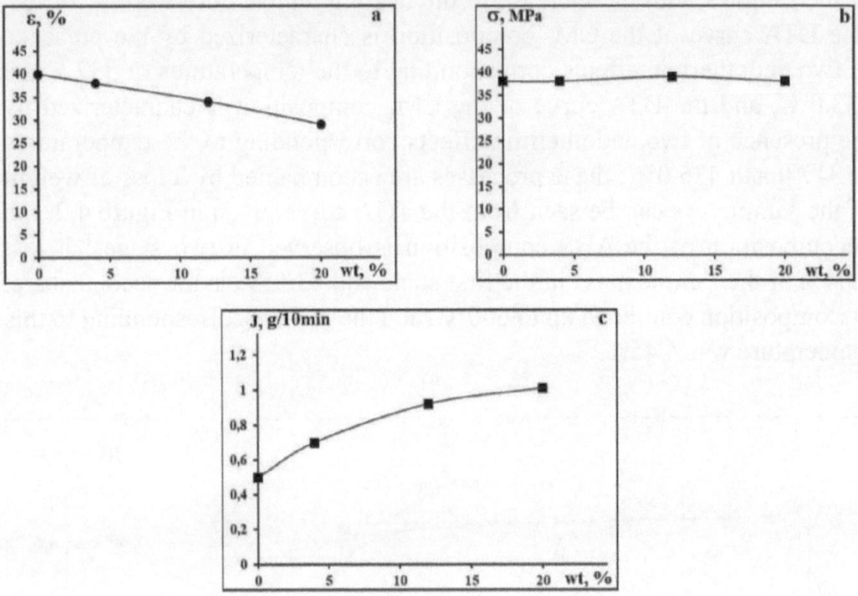

FIGURE 4.1 Dependences of elongation at break (a); ultimate tensile strength (b) melt flow index (c) on the amount of STAlEMMABP (wt%) in ABS CM+ CPAESAMM composite.

Physical-mechanical properties of polymer composite samples are characterized by ultimate tensile strength (σ, MPa) and elongation at break (ε,%), which are more sensitive to the structure and composition of composite materials [17]. Based on the data in Figure 4.1, it can be noted that there is no significant change in the value of the ultimate tensile strength of the polymer composition samples. In the elongation at break of the polymer composition samples, a downward tendency is noticeable with the increase of the amount of antibacterial additive. The MFI of the composite materials containing the copolymer of salicylic acid allyl ester with methyl methacrylate as an antibacterial additive is higher than the corresponding parameter of the original ABS, and the value of this parameter increases proportionally with the increase in the amount of the antibacterial additive.

4.3.1.2 THERMO-PHYSICAL PROPERTIES OF ABS-BASED ANTIBACTERIAL CM

Based on the results of the thermal analysis, a thermogram was constructed (Figures 4.2 and 4.3), on which the thermal effects and the change in the mass of the samples with the increase in the heating temperature were recorded. The DTA curve of the CM_1 composition is characterized by the presence of two endothermic effects corresponding to the temperatures of 332.6 and 433.0°C, and the DTA curve of the CM_3 composition is characterized by the presence of two endothermic effects corresponding to the temperatures of 317.0 and 435.0°C; these processes are accompanied by a loss of weight of the sample. As can be seen from the TGA curve given in Figure 4.2, the decomposition of the ABS composition is observed in two stages: it lost 8.04% of the sample mass in the first stage and 83.55% in the second stage. Decomposition continued up to 600°C, and the residue corresponding to this temperature was 4.45%.

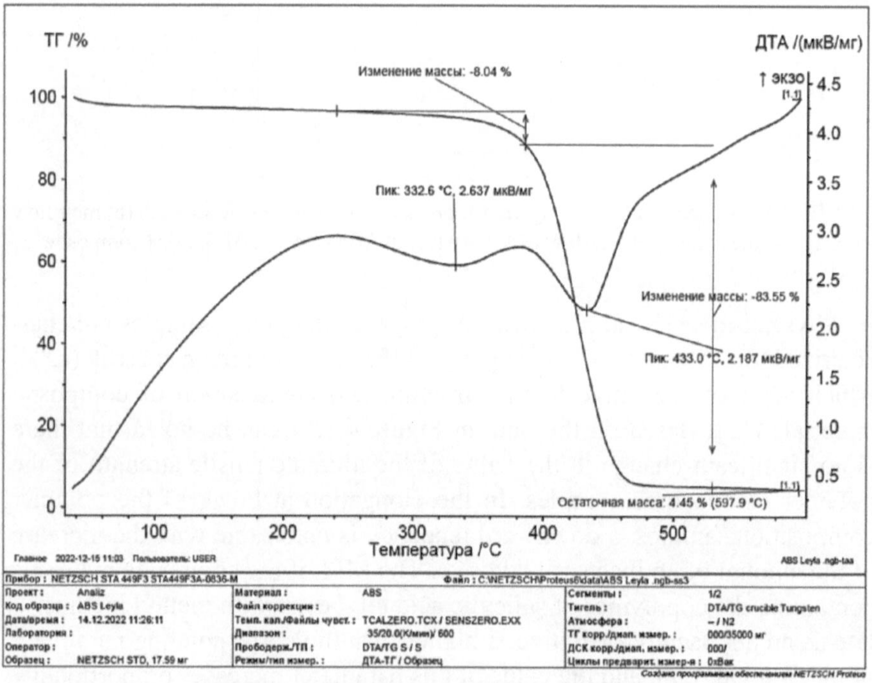

FIGURE 4.2 DTA and TGA curves of antibacterial CM_1.

Figure 4.3 shows the TGA curve of the ABS composite material containing 12 wt% CPAESAMM, which is decomposed in 2 stages: lost 4.28% of the sample mass in the first stage, and 87.86% in the second stage. Decomposition continued up to 600°C, the residue corresponding to this temperature was 6.33%.

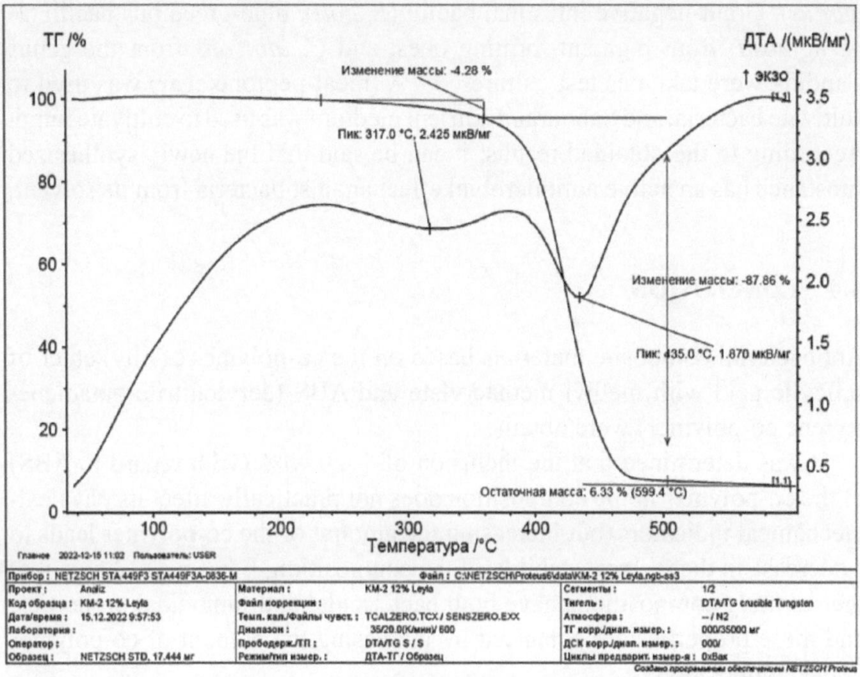

FIGURE 4.3 DTA and TGA curves of antibacterial CM_3.

The thermal stability of the composites was determined based on the T_{10}, T_{20}, T_{50} temperatures at which the samples lost 10, 20, 50 mass% [18] and the amount of the residue corresponding to the temperature of 600°C. The results obtained showed that the heat resistance of the composite material containing the CPAESAMM antibacterial additive is higher than that of the initial ABS composition; that is, the thermal stability of the composition with the antibacterial additive has increased.

4.3.1.3 EFFECT OF ANTIMICROBIAL PROPERTIES

The study of the antimicrobial properties of the synthesized substance was conducted using the serial dilution method. For this, the following dilutions of the 1% solution of the new substance – methyl-ethyl ketone (MEK) in sterile distilled water were performed (1:100, 1:200, 1:400, 1:800), (1, 2, 3, 4). The antimicrobial effect of this substance was studied in comparison with MEK, ethanol, nitrofuran, and rivanol. Golden staphylococci (*S.*

aureus), Gram-negative intestinal bacilli (*E. coli*), blue-green pus bacilli (*P. aeruginosa*) from pigment-forming ones, and *C. albicans* from the genus Candida were taken as test cultures. MPA (meat-peptone agar) was used to cultivate bacteria, and Sabouraud nutrient medium was used to cultivate fungi. According to the obtained results, it can be said that the newly synthesized substance has an active antimicrobial effect against bacteria from its solvent.

4.4 CONCLUSION

Antibacterial composite materials based on the co-polymer of allyl ether of salicylic acid with methyl methacrylate and ABS (acrylonitrile-butadiene-styrene co-polymer) were obtained.

It was determined that the inclusion of 4–20 wt% (with regard to ABS) of the co-polymer in the composition does not practically affect its physical-mechanical indicators, but increasing the amount of the co-polymer leads to a decrease in the thermal stability of the composition. It has been shown that the resulting compositions have both bactericidal and fungicidal properties, and these properties are enhanced by increasing the amount of co-polymer in the composition.

KEYWORDS

- **acrylonitrile-butadiene-styrene copolymer**
- **antibacterial composition material**
- **antimicrobial polymer materials**
- **copolymer of salicylic acid**
- **methyl-ethyl ketone**
- **NETZSCH proteus software**

REFERENCES

1. Rasulzade, N. Sh., Dostueva, V. M., Kahrmanov, N. T., & Arzumanova, N. B. (2020). Rheological features of antimicrobial composites based on polypropylene/magnesium hydroxide mixture and oligopropylene ether of salicylic acid. *Prospective Materials, 2*, 22–29.

2. Aldabaldetrecu, M., Tamayo, L., Alarcon, R., et al. (2018). Stability of antibacterial silver carboxylate complexes against *Staphylococcus epidermidis* and their cytotoxic effects. *Molecules, 23*(7), Article 1629.
3. Lagunin, A., Stepanchikova, A., Filimonov, D., & Poroikov, V. (2000). PASS: Prediction of activity spectra for biologically active substances. *Bioinformatics, 16*(8), 747–748.
4. Shtilman, M. I. (1998). Polymers in biologically active systems. *Soros Educational Journal, 5*, 48–53.
5. Azizov, A. G., Asadov, Z. G., & Akhmedova, G. A. (2009). *Macromonomers*. Baku: Elm.
6. Ito, K., & Kawaguchi, S. (1999). Poly(macromonomers): Homo- and copolymerization. In J. Roovers (Ed.), *Branched Polymers I*. Advances in Polymer Science (Vol. 142). Springer, Berlin, Heidelberg.
7. Plate, N. A., & Slivinskij, U. V. (2002). *Fundamentals of Chemistry and Technology of Monomers*. MAIK Nauka/Interperiodika.
8. Wei, D., Guan, Y., Ma, Q., Zhang, X., Zheng, T., Jiang, H., & Zheng, A. (2012). Condensation between guanidine hydrochloride and diamine/multi-amine and its influence on the structures and antibacterial activity of oligoguanidines. *e-Polymers, 12*(1), 1–10.
9. Rasulzadeh, N. Sh., & İbadov, E. A. (2016). The obtaining of polyethylene macromonomers and methacryloyl salicylates copolymers and research of their properties. *Proceedings of Young Scientists, 14*, 61–64.
10. Rasulzadeh, N. Sh., & İbadov, E. A. (2017). The synthesis and properties of acrylic and methacrylic ether of salicylic acid. *International Journal of Research Studies in Science, Engineering and Technology, 4*(3), 1–3.
11. Clivkin, A. I., Lapenko, V. D., & Bolgov, A. A. (2005). Synthesis of chitosan analogs. *Vestnik VGU. Series: Chemistry, Biology, Pharmacy, 2*, 205–208.
12. Abbasova, L. Sh., Rasulzade, N. Sh., Mammedov, B. A., Muradov, P. Z., & Bakhshaliyeva, K. F. (2022). Obtaining and studying antibacterial properties of the copolymer allyl ether of salicylic acid with methyl methacrylate. *PPOR, 23*(2), 13–20.
13. Balouiri, M., Sadiki, M., & Ibnsouda, S. K. (2016). Methods for *in vitro* evaluating antimicrobial activity: A review. *Journal of Pharmaceutical Analysis, 6*(2), 71–79.
14. Perez-Gavilan, A., de Castro, J. V., Arana, A., et al. (2021). Antibacterial activity testing methods for hydrophobic patterned surfaces. *Scientific Reports, 11*, Article 6675.
15. Jiang, L., Wang, F., Han, F., Prinyawiwatkul, W., No, H. K., & Ge, B. (2013). Evaluation of diffusion and dilution methods to determine the antimicrobial activity of water-soluble chitosan derivatives. *Journal of Applied Microbiology, 114*, 956–963.
16. Berghaus, L. J., Giguère, S., & Guldbech, K. (2015). Comparison of Etest, disk diffusion, and broth macrodilution for *in vitro* susceptibility testing of *Rhodococcus equi*. *Journal of Clinical Microbiology, 53*, 314–318.
17. Kakhramanov, N., Allahverdiyeva, K., Gahramanli, Y., Mustafayeva, F., & Martynova, G. (2023). Physical–Mechanical properties of multifunctional thermoplastic elastomers based on polyolefins and styrene-butadiene elastomer. *Journal of Elastomers & Plastics, 55*(2), 279–302.
18. Mustafayeva, F. A., Kakhramanov, N. T., & Koseva, N. S. (2020). Physicomechanical properties of composites based on the mixture of polyethylenes and the system multifunctional modifier-flame retardant. *New Materials, Compounds and Applications, 4*(1), 28–37.

CHAPTER 5

Preparation and Optical Investigation of Red Luminescent Europium Complexes with Unsaturated Aromatic β-Diketones and Auxiliary Ligands

O. S. BEREZHNYTSKA,[1,2] I. A. SAVCHENKO,[3] O. O. ROHOVTSOV,[1] S. S. SMOLA,[4] N. V. RUSAKOVA,[4] and O. K. TRUNOVA[2]

[1]*V.I. Vernadsky Institute of General and Inorganic Chemistry, NAS of Ukraine Akad., Kyiv, Ukraine*

[2]*National Technical University, Ukraine "Igor Sikorsky Kyiv Polytechnic Institute," 37, Prosp. Peremohy, Kyiv, Ukraine*

[3]*National Taras Shevchenko University, Kyiv, Ukraine*

[4]*A.V. Bogatsky Physico-Chemical Institute NASU, Odessa, Ukraine*

ABSTRACT

A new complex, $Eu(mnfpd)_3 \cdot 2H_2O$, $Eu(mnfpd)_3 \cdot dipy$, $Eu(mnfpd)_3 \cdot Phen$, $Eu(mphpd)_3 \cdot 2H_2O$, and $Eu(mphpd)_3 \cdot Phen$, was synthesized incorporating 2-methyl-5-biphenylpentene-1–3,5-dione (mbphpd) and 2-methyl-5-naphthenpentene-1–3,5-dione (mnfpd) as the primary ligands, along with phenanthroline (Phen) and dipyridyl (Dipy) as the auxiliary ligands. Metal polymers and copolymers with styrene and methyl methacrylate based on the synthesized complexes $[Ln(mbphpd)_3]_n$, $[Ln(mbphpd)_3Phen]_n$, $[Ln(mbphpd)_3Dipy]_n$, $[Ln(mbphpd)_3MMA]_n$, and $[Ln(mbphpd)_3Styrene]_n$ were obtained by the method of radical homo- and copolymerization. The complexes were characterized by analytical and spectroscopic techniques, with special emphasis on photoluminescence properties. The contribution

Advanced Topics in Polymer Chemistry and Materials Science: Tamara Tatrishvili and Marc Jean Médard Abadie (Eds)

of each ligand to the photoluminescence of Eu(III) has been analyzed and discussed. The relationship between complex structure and the intensity of fluorescence emission was established.

5.1 INTRODUCTION

In recent decades, interest in lanthanide coordination compounds has been exponential. This is due to both fundamental and applied aspects. The first is related to the establishment of the relationship between composition, structure, properties, and structural features, and the second is related to the use of luminescent coordination compounds as precursors of electroluminescent materials in organic light-emitting diodes [1–4], liquid crystal displays [5], sensors [6], molecular thermometers [7, 8], and fluorescent labels in biology and analytical chemistry [9, 10].

Unfortunately, non-radiative energy losses in compounds known today lead to the deterioration of luminescent characteristics. For this reason, the question of targeted synthesis and research of lanthanide-containing complexes with β-diketones containing unsaturated substituents in chelate rings arises. The introduction of various substituents into the chelate fragment makes it possible to select optimal ligand systems, which in the future will enable the achievement of high values of the intensity of 4f-luminescence, both due to the high concentration of emitting centers and due to the minimization of emission quenching processes. The effective transfer of energy from organic ligands, which is observed in unsaturated β-diketonates of lanthanides, allows for a more detailed analysis of the influence of the composition of the complexes on the luminescent characteristics, the uniformity of the distribution of emitting centers, and the processes of concentration quenching. Thus, the development of methods for the synthesis of new coordination compounds of lanthanides with modified β-diketones, which contain unsaturated substituents in the α-positions of the chelate ring, as components for creating materials with high luminescent characteristics, is an urgent task of modern chemical science. It should be noted that the use of low-molecular-weight complexes has a number of disadvantages due to the aggregation or crystallization of the latter in the process of obtaining films. In this regard, it is of interest to create metal-polymer systems based on monomer complexes. There are several ways to solve this problem: intercalation of coordination compounds in polymer matrices, homopolymerization of metal complexes containing terminal unsaturated fragments, and copolymerization with industrial monomers.

The introduction of emission materials based on lanthanide complexes and polymers increases their chemical and thermal stability and also optimizes the process of obtaining films of different thicknesses depending on the composition of the polymer system and the solvent.

Among the coordination compounds of lanthanides as precursors of luminescent materials, the first positions are occupied by β-diketones because they form chemically and thermodynamically stable complexes with lanthanide ions. Classical β-diketonate complexes (acetylacetonates, benzoyl acetylacetonates, dipivaloylmethanates, fluorine-substituted compounds) are already quite well studied, so there is a need for the synthesis of new ligand systems. Thus, in works [11–13], the coordination chemical and luminescent properties of carbacylamidophosphates and sulfonylamidophosphates, which are structural analogs of β-diketones with P,N-heterosubstituted atoms and have certain advantages, were synthesized and investigated. In particular, the replacement of carbon atoms with a nitrogen atom excludes high-frequency C–H oscillations, and the replacement of one of the carbonyl groups with a phosphoryl group in the ligands leads to a decrease in the vibrational energy of the fragments coordinated to the Ln^{3+} ion, which significantly reduces the multiphonon quenching of lanthanide luminescence in coordination compounds [14, 15].

The compounds based on europium and terbium are the most studied to date, which is due to significantly higher luminescence quantum yields compared to similar samarium compounds [16, 17].

Earlier studies of β-diketonates of lanthanides with unsaturated substituents in chelate rings showed that they can be used as precursors of organic layers in OLED devices (our articles). Detailed studies of polymer and copolymer systems based on them were conducted. Thus, in order to increase the intensity of luminescence, water molecules in the closest coordination environment of the lanthanide ion were replaced by a phenanthroline molecule, and copolymers with styrene and methyl methacrylate were obtained at different ratios of complex to industrial monomer (styrene, methyl methacrylate). A copolymer with styrene at a complex to styrene ratio of 5:95 was described earlier [18]. In connection with the manufacturability and higher thermal stability of copolymers compared to monomers and metal-polymer complexes, the synthesis and research of copolymers with the same metal chelate to monomer ratio is of interest.

According to the nature of luminescence, Ln(III) can be divided into two groups: those that exhibit luminescence in the visible region and in the near-IR range. The set of luminescence wavelengths of each of the lanthanide ions is preserved regardless of the choice of ligand. However, despite this, the choice of ligand is very important: it determines whether the complex will have effective luminescence, whether the energy transfer from the ligand to the central ion will be insignificant, whether the complex will be stable over time when used in real devices, whether it is safe for use with biological objects, and whether it will be soluble in the chosen solvent, which dictates the possible application, and so on. Direct photoexcitation of lanthanide ions is inefficient due to their weak ability to absorb light from their own transitions inside the 4f shell. This problem can be solved by transferring excitation energy to lanthanide ions from organic molecules (so-called "photoantennas"). The choice of ligand is very important because energy must not only be transferred from the central atom to the ligand but also transferred back. For high luminescence efficiency of the complex, phosphorescence and fluorescence of the ligand should be minimal. Therefore, when choosing ligands, it is necessary to take into account two factors: – the energy of the resonance level should be close to, but slightly lower than the energy of the triplet level, so that the probability of transition from the triplet level of the ligand to the resonance level of REE is high; – the probability of non-radiative deactivation of the resonance level should be small compared to the probability of a radiative transition. At the same time, the difference between the triplet level of the ligand and the singlet level of the lanthanide ion should be in the range of 2500–3500 cm^{-1}.

Recently, europium complexes have attracted more interest in organic light-emitting diodes for their saturated red emission [19, 20]. Also, several europium complexes have been applied as red emitters in electroluminescent devices [21–22].

Synthesis and investigation of the properties of polymer compounds, in particular lanthanum-containing ones, based on macromolecular metal chelates is one of the new and promising areas of coordination chemistry. Lanthanum-containing matrices encapsulated in polyvinyl alcohol (PVA) show better photostability compared to pure complexes, with improved photoluminescence efficiency [10, 16]. Complexes with excitation energy transfer deserve the greatest attention, as they allow us to realize sensitized fluorescence of lanthanides and increase the radiation intensity by several orders of magnitude. Such compounds include hybrid materials obtained by grafting lanthanide complexes onto the surface of aerosol or other silicon-containing compounds; compounds obtained by intercalation of lanthanide salts in polymer matrices of polyvinylcarbazole (PVK), polystyrene (PS), or polymethyl methacrylate (PMMA); copolymers obtained by copolymerization of complexes and polymers; and polymeric metal chelates obtained by polymerization of monomer complexes.

Copolymerization is of great practical importance, as it allows the purposeful synthesis of polymers with specified properties. By changing the composition and ratio of monomer units in the copolymer, it is possible to influence such characteristics of materials as heat resistance, elasticity, strength, transparency, electrophysical properties, resistance to solvents, etc. The paper presents the results of the synthesis of ligands, mono- and heteroligand metal complexes, and Eu(III) metallopolymers, as well as their copolymers with styrene and methyl methacrylate. The composition and structure of the compounds were determined, and their luminescent properties, morphology, and dispersion of nanomaterials obtained on their basis were investigated. The study established the influence of the chemical nature of ligands and additional ligands on the spectral-luminescence characteristics of lanthanide compounds.

5.2 EXPERIMENTAL METHODS AND MATERIALS

5.2.1 SYNTHESIS OF LIGANDS AND COMPLEXES

Ligands were synthesized according to previously described methods [23]. As starting compounds are chosen β-diketonate ligands (2,5-dimethylheptene-1–3,5-dione (dmhpd), (2,6-dimethyloctene-1–3,5-dione (dmokd), (2-methyl-5-phenylpentene-1–3,5-dione (mphpd) (2-methyl-5-biphenyl pentene-1–3,5-dione (mbphpd) 2-methyl-5-naphthenpentene-1–3,5-dione (mnfpd) are shown in Figure 5.1.

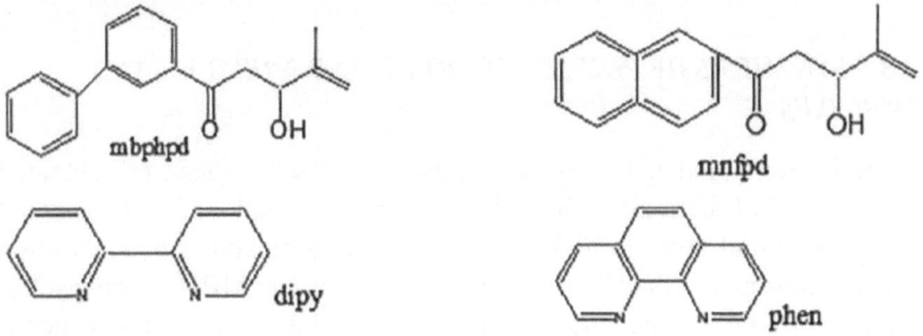

FIGURE 5.1 The structure of the ligands used in the work.

The synthesis of Eu(III) metal complexes based on 2-methyl-5-biphenyl-pentene-1–3,5-dione (mbphpd), 2-methyl-5-naphthenpentene-1–3,5-dione (mnfpd) was carried out by the interaction of aqueous solutions salts of the corresponding metals with an aqueous solution of the sodium salt of the ligand at a molar ratio of reagents of 1:3 and a slight excess of the ligand (pH 8–9) at room temperature.

$$Ln^{3+} + 3NaL \rightarrow LnL_3 + 3Na^+$$

$$Ln\,(III) = Eu$$

The resulting precipitates of the complexes were separated from the mother liquor by centrifugation, washed five times with water, and dried in a vacuum desiccator over anhydrous $CaCl_2$. The obtained complexes had a yellow color, which is due to the coloring of the ligand.

5.2.2 SYNTHESIS OF MIXED LIGAND COMPLEXES

The synthesis of the complex, in which phenanthroline (Phen) and dipyridyl (Dipy) were used as the additional ligands, was carried out in an alcoholic solution at a ratio of $Ln(\beta\text{-dik})_3$:Phen = 1:1. The solution was left for several hours until equilibrium was established and complexation was completed. The precipitate of the complex, which formed after 3–5 hours, was filtered off and washed with ethanol.

$$LnL^1_3{\cdot}nH_2O + L^2 \rightarrow LnL^1_3{\cdot}L^2 + nH_2O$$

$$Ln(III) = Eu(III).$$

$$L^1 = \text{mbphpd, mnfpd}\quad L^2 = \text{phen, dipy}$$

5.2.3 SYNTHESIS OF POLYMERIC MONO- AND MIXED LIGAND COMPLEXES

Polymeric mono- and mixed-ligand complexes of lanthanides were obtained by the method of thermally initiated radical polymerization. Polymerization was carried out at 80°C in a DMF solution with a monomer concentration (β-dicarbonyl metal complex) of 0.3 mol/dm^3 and an AIBN (azobisisobutyronitrile) initiator concentration of 0.003 mol/dm^3. The obtained metallopolymers were precipitated from solutions with methanol, reprecipitated into methanol and then dried at 20°C overnight.

Copolymers were synthesized by free-radical polymerization in DMF. The copolymerization was carried out in a 10 wt.% DMF solution of monomers with AIBN as a free radical initiator (1 wt.% with respect to the mass of monomers) at 80°C for more than 8 hours in a thermostat. The copolymerization was conducted at a ratio of metal complex to industrial monomer (styrene, methyl methacrylate) of 50:50 mass%, respectively. The polymerization mixture was poured into methanol. The solid precipitate was filtered, dissolved in DMF, reprecipitated into methanol, and then dried at 20°C overnight. Polymer powders had a yellowish color, and copolymers with methyl methacrylate had a glassy structure.

5.2.4 PHYSICOCHEMICAL METHODS

The complexes were studied using a number of physicochemical methods, including thermal analysis, infrared (IR), electronic absorption, diffuse reflectance, and luminescent spectroscopies.

The method of IR spectroscopy was used to establish the method of coordination of lanthanide ions to ligands. IR spectra were recorded on a Spectrum BX II FT–IR spectrophotometer (Rerkin-Elmer) in the range from 400 to 4000 cm^{-1} in a KBr tablet.

The method of electronic absorption spectroscopy (EAS) and diffuse reflection spectroscopy (DRS) was used to establish the symmetry and structure of the obtained complexes. EAS and DRS were recorded on a Shimadzu UV-3600 UV-VIS-IR spectrophotometer in the 250–1500 nm range. Spectra of excitation, molecular luminescence, and 4f luminescence of Ln(III) ions in complexes (solid samples and solutions (10–3 M in $CHCl_3$)) were recorded on a spectrofluorimeter Fluorolog FL 3–22, Horiba Jobin Yvon (He-lamp 450 W), equipped with an R928P PV array (Hamamatsu, Japan) for measurements both at room temperature and at 77 K, using an OS 11 light filter with their subsequent adjustment taking into account the radiation distribution of the xenon lamp and the PV array sensitivity.

Powder microphotographs were taken on a Hitachi H-800 scanning electron microscope (TEM) and a Tescan Mira 3 LMU scanning electron microscope (SEM).

The integrated luminescence intensity (I_{lum}) was measured based on the area under the strip contour. The value of the relative quantum yield φ4f (measurement error ± 20%) of the luminescence of lanthanide ions in the investigated complexes was calculated according to the method.

5.3 RESULTS AND DISCUSSION

5.3.1 STUDY OF THE COMPOSITION AND STRUCTURE OF β-DIKETONATE COMPLEXES OF LANTHANIDES

The monomeric mononuclear $EuL_3.2H_2O$; polymeric mononuclear $-[EuL_3]_n$, monomeric heteroligand complexes with phenanthroline and dipyridyl – EuL_3Phen, EuL_3.Dipy, copolymers $[EuL_3.MMA]_n$ $[EuL_3.Styrene]_n$ where L = 2-methyl-5-biphenylpentene-1–3,5-dione (mbphpd), 2-methyl-5-naphthen-pentene-1–3,5-dione (mnfpd) were synthesized.

The coordination mode of ligands to lanthanide ions was determined by IR spectroscopy.

The method of IR spectroscopy allows us to establish the method of coordination of functional groups of ligands to metal ions. β-Diketones are bidentate ligands, and complex formation with them occurs through the formation of a covalent bond between the enol oxygen atom and a coordination bond with the oxygen of the carbonyl group with a metal ion. The band of valence vibrations of the M-O bond is located in the region of 400 cm^{-1} [23]. The presence of these bands in the IR spectra of the complexes indicates both complex formation and the covalent nature of the M-O bond. Low intensity, splitting and overlap of bands *v* (M-O) and δhel cycle does not allow to make correlations between the frequency of the connection and its strength, so they can be used only for the purpose of identifying the given connection (Table 5.1).

In the IR spectra of the synthesized lanthanide complexes, the bands of valence vibrations ν(CO) and ν(CC) are in the range of 1500–1700 cm^{-1}, which indicates chelate coordination of the ligand to the central metal ion. In addition, in the region of 1620–1680 cm^{-1}, there is a medium-intensity band of valence vibrations of the C=C double bond, the high intensity of this band, or its splitting may be due to the presence of an additional valence vibration of the C-Ar bond, since in the molecules acetylacetonates, one methyl group is replaced by a biphenyl or naphthenic group.

TABLE 5.1 Characteristic Oscillation Frequencies (cm^{-1}) in the IR Spectra of Some β-Diketonate Complexes of Lanthanides

Complex	n (Ln-O) +dchat. ring (+ n (Ln-N)	nas(C-C)	ns(C-O)	ns(C=C)
Eu $(mnfpd)_3 \cdot 2H_2O$	420, 438, 460, 475, 540	1555	1590	1640
Eu $(mnfpd)_3$·dipy	415, 440, 450, 472, 556	1556	1600	1660
Eu $(mnfpd)_3$·phen	417, 420, 435, 450, 468, 555	1555	1600	1655

Valence vibrations of the conjugated system ν(C=C-C-O) appear in the IR spectra of lanthanide complexes in the form of bands at 1560–1600 cm^{-1}, which indicates the formation of the enol form of β-dicarbonyl compounds stabilized by a hydrogen bond. A set of absorption bands in the range of 1420–1520cm^{-1} characterizes the system of valence ν(C=O) and deformation δ(C-H) vibrations. Bands with a frequency of 1420 and 1450 cm^{-1} correspond to the sum ν(C-C=C-O) + δ(OH) + δ(C-H) + ν(C-CH_3) [24]. In the IR spectra of all compounds there are bands of 996–1068 cm^{-1}, which are characteristic of deformation vibrations of methyl groups. In the region 937–980 cm^{-1} there are valence vibrations of the bond (C-CH_3). In the region 2983–2987 cm^{-1} there are fluctuations of CH and hydrogen bonds [25, 26].

In the region 3060–3400 cm^{-1} for all synthesized complexes there is a broad band that corresponds to valence vibrations of adsorbed and coordinated water molecules. The position of this band does not allow for distinguishing the nature of the water bond in the complex because the formation of hydrogen bonds and the coordination of water molecules to the metal contribute to a decrease in the valence vibration frequency ν(OH). But in the IR spectra of β-dicarbonyl complexes, bands of deformation oscillation δ(MOH) are observed for all types of bound water at a frequency of 1100 cm^{-1}. The presence of water molecules coordinated by the metal cation in the composition of the complex is evidenced by the appearance of vibrational oscillations in the IR spectra, which are types of deformational oscillations. It was established that the pendulum oscillations are most sensitive to the nature of the bond between water molecules and the metal ion: the more covalent the M–O bond, the higher the frequency of the pendulum oscillations of coordinated water molecules; as a rule, the band of these oscillations is in the region of 600–900 cm^{-1}, fan oscillations are in the region of 500–600 cm^{-1}. In the region of higher frequencies, the bands 760–780 cm^{-1} correspond to π(C-H) vibrations, and 650–670 cm^{-1} correspond to deformation vibrations of the chelate ring formed by hydrogen bonds. In the IR spectra of mixed ligand complexes, in addition to the main bands characteristic of β-dicarbonyl fragments, there are bands characteristic of vibrations of donor ligand molecules and valence vibrations of the M-N bond. The band of valence vibration of the C-N bond, which for pure phenanthroline and dipyridyl is observed at 1560 cm^{-1}, overlaps with the bands ν_{as}(C⩦C) and ν_{as}(C⩦O), since during the coordination of the nitrogen-containing base there is a displacement of ν (C-N) in high-energy region of the spectrum. The oscillations of the C = N ring of free phenanthroline (1420–1430 cm^{-1}) and the CH bond are shifted in the complexes by 20–30 cm^{-1} due to the coordination of phenathroline or dipyridyl to the lanthanide, due to the redistribution of electron density in the molecule. However, the position of the bands characteristic of bidentately coordinated dicarbonyl molecules remains unchanged. In the region of 445–475 cm^{-1} there are bands of Ln-N valence vibration that overlap with Ln-O vibrations, which additionally indicates the formation of a chemical bond of the lanthanide ion with the donor molecules of phenathroline and dipyridyl.

In general, the shape and position of the bands in the IR spectra of all synthesized β-diketonates are similar and indicate the bidentate-cyclic coordination of β-dicarbonyl ligands to the metal cation. The general appearance of the spectrum and the position of the main characteristic bands of mixed-ligand complexes are similar to those of monoligand complexes, which indicates a similarity in structural complexes, namely, the replacement of water molecules in the nearest coordination environment of the lanthanide ion by the donor molecule of phenanthroline.

The hydrate composition of the synthesized complexes was determined by the DTA method. The general appearance of the thermograms and the temperature interval practically do not differ from the previously studied complexes $Ln(mphpd)_3 \cdot 2H_2O$ $Ln(mphpd)_3 \cdot Phen$ [21, 23, 24]. The mass loss curves for β-diketonate complexes containing unsaturated substituents in the α-position for a number of lanthanides practically do not differ. For monoligand complexes in the region of 120–125°C on the DTA curve, an endo effect with a mass loss of 3.8–4.1% is observed, which corresponds to the cleavage of two coordinated water molecules. A weak endo effect in the range of 190–200°C corresponds to the melting temperature of the complexes; the observed mass loss is insignificant (up to 2.5%). The exothermic effect at 240°C corresponds to the cleavage of one ligand molecule (Δm = 25–26%), which is typical for β-diketonate complexes that contain unsaturated substituents in chelate rings. The process of complete decomposition of the complex begins at 330–340°C. The total mass loss in the studied temperature range is 65–78%.

Comparing the data of IR spectroscopy and DTA, the schematic structure of the obtained complexes can be depicted as follows (Figure 5.2):

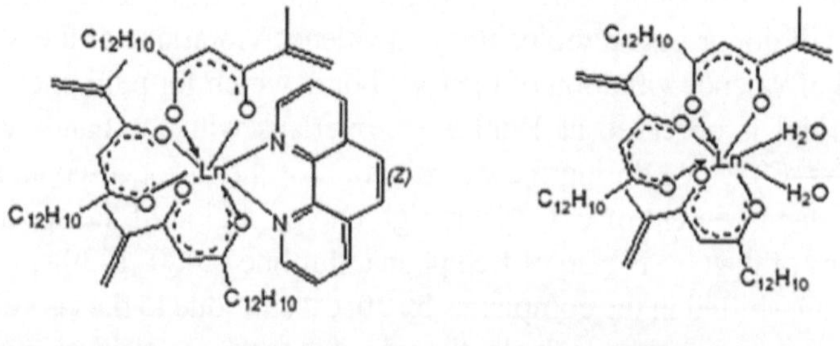

FIGURE 5.2 The method of coordination of ligands to the central ion of metals, Ln (III)= Eu.

The method of electron spectroscopy makes it possible to estimate the geometry, the coordination number of the central atom, and to establish the symmetry of the nearest coordination environment of the Ln(III) ion based on the detailed spectra. Taking into account the fact that in some cases, the composition of the complexes may change upon dissolution, the diffuse reflection spectra of the synthesized complexes were recorded. For lanthanides, the symmetry of the coordination polyhedron mainly depends on the ion-dipole electrostatic interaction. The difference between the energies in different geometric polyhedra is quite insignificant and is determined by the packing density. So, for an eight-coordinated complex, the most common polyhedra are a square antiprism and a dodecahedron.

The shape and position of the bands in the ESD of complexes in the area of super sensitivity, along with the analysis of literary sources [12–14], allow us to assume tetragonal symmetry for all synthesized monomeric and metallopolymeric compounds, with the coordination polyhedron being a square antiprism. The slightly different nature of displacement, splitting, and the ratio of spectral line intensities is due to the deformation of the coordination polyhedron. This is attributed to the different geometrical structures of the starting ligands and the steric difficulties arising from the coordination of an additional molecule of phenanthroline or dipyridyl.

Thus, changing the substituents in the ligands (biphenyl, naphthenic) does not lead to a significant change in the geometry of the coordination polyhedron and the lengths of bonds, not only in the chelate fragment but also in the Ln-O bond.

5.3.2 LUMINESCENT PROPERTIES OF MONOMERIC AND COPOLYMER COMPLEXES OF LANTHANIDES

One of the most common compounds tested as precursors of emissive materials is europium coordination compounds [11, 12, 20, 23]. Previous studies [27–29] have shown that europium β-diketonate complexes with unsaturated substituents exhibit high emission intensity. However, when obtaining materials, it is more appropriate to use polymer or copolymer systems, which is due to the aggregation of low molecular weight substances on the substrate. Therefore, europium polymeric coordination compounds were synthesized on the basis of mbphpd, which showed the best results for mphpd. The luminescence spectra of all compounds are similar, which indicates a similar structure of coordination polyhedra. Intense red luminescence is observed for both complexes. In the luminescence spectra, bands with a maximum in the region of 612–616 nm have the greatest intensity, which corresponds to an electro-dipole transition $^5D_0 \rightarrow ^7F_2$. In addition, bands of much lower intensity in the spectra correspond to the $^5D_0 \rightarrow ^7F_0$ (577–578nm) transition, the magnetic-dipole transition $^5D_0 \rightarrow ^7F_1$ (589–593) and the electric-dipole transitions $^5D_0 \rightarrow ^7F_3$ (648–653nm), $^5D_0 \rightarrow ^7F_4$, (691–700nm). As can be seen from Figure 5.3(a, b), the maximum intensity of the emission is characteristic of phenanthroline complexes of europium, both monomeric and metallopolymeric. The emission intensity of these complexes does not differ significantly. This fact can have two reasons: as a result of the polymerization process, the chain was broken and polymers with a low degree of polymerization were formed, or in a complex polymer structure, due to the presence of bulky ligand molecules and a rather bulky substituent in the diketone, the emitting centers are shielded. This phenomenon causes a decrease in emissions.

However, as can be seen from Figure 5.3b, in the case of europium metallopolymers, the maximum intensity of emission is characteristic of metal complexes containing aromatic substituents. This fact is due to the optimal

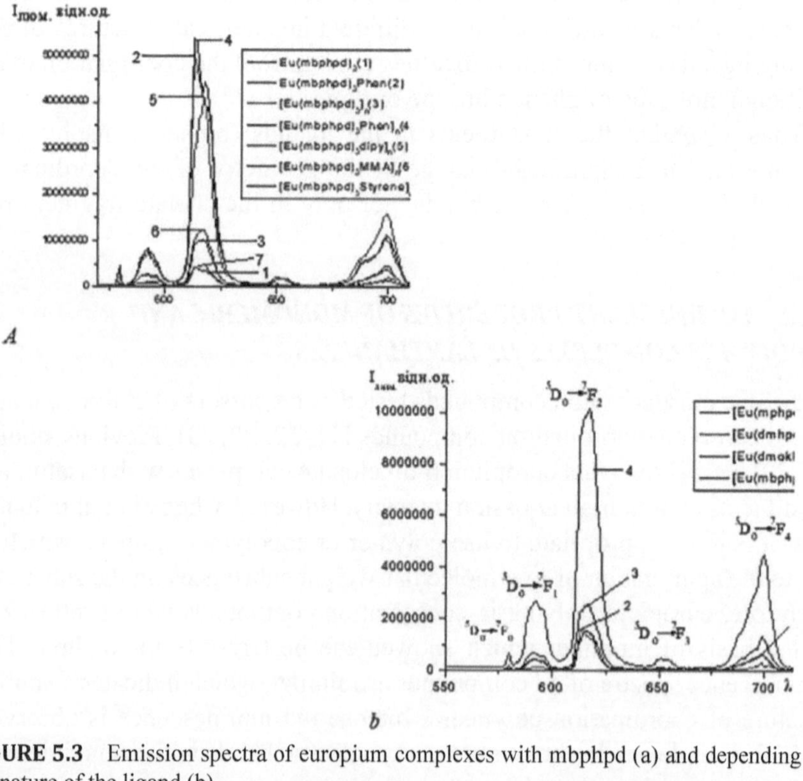

FIGURE 5.3 Emission spectra of europium complexes with mbphpd (a) and depending on the nature of the ligand (b).

difference in the energies of the triplet level of the ligand and the singlet level of europium. However, the introduction of additional bulky molecules to the metal complexes still has a negative effect on the emission characteristics.

The confirmation of the negative influence of the steric factor is a significant decrease in the emission intensity of the copolymer with styrene compared to the copolymer with methyl methacrylate. Therefore, bulky aromatic substituents do not act as antennas or luminescence amplifiers in all cases; quite often, they exhibit a quenching shielding effect. Taking into account the expediency of using polymeric systems as emission materials, Figure 5.4 (a, b) shows the spectra of the monoligand metallopolymer and copolymers. As can be seen from figure 5.4, the shape and position of the emission bands in all compounds are identical, which indicates a similar structure of the coordination polyhedron, since luminescence spectroscopy is quite sensitive to changes in symmetry.

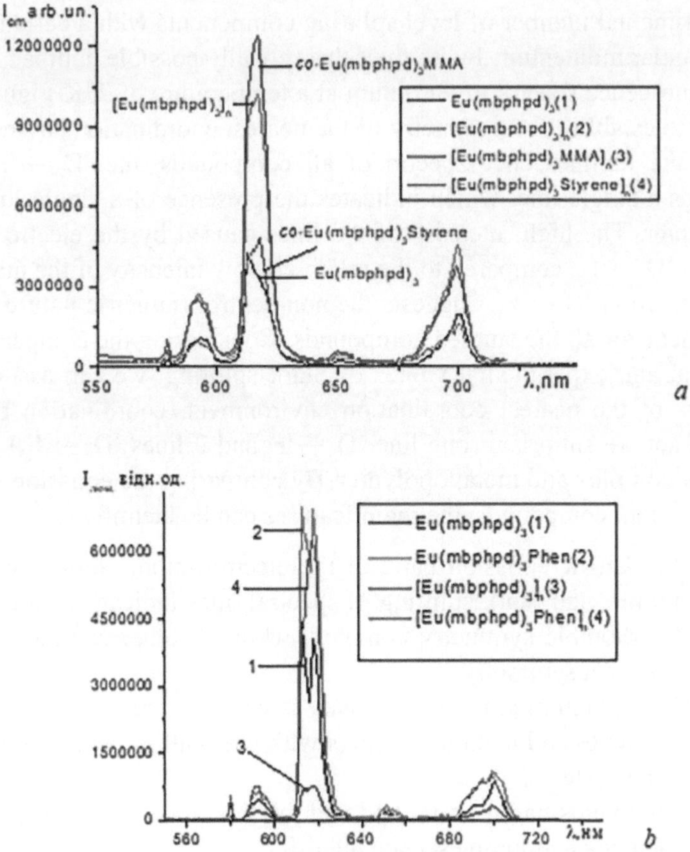

FIGURE 5.4 Emission spectra of europium compounds at 298K (a) and 77 (b).

The maximum emission intensity is characteristic of the copolymer with methyl methacrylate (three times higher than that of the monomer complex), which is practically commensurate with the metallopolymer $[Eu(mbphpd)]_n$ (2.5 times higher than that of the monomer). Since the emission intensity is only a qualitative characteristic, it is appropriate to compare the lifetime of the excited state for all synthesized compounds.

The closest environment has a weak influence on the position of the levels of the $4f^n$ electronic configuration of lanthanide ions, however, under the influence of the crystal field of the environment, these levels due to the Coulomb and spin-orbit interaction are split. Based on the analysis of the splitting values, the symmetry of the surroundings of the luminescence center can be determined. The type of symmetry for europium complexes can be qualitatively assessed without determining the parameters of the crystal field based on experimental data on Stark splitting values, but only by comparing the experimental number of level splitting components with a certain value of the angular momentum J with their theoretically possible number.

Luminescence spectra of europium at a temperature of 77K Figure 5.4b allow us to establish the symmetry of the nearest coordination environment [30]. In the luminescence spectra of all compounds, the $^5D_0 \rightarrow {}^7F_0$ band appears as a single line, which indicates the presence of a single luminescence center. The high intensity of the lines caused by the electro dipole transition $^5D_0 \rightarrow {}^7F_2$, compared to the relatively low intensity of the magnetic dipole transition $^5D_0 \rightarrow {}^7F_1$, suggests the non-centrosymmetric nature of this environment for all the studied compounds. Considering the comparison of theoretical and experimental values of Stark splitting, we can assume C_{4v} symmetry of the nearest coordination environment, coordination polyhedron is a square antiprism (one line $^5D_0 \rightarrow {}^7F_0$ and 2 lines $^5D_0 \rightarrow {}^7F_2$) for the monomer complex and metallopolymer. By comparing the emission spectra of all europium compounds, the main features can be identified:

1. The narrow emission band of the electron-dipole transition, short lifetime, and Stark splitting of spectral lines indicate the formation of a rhombic symmetry complex, while all other complexes have tetragonal symmetry.
2. The maximum emission intensity is characteristic of mixed ligand complexes and metallopolymers with phenanthroline for all studied ligand systems.
3. Compounds based on mphpd and mbphpd demonstrate the highest efficiency and luminescence intensity.

As can be seen from the presented results (Table 5.2), the emission intensity for all synthesized compounds is quite high, which indicates the potential for using europium β-diketonate complexes as precursors of luminescent materials.

TABLE 5.2 Excited State Lifetime of Europium Complexes

	Complex	τEu, μs	
		λex.=360nm	λex.=275 nm
1	$Ln(mphpd)_3$	575	680
2	$Ln(mbphpd)_3$	277	480
3	$Ln(mbphpd)_3dipy$	560	750
4	$Ln(mbphpd)_3Phen$	921	1180
5	$Ln(dmhpd)_3$	165	
6	$Ln(dmhpd)_3dipy$	231	
7	$Ln(dmhpd)_3Phen$	308	
8	$Ln(dmokd)_3$	230	
9	$Ln(dmokd)_3Phen$	837	1120
10	$[Ln(mphpd)_3]_n$	650	
11	$[Ln(mbphpd)_3]_n$	1170	
12	$[Ln(mbphpd)_3Phen]_n$	1580	
13	$[Ln(mbphpd)_3dipy]_n$	811	950
14	$[Ln(mbphpd)_3MMA]_n$	839	961
15	$[Ln(mbphpd)_3Styrene]_n$	676	681
16	$[Ln(dmhpd)_3]_n$	290	
17	$[Ln(dmhpd)_3Phen]_n$	755	
18	$[Ln(dmokd)_3]_n$	380	
19	$[Ln(dmokd)_3Phen]_n$	1020	1321

It is known that the maximum lifetime for europium complexes is 500 μs. As can be seen from table 5.2, the lifetime of almost all europium complexes exceeds this indicator. This fact can be explained based on the energy values of singlet and triplet levels of ligands and resonance levels of lanthanide ions.

The energy of the resonance level of europium is EsEu $=17350cm^{-1}$, and the energies of the triplet levels of the ligands from which the transition occurs are within $18900–21000cm^{-1}$. The optimal difference in energy between these levels of $2000–3500\ cm^{-1}$ is the key to a high lifetime of the excited state and, accordingly, intense emission. For example, E^T (dmhpd) $= 21000cm^{-1}$, i.e., there is practically no energy gap between the levels of the ligand and metals, and as a result, the lifetime of the excited state of these complexes is low. The introduction of phenathroline into the complex, which replaces water molecules in the close coordination environment of the lanthanide ion, eliminates the additional quenching effect of water molecules, so the lifetime is somewhat higher for mixed-ligand complexes.

5.3.3 DISPERSION AND MORPHOLOGY OF THE INVESTIGATED SYSTEMS

To analyze the surface morphology, SEM microphotographs of the synthesized complexes and copolymers were recorded (Figure 5.5). If the micrographs of monomer powders resemble ordinary salts, then in the micrographs of polymer systems, we can observe a certain morphology. At the same time, styrene-based copolymers are characterized by the formation of fairly large spherical particles; their structure is closer to a mesh-layered one. In general, regardless of the nature of the metal and the structural features of the ligand, the general appearance of the micrographs of monomer complexes and copolymers based on the same PMs is practically the same. It should be noted that copolymers are characterized by a more ordered structure. For europium (III) metallopolymers, the fluorescence of the ligand is proportional to the emission of the metal complex, which significantly increases the number of non-radiative losses and, as a result, reduces the intensity and efficiency of the process (Figure 5.5).

From the figures, the highest emission intensity and lifetime are characteristic of $[Eu(dmokd)_3]_n$, which indicates the influence of morphology on the emission properties.

In this way, the studies carried out indicate significant differences in morphology and dispersion for monomeric, polymeric, mixed-ligand, and copolymer systems. If monomeric complexes have a loose structure, as in the vast majority of amorphous compounds, then metallopolymers mainly form spherical-ellipsoidal particles, the size of which depends on the nature of the substituents in the ligand. Styrene-based copolymers are characterized by a similar morphology, although the particles are much larger in size, and their agglomeration is noticeable. In general, the main difference between polymer and copolymer systems is their dispersion; the particle size in copolymers is higher than in homopolymers, which is due to agglomeration processes. A somewhat different picture occurs in mixed-ligand systems. In the case of monomer complexes with phenanthroline and bipyridyl, the

FIGURE 5.5 Photomicrographs of powders $[Eu(dmokd)_3]_n$ (a), $[Eu(dmokd)_3Phen]_n$ (b), co- $Eu(dmhpd)_3$Styrene (c).

surface morphology resembles polymers, but the particle size is somewhat larger. Polymerization processes completely change the morphological structure, which acquires a layered heterogeneous structure from isolated spherical particles. This may be a consequence of the formation of polymers with a low molecular weight due to steric difficulties, and heterogeneous low molecular weight metallopolymers, which correlate quite well with the symmetry of the complexes. The symmetry of mixed-ligand metallopolymers is lower than that of monomeric mixed-ligand complexes.

5.4 CONCLUSION

In this work, Eu(III) coordination compounds based on 2-methyl-5-biphenyl-pentene-1–3,5-dione (mbphpd), 2-methyl-5-naphthenepentene-1–3,5-dione (mnfpd), mixed ligand complexes with phenanthroline and bipyridyl, as well as polymers and copolymers with styrene and methyl methacrylate based on them, were obtained.

The composition, method of coordination, and structure of coordination compounds and copolymers were established by IR, ESD, ESP, and thermal analysis methods. It is shown that all coordination compounds are characterized by a coordination number of 8, which corresponds to the coordination polyhedron of a square antiprism. During polymerization, the structure of the coordination polyhedron remains unchanged, and polymer and copolymer complexes are characterized by higher thermal stability than monomer complexes.

1. The luminescent properties of all synthesized compounds were studied, and a number of regularities were established.
2. The luminescence intensity of metallopolymers is higher for almost all complexes than for monomeric complexes. The most significant increase in emission from monomer to polymer is observed for europium (III) complexes with aromatic substituents.
3. The luminescence intensity of monomeric adducts with phenanthroline and dipyridyl for all synthesized mixed-ligand complexes is higher than for monoligand complexes. This phenomenon is due to the displacement of water molecules in the closest coordination environment of the lanthanide ion by the phenanthroline molecule and the antenna effect characteristic of phenanthroline. Moreover, for ions emitting in the visible region of the spectrum, the intensity of luminescence increases tenfold, while for ions emitting in the IR region of the spectrum, it is 5–10 times the maximum. A significant increase in emission is characteristic of complexes based on dmokd and dmhpd.
4. As a result of shielding by bulky phenanthroline molecules and substituents in chelate rings (phenyl, biphenyl, and naphthyl), emission decreases in mixed-ligand polymer systems.
5. The luminescence intensity of copolymers is determined by their structure, in the case of aliphatic substituents in the ligand, the highest emission efficiency is shown by copolymers with methyl methacrylate, in the case of aromatic ones – with styrene. Accordingly, in the series $Ln(\beta\text{-dik})_3]_n$ > $[Ln(\beta\text{-dik})_3]_n[Styrene]_m$ > $[Ln(\beta\text{-dik})_3]_n[MMA]_m$,

where β-dik = mphpd, mbphpd, the emission intensity decreases, and if β-dik = dmokd, dmhpd, then a decrease is observed in the series $Ln(\beta\text{-dik})_3]_n[MMA]_m > [Ln(\beta\text{-dik})_3]_n > [Ln(\beta\text{-dik})_3]_n[Styrene]_m$.

6. The luminescence intensity of metallopolymers $[Ln(mphpd)_3]_n$, $[Ln(mbphpd)_3]_n$ is significantly higher (several orders of magnitude) than that of monomers, and $[Ln(dmokd)_3]_n$, $[Ln(dmhpd)_3]_n$ exceeds the intensity of monomer complexes no more than 2 times. This phenomenon is caused by rapid chain breakage during polymerization and, accordingly, the formation of oligomeric disordered structures of low molecular weight, in addition, the formation of several oligomeric forms of complexes, which is confirmed by the luminescence spectra.
7. Particle dispersion and surface morphology were determined. Monomeric systems are characterized by the formation of loose amorphous structures. Metallopolymers are characterized by the formation of nanoscale systems with predominantly spherical particles, the size of which depends on the nature of the substituents in the ligand. The morphology of copolymer systems depends on the industrial monomer. In the case of styrene, a morphology similar to polymer systems is observed, but the formed nanoparticles are much larger, which is due to their agglomeration. The formation of layered structures is characteristic of methyl methacrylate copolymers. The influence of morphology on the luminescent properties of the synthesized compounds is shown. The more orderly and less agglomerated the system is, the higher the emission characteristics.

KEYWORDS

- **β-diketones**
- **complexes**
- **europium**
- **lanthanides**
- **metallopolymers**
- **monoligand complexes**
- **photoluminescence metal (co)polymer**

REFERENCES

1. Martins, J. P., Martín-Ramos, P., Coya, C., Álvarez, A. L., Pereira, L. C., Díaz, R., Martín-Gil, J., & Ramos Silva, M. (2014). Lanthanide tetrakis-β-diketonate dimers for solution-processed OLEDs. *Materials Chemistry and Physics, 147*, 1157–1164.
2. Lima, P. P., Paz, F. A. A., Brites, C. D. S., Quirino, W. G., Legnani, C., Costa e Silva, M., Ferreira, R. A. S., Junior, S. A., Malta, O. L., & Cremona, M. (2014). White OLED based on a temperature sensitive Eu3+/Tb3+ β-diketonate complex. *Organic Electronics, 15*, 798–808.
3. Anuj, D., Kapeesha, N., Anjli, H., Devender, S., Parvin, K., Sumit, K., & Rajender, S. M. (2023). Luminous lanthanide diketonates: Review on synthesis and optoelectronic characterizations. *Inorganica Chimica Acta, 550*, 121406.
4. Savchenko, I., Berezhnytska, A., Trunova, E., Rusakova, N., Fedorov, Ya., & Grozduyk, G. (2016). Polycomplexes based on unsaturated β-diketones and rare earth elements for optoelectronics. *Molecular Crystals and Liquid Crystals, 640*, 1–14.
5. Bakker, B. H., Goes, M., & Hoebe, N. (2000). Luminescent materials and devices: Lanthanide azatriphenylene complexes and electroluminescent charge transfer systems. *Coordination Chemistry Reviews, 208*(1), 3–16.
6. Pettinari, C., Drozdov, A., & Belousov, Yu. (2022). Coordination compounds of lanthanides as materials for luminescent turn-off sensors. In *Rare Earth Elements–Emerging Advances, Technology Utilization, and Resource Procurement*. https://doi.org/10.5772/intechopen.109011.
7. Brites, C. D. S., Millan, A., & Carlos, L. D. (2016). Lanthanides in luminescent thermometry. In *Handbook on the Physics and Chemistry of Rare Earths* (Vol. 49, pp. 339–427). Elsevier Science B.V.
8. Rocha, J., Brites, C. D. S., & Carlos, L. D. (2016). Lanthanide organic framework luminescent thermometers. *Chemistry–A European Journal, 22*, 14782–14795.
9. Charbonnière, L. J. (2011). Luminescent lanthanide labels. *Current Inorganic Chemistry, 1*, 2–16.
10. Comby, S., & Bünzli, J. C. G. (2007). Lanthanide near-infrared luminescence in molecular probes and devices. In K. A. Gschneidner Jr., J. C. G. Bünzli, & V. K. Pecharsky (Eds.), *Handbook on the Physics and Chemistry of Rare Earths* (Vol. 37, pp. 217–470). Elsevier B.V.
11. Pei, H., Huihui, W., Shenggui, L., Jianxin, S., Gang, W., & Menglian, G. (2009). Effect of different alkyl groups at the N-position on the luminescence of carbazole-based diketonate europium(III) complexes. *The Journal of Physical Chemistry A, 113*, 12885–12890.
12. Olyshevets, I., Kariaka, N., Znovjyak, K., Gerasimchuk, N., Lindeman, S., Smola, S., Seredyuk, M., & Sliva, T. Yu. (2020). Synthesis and characterization of anionic lanthanide(III) complexes with a bidentate sulfonylamidophosphate (SAPh) ligand. *Inorganic Chemistry, 59*, 76–85.
13. Olyshevets, I., Ovchynnikov, V., Kariaka, N., Dyakonenko, V., Shishkina, S., Sliva, T., Ostrowska, M., Jedyńczuk, A., Gumienna-Kontecka, E., & Amirkhanov, V. (2020). Lanthanide complexes based on new bis-chelating carbacylamidophosphate (CAPh) scorpionate-like ligand. *RSC Advances, 10*, 24808–24816.
14. Amirkhanov, V. M., Ovchynnikov, V. A., & Glowiak, T. (1997). Crystal and molecular structures of N,N'-diphenyl-N"-trichloroacetylphosphorictriamide and

N,N,-tetraethyl-N"-benzoylphosphorictriamide: The effect of various substituents on the structural parameters of the [C(O)N(H)P(O)] moiety. *Zeitschrift für Naturforschung B, 52*(11), 1331–1336.

15. Berezhnytska, O., Savchenko, I., Rohovtsov, O., Smola, S., Fedorov, Ya., & Trunova, O. (2021). Luminescent properties of complexes and polymers of Sm(III). *Optical Materials, 120*, Article 111492. https://doi.org/10.1016/j.optmat.2021.111492.
16. Saman, Q. M., Mudhafar, M. A., Md. Rahim, S., & Kasim, F. A. (2017). Plasmon-enhanced luminescence of samarium-doped sodium tellurite glasses embedded with gold nanoparticles: Judd-Ofelt parameter. *Journal of Luminescence, 190*, 468–475.
17. Savchenko, I., Berezhnytska, O., Trunova, O., Fedorov, Ya., Smola, S., & Rusakova, N. (2019). Monomer and metallopolymer compounds of Tb(III) as precursors for OLEDs. *Applied Nanoscience, 9*, 787–793.
18. Berezhnytska, O. S., Savchenko, I. O., Smola, S. S., & Ivakha, N. B. (2013). Synthesis and characterization of copolymers of lanthanide complexes with styrene. In *Fr.-Ukr. J. of Chem., Proc. of VII Intern. Conf. in Chem.* (pp. 94–99).
19. Al-Busaidi, I. J., Ilmi, R., Zhang, D. Y., Dutra, J. D. L., Oliveira, W. F., Al Rasbi, N. K., Zhou, L., Wong, W. Y., Raithby, P. R., & Khan, M. S. (2022). Synthesis and photophysical properties of ternary beta-diketonate europium(III) complexes incorporating bipyridine and its derivatives. *Dyes and Pigments, 197*, 109879.
20. Ilmi, R., Khan, M. S., Li, Z., Liang, Z., Wong, W., Marken, F., & Raithby, P. R. (2019). Utilization of ternary europium complex for organic electroluminescent devices and as a sensitizer to improve electroluminescence of red-emitting iridium complex. *Inorganic Chemistry, 58*(13), 8316–8331.
21. Ilmi, R., Sun, W., Dutra, J. D. L., Al-Rasbi, N. K., Zhou, L., Qian, P., Wong, W., Raithby, P. R., & Khan, M. S. (2020). Monochromatic red electroluminescence from a homodinuclear europium(III) complex of a β-diketone tethered by 2,2′-bipyrimidine. *Journal of Materials Chemistry C, 8*, 9816–9827.
22. Ilmi, R., Anjum, S., Haque, A., & Khan, M. S. (2019). A new brilliant red-emitting Eu(III) ternary complex and its transparent flexible and photostable poly(urethane) hybrid thin film for optoelectronic applications. *Journal of Photochemistry and Photobiology A: Chemistry, 383*, 111968.
23. Berezhnytska, O. S., Savchenko, I. O., Ivakha, N. B., Trunova, O. K., Smola, S. S., & Zheleznova, L. I. (2018). Near-infrared electroluminescence polymeric systems containing β-diketones and lanthanides as emitters for organic light-emitting diodes. *Molecular Crystals and Liquid Crystals, 670*, 48–60.
24. Nakamoto, K. (1991). *Infrared Spectroscopy of Inorganic and Coordination Compounds* (p. 536). Mir.
25. Savchenko, I., Berezhnytska, O. S., Fedorov, Ya., Smola, S., & Trunova, O. (2018). Luminescent properties of new polymer metal complexes based on β-diketones and REE. *Molecular Crystals and Liquid Crystals, 673*, 48–60.
26. Berezhnytska, O., Savchenko, I., Denysova, Z., Rusakova, N., Fedorov, Ya., Veligura, L., Rogovtsov, O., & Trunova, E. (2014). The new nanosized system on the basis of Eu(III) complexes as precursors for organic electroluminescent diodes. *Molecular Crystals and Liquid Crystals, 590*, 58–65.
27. Savchenko, I., Berezhnytska, A., Ivakha, N., & Trunova, E. (2015). New nanosized systems based on lanthanide diketonate complexes for OLEDs. In *Nanocomposites, Nanophotonics, Nanobiotechnology, and Applications* (pp. 85–94). Springer.

28. Devi, R., Bala, M., Khatkar, S. P., Taxak, V. B., & Boora, P. (2016). Investigations of luminescent behavior and intramolecular energy transfer mechanism of europium(III) complexes with fluorinated β-ketoester ligand. *Journal of Fluorine Chemistry, 181*, 36–44.
29. Savchenko, I. O., Berezhnytska, A. S., Fedorov, Ya. V., & Trunova, E. K. (2014). Copolymers of rare earth elements complexes with unsaturated β-diketones and N-vinylcarbazole for OLEDs. *Molecular Crystals and Liquid Crystals, 590*, 49–57.
30. Hooda, A., Singh, D., Nehra, K., Dalal, A., Kumar, S., Malik, R. S., Kumar, R., & Kumar, P. (2023). Photophysical characteristics of Eu(III) 1,3-diketonates with substituted 1,10-phenanthroline auxiliary moieties: Experimental and theoretical approach. *Journal of Molecular Structure, 1282*, 135200.

CHAPTER 6

Alternating Donor–Acceptor Copolymers as Through-Space Charge-Transfer Thermally Activated Delayed Fluorescence Emitters for Solution-Processable OLEDs: Synthesis, Theoretical Study, and Application

GEORGE K. BELOUSOV,[1,2] ALIAKSEI A. VAITUSIONAK,[1] JUOZAS VIDAS GRAZULEVICIUS,[3] and SERGEI V. KOSTJUK[1,2]

[1]*Research Institute for Physical Chemical Problems, Belarusian State University, Minsk, Belarus*

[2]*Department of Chemistry, Belarusian State University, Minsk, Belarus*

[3]*Department of Polymer Chemistry and Technology, Kaunas University of Technology, Kaunas, Lithuania*

ABSTRACT

A series of alternating donor-acceptor copolymers with M_n(SEC) = 7300-9900 g/mol and moderate polydispersity (Đ ≤ 1.9) were synthesized via RAFT-copolymerization of four donor styrene-type monomers with acceptor maleimide-type ones using 2-(dodecylthiocarbonothioylthio)-2-methylpropionic acid as a chain-transfer agent. The obtained alternating copolymers, being characterized by high thermal stability (T_{d5} = 402–417°C), exhibited aggregation-induced emission via through-space charge-transfer (TSCT) mechanism. A computational study was performed for an alternating copolymer with the most outstanding TSCT in order to reveal features of

Advanced Topics in Polymer Chemistry and Materials Science: Tamara Tatrishvili and Marc Jean Médard Abadie (Eds)

electron transitions responsible for thermally activated delayed fluorescence. A host-free, solution-processable TADF OLED was fabricated using the synthesized alternating copolymer as a green emitter. It showed good stability of electroluminescence spectra at different voltages and reached an external quantum efficiency (EQE) of 7.84%.

6.1 INTRODUCTION

Today, organic light-emitting diodes (OLEDs) are designated as one of the most promising electroluminescence devices, especially for such difficult tasks as creating flexible and large-scale displays [1–6]. According to Precedence Research analytics [7], the global OLED market size was estimated at USD 45.63 billion in 2022 and is expected to be worth around USD 214.8 billion by 2030, poised to grow at a compound annual growth rate of 21.37% over the projection period from 2022 to 2030 (Figure 6.1a). OLEDs are multilayer compositions, consisting of emissive and other auxiliary layers sandwiched between anode and cathode (Figure 6.1b) [8]. The main trend of the last decade in OLED fabrication is the transition from low molecular weight compounds to polymers [9]. This trend is associated with two reasons: first of all, the transition to high molecular weight compounds allows the use of solution (or "wet") fabrication methods, such as spin-coating, roll-to-roll coating, and inkjet printing, which are far cheaper and simpler than vacuum deposition, which is traditionally applied for low molecular weight compounds; secondly, polymers are characterized by higher morphological and thermal stability than usual organic compounds, which makes the service life of polymer-based devices longer [9–22].

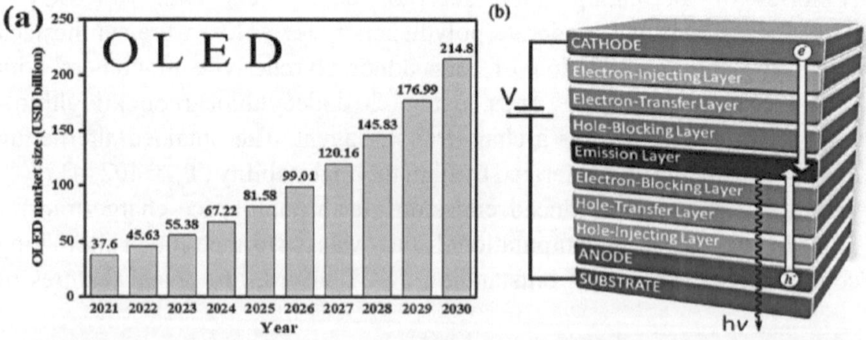

FIGURE 6.1 (a) The estimation and expectation of the global OLED market size (reproduced from reference 7). (b) Complex structure of OLED.

All OLEDs are based on a simple principle: under external voltage, electrons generated in the cathode and holes generated in the anode meet each other in the emissive layer, producing excitons, which then turn into light quanta (Figure 6.1b) [8, 19]. However, depending on the emitter material, the mechanism of electroluminescence may differ (Figure 6.2). First-generation OLEDs are based on fluorescence (Fl), allowing for the harvesting of only singlet excitons for luminescence [23]. Devices of the second and third generations are able to harvest both singlet and triplet exciton energies through phosphorescence (Ph) and thermally activated delayed fluorescence (TADF), respectively [24–26]. According to spin statistics, 75% of generated excitons are triplet, and only 25% are singlet; that's why first-generation OLEDs have a limitation in internal quantum efficiency (IQE ≤ 25%), while for second and third generations, IQE could reach 100%. It should be noted that phosphorescent emitters, in contrast to TADF ones, often contain heavy metals in their structure to ensure sufficient spin-orbit coupling, allowing the photophysical system to undergo intersystem crossing (ISC) [25, 27, 28]. This fact makes the fabrication of second-generation OLEDs expensive and less eco-friendly, while third-generation devices are more preferred.

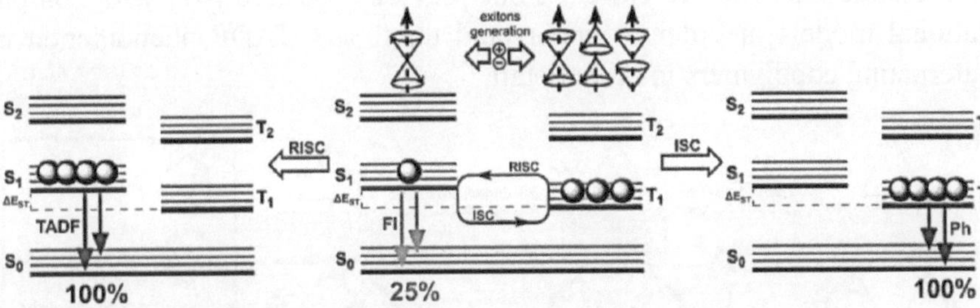

FIGURE 6.2 Possible electroluminescence mechanism: TADF – thermally activated delayed fluorescence; Fl – fluorescence; Ph – phosphorescence; ISC – intersystem crossing; RISC – reversible intersystem crossing.

One of the most progressive approaches to achieving effective TADF emission is the design of compounds with a donor-acceptor structure capable of through-space charge transfer (TSCT) [29–38]. Once again, in this case, polymers prove to be the most attractive materials, as they provide a huge amount of design variability. In recent years, a large number of articles devoted to the synthesis of TSCT TADF polymers have been published [9, 39–44], and there are two main strategies among them (Figure 6.3). The first one is the synthesis of a monomer with complete TADF properties and its further (co)polymerization (Figure 6.3a). Such a monomer must have in its structure donor and acceptor moieties, which are located geometrically close but chemically separated by some spacer [39, 40]. The second strategy implies the synthesis of a monomer with a side donor group as well as a monomer with a side acceptor group and their further copolymerization (Figure 6.3b) [39, 41–43]. Favorable orientation of donor and acceptor groups leading to their interaction ensures the TSCT TADF effect of the resulting copolymer. Moreover, aggregation-induced emission (AIE) is often observed for such copolymers [42, 44]. This fact is associated with the closer arrangement of donor and acceptor moieties relative to each other during aggregation, which makes favorable interaction more likely. That is why TSCT TADF emission of donor-acceptor copolymers can increase significantly during the transition from solution to solid polymer film. However, in both strategies, it is necessary to correctly select donor and acceptor fragments to achieve the most efficient TSCT TADF emission, taking into account their energetic characteristics.

In our present study, we propose an approach to the synthesis of through-space charge-transfer thermally activated delayed fluorescence alternating donor-acceptor copolymers, based on donor styrene-type monomers and benzophenone maleimide-type monomers, to fabricate non-doped solution-processible OLEDs. We enriched our previous research [41] with computational models in order to understand the TSCT TADF phenomenon of alternating copolymers in more detail.

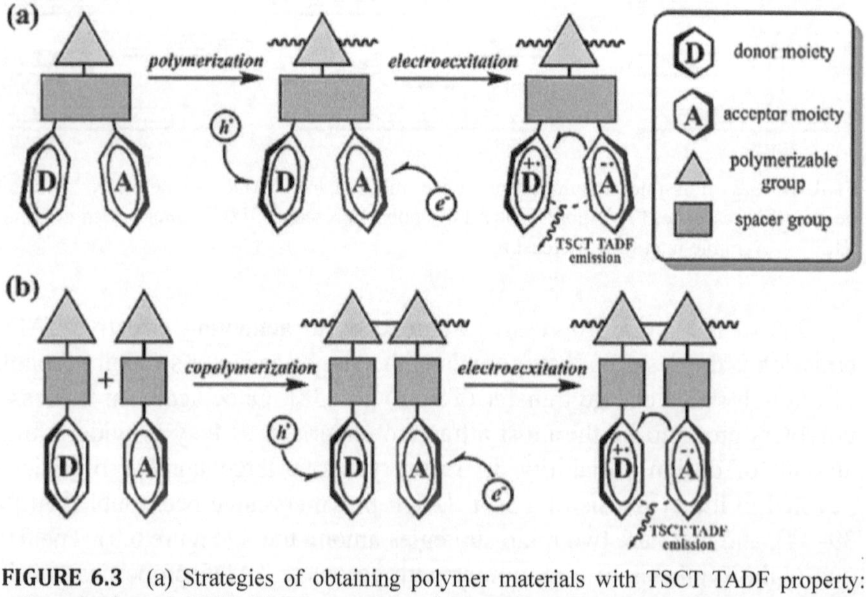

FIGURE 6.3 (a) Strategies of obtaining polymer materials with TSCT TADF property: polymerization of TSCT TADF monomer; (b) copolymerization of donor and acceptor monomers.

6.2 EXPERIMENTAL METHODS AND MATERIALS

6.2.1 *MATERIALS*

Cyclohexanone (Sigma-Aldrich, ≥ 99%) was dried overnight with CaH_2 and then distilled from CaH_2 under reduced pressure. THF (Sigma-Aldrich, ≥ 99%) was treated with KOH and distilled twice from Na under an inert atmosphere. Toluene (Sigma-Aldrich, ≥ 99%) was refluxed with Na and distilled twice from Na under an inert atmosphere. Acetic anhydride was refluxed with AcONa (Sigma-Aldrich, ≥ 99%) and then distilled from AcONa under reduced pressure 2-(Dodecylthiocarbonothioylthio)-2-methylpropionic acid (DMP) was synthesized according to the published procedure [45]. 2,2'-Azobisisobutyronitrile (AIBN) (Sigma-Aldrich, ≥ 98%) was recrystallized from ethanol. 3,6-Dimethoxy-9H-carbazole (Fluorochem, 97%), 9H-carbazole (Sigma-Aldrich, ≥ 95%), 10H-phenoxazine (Sigma-Aldrich, 97%), 9,10-dihydro-9,9-dimethylacridine (TCI, 97%), 2-aminobenzophenone (Sigma-Aldrich, 98%), 1-bromo-4-vinylbenzene (Sigma-Aldrich, 97%), 1-bromo-3-vinylbenzene (Sigma-Aldrich, 97%), tris(dibenzylideneacetone) dipalladium(0) ($Pd(dba)_3$) (Sigma-Aldrich, 97%), 2-dicyclohexylphosphino-2′,4′,6′-triisopropylbiphenyl (X-Phos) (Sigma-Aldrich, 98%), sodium tert-butoxide (t-BuONa) (Sigma-Aldrich, 97%), $CDCl_3$ (Euriso-top®), $CHCl_3$ (Sigma-Aldrich, 99.5%), methanol (Sigma-Aldrich, 99.9%), and ethanol (Sigma-Aldrich, 96%) were used as received.

6.2.2 *INSTRUMENTATION*

^1H (400 MHz) NMR spectra of the synthesized monomers were recorded in $CDCl_3$ at 25°C on a Bruker Avance III. Mass spectra were obtained by electron impact mass spectrometry (EI-MS) on a GSMS-QP2010 Plus. Size exclusion chromatography (SEC) was performed on an Ultimate 3000 Thermo Scientific apparatus with an Agilent PLgel 5 μm MIXED-C (300×7.5 mm) column and one precolumn (PLgel 5 μm guard 50×7.5 mm) thermostated at 30°C. Detection was achieved by a differential refractometer (thermostated at 35°C). Tetrahydrofuran (THF) was eluted at a flow rate of 1.0 ml min^{-1}. The calculation of molecular weights and polydispersity was carried out using polystyrene standards (Polymer Labs, Germany). Differential scanning calorimetry (DSC) and thermogravimetric analysis (TGA) measurements were carried out using a Netzsch STA (Simultaneous thermal analysis) 449 F3 device at a heating rate of 20°C min^{-1} under nitrogen flow. UV–vis and photoluminescence spectra of dilute solutions were recorded with the

PerkinElmer Lambda 25 UV-Vis Spectrometer and Edinburgh Instruments FLS980 Spectrometer, respectively. For these measurements, the dilute solutions of the investigated compounds were prepared by dissolving them in a spectral grade THF at 10^{-4} M concentration. Edinburgh Instruments FLS980 spectrometer and PicoQuant LDH-DC-375 laser (wavelength 374 nm) as the excitation source were used for recording photoluminescence (PL) decay curves. Photoluminescence quantum yields (PLQY) of THF solutions were recorded by using an integrating sphere (inner diameter of 120 mm).

6.2.3 DEVICE FABRICATION

Electroluminescent devices were fabricated using wet and vacuum techniques. Functional materials poly(2,3-dihydrothieno-1,4-dioxin)-poly(styrene sulfonate) (PEDOT:PSS), diphenyl-4-triphenylsilyl-phenylphosphine oxide (TSPO1), and 2,2,'2"-(1,3,5-benzinetriyl)-tris(1-phenyl-1H-benzimidazole) (TPBi) were purchased from Sigma Aldrich and used as received. Tris(4-formylphenyl)amine tri(N-4-vinylbenzyl-N-phenylhydrazone) (VM35) was synthesized according to the previously published procedure [46]. The device was fabricated using glass substrates with pre-patterned indium-tin-oxide (ITO) anodes (with a sheet resistance of 15 Ω/sq) from Ossila (with the possibility to deposit four OLEDs on a substrate). The substrate was cleaned in different solvents and in a UV ozone cleaner from Ossila. The PEDOT:PSS layer was spin-coated in the air at 3000 rpm for 3 minutes and dried at 120 degrees for 40 minutes. The VM35 layer was spin-coated at 3000 rpm in a glove box from a chlorobenzene solution (8 mg/mL) in an inert atmosphere. The VM35 layer was dried at 100°C for 30 minutes, followed by annealing at 200°C for 45 minutes to achieve cross-linking [46]. As a doping-free light-emitting layer, P4 was spin-coated at 3000 rpm from THF (4 mg/mL) and dried at 70°C for 30 minutes. The device structure was finalized by thermal deposition of TPBi/LiF/Al at vacuum ca. 2×10^{-6} mBar. Certificated photodiode PH100-Si-HA-D0 in combination with the PC-Based Power and Energy Monitor 11S-LINK (from STANDA), Keithley 2400C source meter, and Avantes AvaSpec-2048XL spectrometer were used to record electroluminescent characteristics.

6.2.4 SYNTHESIS OF MONOMERS

All of the monomers M1–5 were synthesized according to the procedures described earlier in our previous researches (Figure 6.4a) [42, 47].

3,6-Dimethoxy-9-(4-vinylphenyl)-9H-carbazole (M1): the yield of yellow crystals 41% (1.8 g); m.p. = 76°C. MS (EI, 70 V), m/z (%): 329 (M^+, 100). ^{1}H NMR(400 MHz, $CDCl_3$, δ, ppm): δ: 7.61 (d, *J* = 8.4 Hz, 2H), 7.56 (d, *J* = 2.4 Hz, 2H), 7.51 (d, *J* = 8.4 Hz, 2H), 7.35 (d, *J* = 8.9 Hz, 2H), 7.04 (dd, *J* = 8.9 Hz, *J* = 2.5 Hz, 2H), 6.82 (dd, *J* = 17.6 Hz, *J* = 10.9 Hz, 1H), 5.84 (d, *J* = 17.6 Hz, 1H), 5.35 (d, *J* = 11.0 Hz, 1H), 3.95 (s, 6H).

3,6-Dimethoxy-9-(3-vinylphenyl)-9H-carbazole (M2): the yield of brown solid 85% (3.7 g); m.p. = 62°C. MS (EI, 70 V), m/z (%): 329 (M^+, 100). ^{1}H NMR(400 MHz, $CDCl_3$, δ, ppm): δ: 7.58–7.51 (m, 4H), 7.44–7.33 (m, 2H), 7.33 (d, J = 8.8 Hz, 2H), 7.04 (dd, J = 11.8 Hz, J = 4.6 Hz 2H), 6.79 (dd, J = 17.6, 10.8 Hz, 1H), 5.81 (d, J = 17.6 Hz, 1H), 5.34 (d, J = 10.8 Hz, 1H), 3.95 (s, 6H).

10-(4-Vinylphenyl)-10H-phenoxazine (M3): the yield of white crystals was 80% (5.6 g); m.p. = 162°C. MS (EI, 70 V), m/z (%): 285 (M^+, 100). ^{1}H NMR (400 MHz, $CDCl_3$, δ, ppm): 5.36 (d, 1H, CH = CH_2), 5.83 (d, 1H, CH = CH_2), 5.95 (d, 2H, Ar), 6.55–6.66 (m, 4H, Ar), 6.69 (d, 2H, Ar), 6.79 (dd, 1H, CH = CH_2), 7.30 (d, 2H, Ar), 7.62 (d, 2H, Ar).

9,9-Dimethyl-10-(4-vinylphenyl)-9,10-dihydroacridine (M4): the yield of white crystals was 73% (5.3 g); m.p. = 130°C. MS (EI, 70 V), m/z (%): 296 ($[M - CH_3]^+$, 100). ^{1}H NMR (400 MHz, $CDCl_3$, δ, ppm): 1.69 (s, 6H, CH3), 5.37 (d, 1H, CH = CH_2), 5.86 (d, 1H, CH = CH_2), 6.30 (d, 2H, Ar), 6.84 (dd, 1H, CH = CH_2), 6.89–7.00 (m, 4H, Ar), 6.30 (d, 2H, Ar), 7.45 (d, 2H, Ar), 7.66 (d, 2H, Ar).

1-(2-Benzoylphenyl)pyrrole-2,5-dione (M5): 2.0 g of yellow crystals (71% yield); m.p. = 172°C. MS (EI, 70 V), m/z (%): 277 (M^+, 100). ^{1}H NMR (400 MHz, CDC13, δ, ppm): 6.65 (s, 2H, CH = CH), 7.38 (d, 1H, Ar), 7.42 (t, 2H, Ar), 7.49–7.57 (m, 2H, Ar), 7.64 (d, 1H, Ar), 7.68 (t, 2H, Ar), 7.73 (d, 2H, Ar).

6.2.5 COPOLYMERIZATION PROCEDURE

Copolymerization was carried out under a dry argon atmosphere in a Schlenk tube. Liquid reagents were transferred to the reactor via dry syringes against a continuous argon flow. In a typical polymerization experiment, 115 μL of a solution of DMP in cyclohexanone (30 mM) and 115 μL of a solution of AIBN in cyclohexanone (10 mM) were sequentially added to a crystalline mixture of M1–4 (115 mmol) and M5 (31.9 mg, 115 mmol). After three freeze-pump-thaw cycles, the reaction was started by immersing the Schlenk tube into an oil bath preheated to 70°C. After a predetermined time, the whole solution was withdrawn and poured into an excess of ethanol. The precipitated polymers were separated from the solution by centrifugation and then dried in a vacuum at 50°C. Before analysis, copolymers were reprecipitated twice from $CHCl_3$ into methanol. Monomer conversions were determined gravimetrically.

6.3 RESULTS AND DISCUSSIONS

6.3.1 SYNTHESIS OF MONOMERS AND THEIR ALTERNATING COPOLYMERIZATION

In the first step of the study, we synthesized four donor styrene-type monomers (M1–4) using Buchwald-Hartwig C-N coupling, and an acceptor maleimide-type monomer (M5) from the corresponding amine, utilizing maleic anhydride to form the maleimide ring. All of the monomers were obtained with tolerable yields (40–80%) by simple methods (Figure 6.4a). Their chemical structures were confirmed by ^{1}H NMR spectroscopy and mass spectrometry.

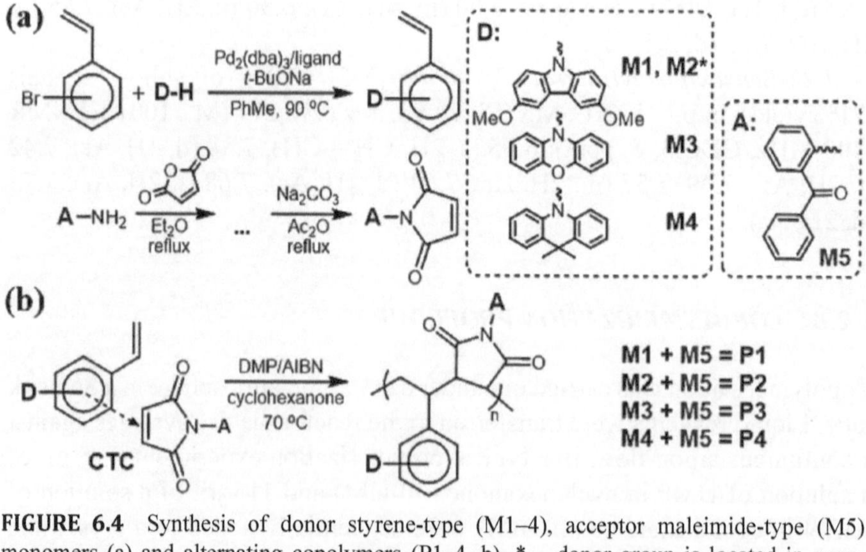

FIGURE 6.4 Synthesis of donor styrene-type (M1–4), acceptor maleimide-type (M5) monomers (a) and alternating copolymers (P1–4, b). * – donor group is located in *para*-position for M1, M3, M4 and in *meta*-position for M2. CTC – charge-transfer complex.

Further, monomers M1–4 were copolymerized with M5 using the DMP/AIBN initiating system (Figure 6.4b). RAFT polymerization with DMP as a chain-transfer agent was chosen because the method allows for the synthesis of alternating copolymers based on styrene-type and maleimide-type monomers in a controlled fashion [48–51]. The polymers (P1–4) were obtained with quantitative conversions, molecular weights in the range of 7300–9900 g/mol, and moderate polydispersity (Đ ≤ 1.9). The deviation in values between M_n(SEC) and M_n(th.) (Table 6.1) could be explained by the difference in hydrodynamic volumes of the obtained copolymers and polystyrene, which was used during SEC analysis as a standard [52]. As it is known, maleimide-type monomers are not able to be homopolymerized by a radical mechanism [53–55], but they can be copolymerized with some kinds of donor monomers like styrene derivatives or vinyl esters. According to the literature, this is associated with the generation of a charge-transfer complex (CTC) from donor and acceptor monomers, which can be polymerized as one unit, leading to the formation of an alternating copolymer. The high conversions of the synthesized copolymers (P1–4) allow us to assume that P1–4 had an alternating structure. The thermal stability of all the synthesized alternating copolymers was then estimated by TGA. According to the data presented in table 6.1, all copolymers showed high thermal stability: the values of the temperature at which a 5% loss of weight is observed (T_{d5}) were in the range of 402–417°C.

TABLE 6.1 RAFT Copolymerization of M1-M4 with M5 Using the DMP/AIBN Initiating System[a]

Copolymer	Conv.,%	Mn(th.)[b], g/mol	M_n(SEC), g/mol	Đ	T_{d5}[c], °C
P1	100	12.300	9.900	1.73	402
P2	87	10.700	7.500	1.91	417
P3	100	11.400	8.900	1.65	404
P4	100	12.000	7.300	1.85	417

[a] Polymerization conditions: T = 70°C; cyclohexanone medium; $[M]_0$ = 1M ($[M1–4]_0$ = $[M5]_0$ = 0.5 M), $[AIBN]_0$ = 5 mM, $[DMP]_0$ = 15 mM. [b] M_n(th.) was calculated according to the following equation: $M_n(th.) = [M]_0 \cdot (M(M1–4) + M(M5)) \cdot conv./\ 2[I]_0$, where $[I]_0 = [DMP]_0 + 2[AIBN]_0$. [c] 5% weight loss was determined by TGA in a nitrogen atmosphere at a heating rate of 20°C/min.

UV-vis absorption spectra of THF solutions of the monomers and alternating copolymers exhibit several bands in the region between 200 and 360 nm, which can be attributed to the lowest π-π* transition of the aromatic rings (Table 6.2). Interestingly, the absorption spectrum of the mixture of donor and acceptor monomers (M4 and M5) roughly resembles the absorption spectra of the single donor monomer (M5). A similar situation was observed for photoluminescent spectra of THF solutions (Figure 6.5a, b): M4+M5 mixture as well as pure M4 show peak with λ_{PL}(max) = 430 nm, while corresponding *alt*-copolymer P4 demonstrate a peaks series with λ_{PL}(max) = 560, 400, 300 nm. This observation indicates the principal difference in electron transitions in both cases.

Then, the photophysical properties of the obtained copolymers (P1–4) were investigated in the solid state (Table 6.2). The UV-absorption spectra of the films of the alternating copolymers were much broader than those of the solutions. The PL spectra of the films of the synthesized copolymers showed predominantly only one broad peak corresponding to TSCT emission, in strong contrast to the PL spectra of the THF solutions (Figure 6.5b, c). In addition, the peaks in the PL spectra of the films of P1–4 displayed a significant blue shift of approximately 24–80 nm in comparison to the PL spectra of the THF solutions (Table 6.2). The largest difference between the emission maxima of the film and of the solution was observed for P3 (~80 nm). Interestingly, the equimolar mixture M4+M5 gives the exciplex emission in the solid state, and the corresponding peak was characterized by a similar form, but was a bit wider and blue-shifted by approximately 20 nm in comparison to the P4 spectrum (Figure 6.5c).

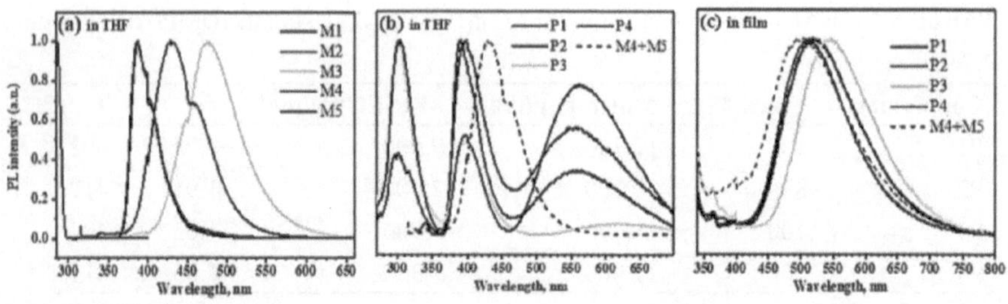

FIGURE 6.5 PL spectra of the synthesized monomers (M1–5) and alt-copolymers (P1–4) in THF solutions and films.

The photoluminescence quantum yields (PLQYs) of the synthesized copolymers in dilute solutions and in the films were estimated in air. PLQY values higher than those listed in table 6.2 are expected for the copolymer solutions in the absence of oxygen. This expectation is supported by the data obtained during the comparison of intensities of copolymer films' PL under air and vacuum. Emission intensity enhancement after deoxygenation was observed for each investigated alternating copolymer (Table 6.2). By overcoming triplet quenching by oxygen after the evacuation of air, triplets are harvested for emissive relaxations via reverse intersystem crossing (RISC), resulting in delayed fluorescence. PLQYs of the films of all of the studied copolymers were found to be higher than those of the corresponding solutions as a result of the AIE effect (Tables 6.2). For example, the solution and film of P3 displayed PLQYs of 0.1% and 2.84%, respectively.

The emission of THF solutions and films of alternating copolymers was further investigated by PL decay measurements (Table 6.2). The prompt fluorescence was observed in the range of 170–500 ns for the films of copolymers, which is ca. 2 orders of magnitude higher than τ_p observed for the solutions of the corresponding copolymers. The films of P1–4 demonstrated delayed fluorescence in the range between 1.37–2.81 μs, while film prepared from the equimolar mixture of M4 and M5 didn't show TADF, confirming the importance of copolymer formation for through-space charge-transfer to occur.

TABLE 6.2 Photophysical Properties of the Synthesized Monomers and *alt*-Copolymers

Systems		λ_{abs}[b] (nm)	λ_{PL}[b] (nm)	τ_p[c] (ns)	τ_d[c] (μs)	PLQE[d],%	Vac/Air[e]	E_g[f], eV
M1	THF	230, 305, 364	387	-	-	-	-	3.31
M2	THF	230, 303, 363	387	-	-	-	-	3.32
M3	THF	234, 318	476	-	-	-	-	3.51
M4	THF	245, 279	429, 459	-	-	-	-	3.95
M5	THF	246	286	-	-	-	-	4.25
P1	THF	234, 295, 355	399, 548	7	0.025	1.00	-	3.34
	film	355	518	172	1.40	2.31	1.14	3.26
P2	THF	239, 295, 354	390, 557	17	0.053	0.30	-	3.34
	film	356	520	193	1.37	2.06	1.13	3.27
P3	THF	227, 240, 307	397, 625	5	0.013	0.10	-	3.57
	film	306	547	225	2.13	2.84	1.14	3.42
P4	THF	242, 272	397, 561	16	0.061	0.86	-	4.03
	film	270	513	501	2.81	3.62	1.21	3.71
M4+M5	THF	252, 284	431, 459	-	-	-	-	3.91
	film	205, 238, 270	501	475	-	-	-	3.90

[a] Measurements were performed at 298 K in 10^{-4} M THF solutions or films obtained by spin-coating. [b] Wavelengths of PL or absorption maxima. [c] Lifetimes of prompt (τ_p) and delayed (τ_d) fluorescence were measured under air. [d] PLQY were measured by integrating sphere under air. [e] Comparison of intensities of copolymer films PL under vacuum and air. [f] Calculated from absorption spectra.

The film of P4 exhibiting the best (among studied alternating copolymers) photophysical characteristics (longest delayed fluorescence, highest PLQYs) was selected for additional investigations at different temperatures. Temperature dependencies of PL spectra and PL decay curves of TSCT TADF alt-copolymers were investigated in more detail in our previous study [42]. TADF of the film of P4 was evidenced by the higher emission intensity recorded at 160 K in comparison to that observed at 78 K. It should be noted that the emission intensity of the film of P4 decreased with the further increase of the temperature from 160 to 300 K. This last observation is quite surprising since, because of TADF, the PL intensity should increase with the increase of the temperature. However, the quenching processes of exciplex emission (TADF exciplex) are intensified with the increasing temperature. The intensification of exciplex emission quenching is mainly related to the intensified molecular movements (rotations/vibrations). In other words, because the distances between the exciplex-forming donor and acceptor moieties are not fixed (e.g., by chemical bonds like in the conventional TADF compounds), quenching processes of exciplex emission are intensified with the increase of temperature. Another possible reason could be related to trapped charges, as demonstrated in a recently published work [56].

To reveal a more detailed understanding of the photoluminescence of P4, a model of the alternating copolymer with a molecular weight of approximately 7000 g/mol was computed using the Merck molecular force field (MMFF94x). The obtained structure of P4 was used to find fragments with the closest orientation of donor and acceptor units relative to each other (Figure 6.6). Among the suitable fragments, a pair of neighboring units (F1) and a pair of units that are not neighbors along the chain (F2) were selected. The both orientations were additionally optimized applying density functional theory (DFT), after which their singlet and triplet excited states were computed using time-dependent DFT. The calculations show that TSCT is presented by "donor unit HOMO → acceptor unit LUMO" transition for both cases under consideration (F1, F2). It should be noted, that TSCT corresponds to S_0-S_1 transition only for F2 orientation, while the same transition for F1 is just local excitation, and TSCT comes from S_0-S_2 transition. Nevertheless, both structures (F1, F2) show relatively small ΔE_{ST} for TSCT: 0.46 eV for F1 and 0.32 eV for F2. Additionally, it should be noted that energy gap (E_g) as well as absorption peak maximum (λ_{abs}) values extracted from the calculations give close agreement with experimental ones, obtained from absorption spectra of P4 film (Table 6.3). All of the observations described above allow us to believe that orientations like F2 are more likely responsible for TSCT TADF properties in *alt*-copolymer P4.

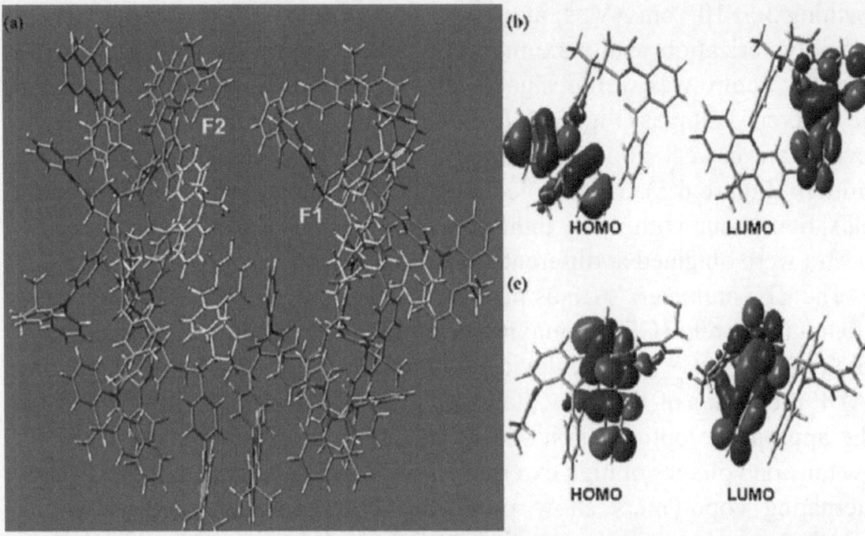

FIGURE 6.6 (a) The model of P4 with molecular weight ≈7000 g/mol, computed using MMFF94x force. The fragments, which were chosen for further TD-DFT calculations (F1, F2), are colored in yellow. The visualization of HOMO and LUMO of (b) F1 and (c) F2 structures, optimized using m062x/6–31+G(d) theory level.

TABLE 6.3 Energy Levels Parameters of P4, Computed Through Simulation F1 and F2 or Measured Directly (Experimental Values)

		ΔE_{ST} (eV)	E_g (eV)	λ_{abs}[a] (nm)
Simulation	F1	0.46	3.68	250
	F2	0.32	3.55	260
Experimental values for P4 film		-	3.71[b]	242, 272

[a] Wavelength of absorption maxima. [b] Calculated from the absorption spectrum.

Aiming to demonstrate applications of the developed alternating copolymers as light-emitting materials for solution-processable TADF OLEDs, P4 was selected as the green emitter. Their electroluminescent (EL) properties were investigated using the device structure ITO/PEDOT:PSS/VM-35 (40 nm)/host-free light-emitting layer (20 nm)/TSPO1 (10 nm)/TPBi (40 nm)/LiF (0.3 nm)/Al (120 nm). The hole-injecting (PEDOT:PSS), hole-transporting (VM-35), and light-emitting (P4) layers were step-by-step spin-coated. Hole-blocking (TSPO1), electron-transporting (TPBi), electron-injecting (LiF) layers, and cathode (Al) were deposited using vacuum technology. Hole-transporting layer of VM35 was selected due to its high hole-mobility reaching 3.5×10^{-3} $cm^2/(V\times s)$ at electric field of 6.4×10^5 V/cm and ability of self-polymerization with maximum polymerization rate at 209°C [46]. The device structure was well designed. The stability of EL spectra recorded at the different voltages (Figure 6.7) confirm this statement. EL spectra of the devices are in very good agreement with the PL spectra of corresponding emitters (Figure 6.5). Thus, hole-electron recombination as well as exciton relaxation occur within the light-emitting layers. Practically the same EL spectra were obtained at different applied voltages.

The EL parameters such as turn-on voltages (V_{on}), maximum brightness, maximum current (CE_{max}), maximum power (PE_{max}) and external quantum efficiencies (EQE_{max}) of the device are presented in Figure 6.7. Outstanding TADF properties of P4 allow the green device to achieve EQE_{max} = 7.84%. The appropriate optimization should be certainly considered for reaching low turn-on voltages or high external efficiencies. Nevertheless, at this stage, alternating copolymers show promising light-emitting, charge-injecting and charge-transporting properties useful for the fabrication of solution-processable electroluminescent devices.

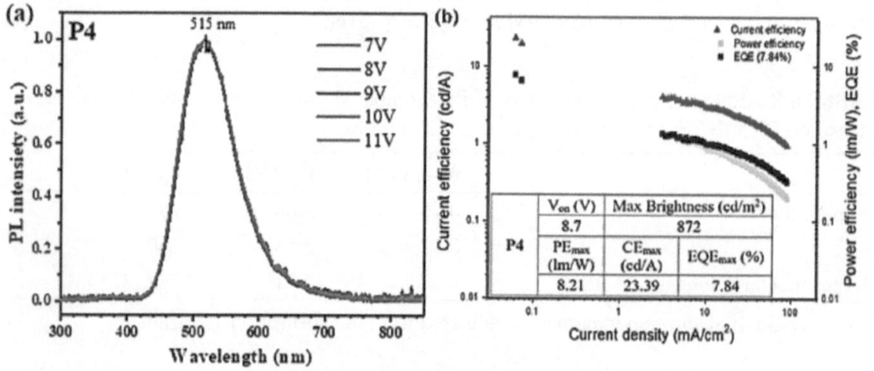

FIGURE 6.7 EL spectra of green OLED based on P4 recorded at the different voltages (a); its EL parameters (b).

6.4 CONCLUSION

In this work, we considered in more detail a new simple and cost-efficient strategy for the synthesis of polymeric TSCT TADF emitters, which was developed in our previous research. The strategy is based on RAFT-mediated alternating copolymerization of styrene-based donor monomers with acceptor maleimide-type monomers, which were obtained using one- or two-step synthesis. The obtained alternating copolymers (M_n(SEC) = 7,300–9,900 g/

mol) are characterized by high thermal stability and moderate polydispersity (Đ ≤ 1.9). All of them displayed aggregation-induced emission via through-space charge-transfer mechanism. The alternating copolymers in solid state demonstrate delayed fluorescence with lifetimes between 1.4 and 2.8 μs, while photoluminescence quantum yields of the solid films of the synthesized alternating copolymers were relatively low (2.1–3.6%) most probably because of non-emissive relaxation of triplets before reverse intersystem crossing. Computational study was performed for an alt-copolymer with the most outstanding TSCT in order to reveal features of electron transitions responsible for thermally activated delayed fluorescence. The calculations allowed us to understand what kind of monomer unit orientations in the polymer chain are suitable for favorable processing of TADF. A green host-free solution-processable TADF OLED was fabricated using the synthesized alternating copolymer with the best TADF properties (P4) as the emitter. It showed good stability of electroluminescence spectra at different voltages and reached suitable external quantum efficiency EQE = 7.84%.

KEYWORDS

- **alternating copolymers**
- **OLED**
- **quantum efficiency**
- **RAFT copolymerization**
- **thermally activated delayed fluorescence**
- **TSCT**
- **TADF polymers**

REFERENCES

1. Tang, C. W., & Vanslyke, S. A. (1987). Organic electroluminescent diodes. *Applied Physics Letters, 51*(11), 913–915.
2. Kulkarni, A. P., Tonzola, C. J., Babel, A., & Jenekhe, S. A. (2004). Electron transport materials for organic light-emitting diodes. *Chemistry of Materials, 16*(22), 4556–4573.
3. Hung, L. S., & Chen, C. H. (2002). Recent progress of molecular organic electroluminescent materials and devices. *Materials Science and Engineering: R: Reports, 39*(3), 143–222.

4. Meyer, J., Gorrn, P., & Riedl, T. (2013). Organic Light-Emitting Diodes (OLEDs): Transparent OLED Displays. In *Woodhead Publishing Limited* (pp. 512–547).
5. Burroughes, J. H., Bradley, D. D. C., Brown, A. R., Marks, R. N., Mackay, K., Friend, R. H., Burns, P. L., & Holmes, A. B. (1990). Light-emitting diodes based on conjugated polymers. *Nature, 347*, 539–541.
6. Jeong, H., Shin, H., Lee, J., Kim, B., Park, Y. I., Yook, K. S., An, B. K., & Park, J. (2015). Recent progress in the use of fluorescent and phosphorescent organic compounds for organic light-emitting diode lighting. *Journal of Photonics Energy, 5*, 057608.
7. Precedence Research. (2022). OLED market analysis. Retrieved from: https://www.precedenceresearch.com/oled-market (accessed on 1 October 2024).
8. Karzazi, Y. (2014). Organic light emitting diodes: Devices and applications. *Journal of Materials and Environmental Science, 5*(1), 1–12.
9. Wei, Q., & Ge, Z. (2019). Thermally activated delayed fluorescent polymers: Structures, properties, and applications in OLED devices. *Macromolecular Rapid Communications, 40*(1), 1–19.
10. Li, C., Harrison, A. K., Liu, Y., Zhao, Z., Dias, F. B., Zeng, C., Yan, S., Bryce, M. R., & Ren, Z. (2022). TADF dendronized polymer with vibrationally enhanced direct spin-flip between charge-transfer states for efficient non-doped solution-processed OLEDs. *Journal of Chemical Engineering, 435*, 134924.
11. Albrecht, K., Matsuoka, K., Fujita, K., & Yamamoto, K. (2015). Carbazole dendrimers as solution-processable thermally activated delayed fluorescence materials. *Angewandte Chemie International Edition, 54*(17), 5677–5682.
12. Wang, J., Liu, C., Jiang, C., Yao, C., Gu, M., & Wang, W. (2019). Solution-processed aggregation-induced delayed fluorescence (AIDF) emitters based on strong π-accepting triazine cores for highly efficient non-doped OLEDs with low efficiency roll-off. *Organic Electronics, 65*, 170–178.
13. Nikolaenko, A. E., Cass, M., Bourcet, F., Mohamad, D., & Roberts, M. (2015). Thermally activated delayed fluorescence in polymers: A new route toward highly efficient solution processable OLEDs. *Advanced Materials, 27*(41), 7236–7240.
14. Kim, H. J., Lee, C., Godumala, M., Choi, S., Park, S. Y., Cho, M. J., Park, S., & Choi, D. H. (2018). Solution-processed thermally activated delayed fluorescence organic light-emitting diodes using a new polymeric emitter containing non-conjugated cyclohexane units. *Polymer Chemistry, 9*(9), 1318–1326.
15. Lee, S. Y., Yasuda, T., Komiyama, H., Lee, J., & Adachi, C. (2016). Thermally activated delayed fluorescence polymers for efficient solution-processed organic light-emitting diodes. *Advanced Materials, 28*(22), 4019–4024.
16. Hu, Y., Cai, W., Ying, L., Chen, D., Yang, X., Jiang, X. F., Su, S., Huang, F., & Cao, Y. (2018). Novel efficient blue and bluish-green light-emitting polymers with delayed fluorescence. *Journal of Materials Chemistry C, 6*(10), 2690–2695.
17. Rao, J., Liu, X., Li, X., Yang, L., Zhao, L., Wang, S., Ding, J., & Wang, L. (2020). Bridging small molecules to conjugated polymers: Efficient thermally activated delayed fluorescence with a methyl-substituted phenylene linker. *American Ethnol., 132*, 1336–1342.
18. Luo, J., Xie, G., Gong, S., Chen, T., & Yang, C. (2016). Creating a thermally activated delayed fluorescence channel in a single polymer system to enhance exciton utilization efficiency for bluish-green electroluminescence. *Chemical Communications, 52*, 2292–2295.

19. Xie, G., Luo, J., Huang, M., Chen, T., Wu, K., Gong, S., & Yang, C. (2017). Inheriting the characteristics of TADF small molecule by side-chain engineering strategy to enable bluish-green polymers with high PLQYs up to 74% and external quantum efficiency over 16% in light-emitting diodes. *Advanced Materials, 29*(16), 1604223.
20. Ren, Z., Nobuyasu, R. S., Dias, F. B., Monkman, A. P., Yan, S., & Bryce, M. R. (2016). Pendant homopolymer and copolymers as solution-processable thermally activated delayed fluorescence materials for organic light-emitting diodes. *Macromolecules, 49*(15), 5452–5460.
21. Li, C., Nobuyasu, R. S., Wang, Y., Dias, F. B., Ren, Z., Bryce, M. R., & Yan, S. (2017). Solution-processable thermally activated delayed fluorescence white OLEDs based on dual-emission polymers with tunable emission colors and aggregation-enhanced emission properties. *Advanced Optical Materials, 5*(1), 1–9.
22. Li, C., Wang, Y., Sun, D., Li, H., Sun, X., Ma, D., Ren, Z., & Yan, S. (2018). Thermally activated delayed fluorescence pendant copolymers with electron- and hole-transporting spacers. *ACS Applied Materials & Interfaces, 10*(7), 5731–5739.
23. Xu, Z., Tang, B. Z., Wang, Y., & Ma, D. (2020). Recent advances in high performance blue organic light-emitting diodes based on fluorescence emitters. *Journal of Materials Chemistry C, 8*(7), 2614–2642.
24. Lee, J. H., Chen, C. H., Lee, P. H., Lin, H. Y., Leung, M., Chiu, T. L., & Lin, C. F. (2019). Blue organic light-emitting diodes: Current status, challenges, and future outlook. *Journal of Materials Chemistry C, 7*(17), 5874–5888.
25. Baldo, M. A., Adachi, C., & Forrest, S. R. (2000). Transient analysis of organic electrophosphorescence. II. Transient analysis of triplet-triplet annihilation. *Physical Review B, 62*(7), 10967–10977.
26. Wei, Q., Fei, N., Islam, A., Lei, T., Hong, L., Peng, R., Fan, X., Chen, L., Gao, P., & Ge, Z. (2018). Small-molecule emitters with high quantum efficiency: Mechanisms, structures, and applications in OLED devices. *Advanced Optical Materials, 6*(18), 1800512.
27. Ganzorig, C., & Fujihira, M. (2002). A possible mechanism for enhanced electrofluorescence emission through triplet-triplet annihilation in organic electroluminescent devices. *Applied Physics Letters, 81*(17), 3137–3139.
28. Leplingard, F., Borne, S., Martinelli, C., Leclère, C., Lopez, T., Guérin, J., Bayart, D., & Vanholsbeeck, F. (2003). FWM-assisted Raman laser for second-order Raman pumping. In *Optics InfoBase Conference Papers* (pp. 431–432). Scimago.
29. Kawasumi, K., Wu, T., Zhu, T., Chae, H. S., Van Voorhis, T., Baldo, M. A., & Swager, T. M. (2015). Thermally activated delayed fluorescence materials based on homoconjugation effect of donor–Acceptor triptycenes. *Journal of the American Chemical Society, 137*(36), 11908.
30. Tsujimoto, H., Ha, D. G., Markopoulos, G., Chae, H. S., Baldo, M. A., & Swager, T. M. (2017). Thermally activated delayed fluorescence and aggregation induced emission with through-space charge transfer. *Journal of the American Chemical Society, 139*(14), 4894.
31. Chen, X. L., Jia, J. H., Yu, R. M., Liao, J. Z., Yang, M. X., Lu, C. Z., & Combining Charge-Transfer Pathways to Achieve Unique Thermally Activated Delayed Fluorescence Emitters for High-Performance Solution-Processed, Non-doped Blue OLEDs. (2017). *Angewandte Chemie International Edition, 56*(48), 15006.

32. Spuling, E., Sharma, N., Samuel, I. D. W., Zysman-Colman, E., & Bräse, S. (2018). (Deep) blue through-space conjugated TADF emitters based on [2.2] paracyclophanes. *Chemical Communications, 54*(70), 9278.
33. Auffray, M., Kim, D. H., Kim, J. U., Bencheikh, F., Kreher, D., Zhang, Q., D'Aléo, A., Ribierre, J. C., Mathevet, F., & Adachi, C. (2019). Dithia [3.3] paracyclophane core: A versatile platform for triplet state fine-tuning and through-space TADF emission. *Chemistry – A European Journal, 14*(6), 1921.
34. Cai, X., Qiao, Z., Li, M., Wu, X., He, Y., Jiang, X., Cao, Y., & Su, S. J. (2019). Purely organic crystals exhibit bright thermally activated delayed fluorescence. *Angewandte Chemie International Edition, 58*(40), 13522.
35. Lin, J. A., Li, S. W., Liu, Z. Y., Chen, D. G., Huang, C. Y., Wei, Y. C., Chen, Y. Y., Tsai, Z. H., Lo, C. Y., Hung, W. Y., Wong, K. T., & Chou, P. T. (2019). Bending-type electron donor–donor–acceptor triad: Dual excited-state charge-transfer coupled structural relaxation. *Chemistry of Materials, 31*(15), 5981.
36. Woon, K. L., Yi, C. L., Pan, K. C., Etherington, M. K., Wu, C. C., Wong, K. T., & Monkman, A. P. (2019). Intramolecular dimerization quenching of delayed emission in asymmetric D–D′–A TADF emitters. *The Journal of Physical Chemistry C, 123*(24), 12400.
37. Zhang, P., Zeng, J., Guo, J., Zhen, S., Xiao, B., Wang, Z., Zhao, Z., & Tang, B. Z. (2019). New aggregation-induced delayed fluorescence luminogens with through-space charge transfer for efficient non-doped OLEDs. *Frontiers in Chemistry, 7*, 199.
38. Li, K., Zhu, Y., Yao, B., Chen, Y., Deng, H., Zhang, Q., Zhan, H., Xie, Z., & Cheng, Y. (2020). Rotation-restricted thermally activated delayed fluorescence compounds for efficient solution-processed OLEDs with EQEs of up to 24.3% and small roll-off. *Chemical Communications, 56*(41), 5957.
39. Shao, S., & Wang, L. (2020). Through-space charge transfer polymers for solution-processed organic light-emitting diodes. *Aggregate, 1*(1), 45–56.
40. Li, Q., Hu, J., Lv, J., Wang, X., Shao, S., Wang, L., Jing, X., & Wang, F. (2020). Through-Space Charge-Transfer Polynorbornenes with Fixed and Controllable Spatial Alignment of Donor and Acceptor for High-Efficiency Blue Thermally Activated Delayed Fluorescence. *Angewandte Chemie, 132*(61), 20349–20357.
41. Shao, S., Hu, J., Wang, X., Wang, L., Jing, X., & Wang, F. (2017). Blue thermally activated delayed fluorescence polymers with nonconjugated backbone and through-space charge transfer effect. *Journal of the American Chemical Society, 139*(51), 17739–17742.
42. Belousov, G. K., Vaitusionak, A. A., Vasilenko, I. V., Ghasemi, M., Andruleviciene, V., Ivanchanka, A., Volyniuk, D., Kim, H., Grazulevicius, J. V., & Kostjuk, S. V. (2023). Through-space charge-transfer thermally activated delayed fluorescence alternating donor–acceptor copolymers for nondoped solution-processable OLEDs. *Macromolecules, 56*(6), 2686–2699.
43. Yu, M., Gao, Y., Ying, A., Li, L., Xie, G., Gong, S., Gao, X., Wang, T., & Yang, C. (2023). Alternating thermally activated delayed fluorescence copolymers featuring through-space charge transfer for efficient electroluminescence. *Macromolecules, 56*(13), 5381–5389.
44. Poisson, J., Tonge, C. M., Paisley, N. R., Sauvé, E. R., McMillan, H., Halldorson, S. V., & Hudson, Z. M. (2021). Exploring the scope of through-space charge-transfer thermally

activated delayed fluorescence in acrylic donor–acceptor copolymers. *Macromolecules, 54*(6), 2466–2476.

45. Lai, J. T., Filla, D., & Shea, R. (2002). Functional polymers from novel carboxyl-terminated trithiocarbonates as highly efficient RAFT agents. *Macromolecules, 43*(1), 122–123.
46. Mimaite, V., Grazulevicius, J. V., Ostrauskaite, J., & Jankauskas, V. (2012). Synthesis and properties of triphenylamine-based hydrazones with reactive vinyl groups. *Dyes and Pigments, 95*(1), 47–52.
47. Bernard, R. S., Andruleviciene, V., Belousov, G. K., Vaitusionak, A. A., Tsiko, U., Volyniuk, D., Kostjuk, S. V., Kublickas, R. H., & Grazulevicius, J. V. (2022). Methoxy-substituted carbazole-based polymers obtained by RAFT polymerization for solution-processable organic light-emitting devices. *European Polymer Journal, 174*, 111323.
48. Perrier, S. (2017). 50th anniversary perspective: RAFT polymerization–A user guide. *Macromolecules, 50*(18), 7433–7447.
49. Boyer, C., Bulmus, V., Davis, T. P., Ladmiral, V., Liu, J., & Perrier, S. (2009). Bioapplications of RAFT polymerization. *Chemical Reviews, 109*(11), 5402–5436.
50. Harrisson, S., & Wooley, K. L. (2005). Shell-crosslinked nanostructures from amphiphilic AB and ABA block copolymers of styrene-alt-(maleic anhydride) and styrene: Polymerization, assembly and stabilization in one pot. *Chemical Communications, 26*, 3259.
51. Abiko, Y., Matsumura, A., Nakabayashi, K., & Mori, H. (2015). Synthesis of sulfur-containing alternating copolymers by RAFT copolymerization of phenyl vinyl sulfides. *Reactive & Functional Polymers, 93*, 170–177.
52. Vaitusionak, A. A., Vasilenko, I. V., Sych, G., Kashina, A. V., Simokaitiene, J., Grazulevicius, J. V., & Kostjuk, S. V. (2020). Atom-transfer radical homo- and copolymerization of carbazole-substituted styrene and perfluorostyrene. *European Polymer Journal, 134*, 109843.
53. Cubbon, R. C. P. (1965). The free radical and anionic polymerization of some N-substituted maleimides. *Polymer, 6*(4), 419–426.
54. Lutz, J. F., Schmidt, B. V. K. J., & Pfeifer, S. (2011). Tailored polymer microstructures prepared by atom transfer radical copolymerization of styrene and N-substituted maleimides. *Macromolecular Rapid Communications, 32*(2), 127–135.
55. Pfeifer, S., & Lutz, J. F. (2007). A facile procedure for controlling monomer sequence distribution in radical chain polymerizations. *Journal of the American Chemical Society, 129*(30), 9542–9543.
56. Chapran, M., Sahalianov, I., Karaush-Karmazin, N. N., Wiosna-Salyga, G., Glowacki, I., Luszczynska, B., Pander, P., & Baryshnikov, G. V. (2023). Electronic structure of exciplexes and the role of local triplet states on the efficiency of thermally activated delayed fluorescence. *ACS Applied Electronic Materials, 5*(3), 1489–1501.

CHAPTER 7

Catalyst Applications in the Petroleum Industry: An Overview

LAURA ESTEFANÍA GARZÓN ROJAS, ABDOLLAH ESMAEILI, AFONSO CESAR RODRIGUES NOGUEIRA, RÔMULO SIMÕES ANGÉLICA, MARYAM DEHGHANI, and CAMILO ANDRES GUERRERO

Universidade Federal do Pará (UFPA), Belem, Brazil

ABSTRACT

The catalysis process is involved in different phases of the processing of several commercial chemical products, as well as in the production of transportation fuels. Catalysts accelerate chemical reactions by facilitating the formation of products from reactant bonds, remaining unchanged themselves and available for subsequent reactions, creating a cyclic process. Petroleum refining heavily employs catalysis in alkylation, catalytic cracking, naphtha reforming, and steam reforming to produce fuels like gasoline, diesel, and kerosene. Catalytic oxidation, often using oxygen, produces significant chemicals, while large-scale reduction, such as hydrogenation, generates various chemical products including polymers like polyethylene and polyesters. Catalytic processes are vital in the oil, gas, and petrochemical industries of developed nations, with increasing importance due to their role in fuel production. Platinum catalysts activate through acid-base interactions, influencing reactions in processes like reforming, dehydrogenation, and hydrogenation. Catalytic cracking transforms heavy fractions into valuable materials, with a final product including light gases, gasoline, and diesel. Alkylation, a crucial refining process, involves adding an alkyl group to an organic compound using catalysts at low temperatures. Although the alkylation reaction can occur without a catalyst at high temperature and pressure,

Advanced Topics in Polymer Chemistry and Materials Science: Tamara Tatrishvili and Marc Jean Médard Abadie (Eds)

industrial alkylation is optimized at lower temperatures and with catalysts. While occupying a smaller volume compared to oil industry processes, catalytic processes yield diverse products, including polyethylene, propene, and cyclohexane. This paper focuses on the manifold applications of catalysts in the oil and gas sector, highlighting their vital role in various production processes and end products.

7.1 INTRODUCTION

Catalysts are substances that increase the speed of chemical reactions but are not consumed in the reaction. Their significance spans both industrial and biological realms, facilitating faster reactions for cost-effective product generation; these reactions exhibit lower energy barriers compared to uncatalyzed ones, resulting in higher reaction rates at identical temperatures. Using a catalyst is the most suitable approach to accelerate chemical reactions efficiently, as raising temperatures significantly increases reaction speed but is uneconomical due to energy consumption and can lead to material decomposition. Catalysts play a key role in addressing contemporary challenges related to the environment, including atmospheric and energy concerns. Their influence extends to vital domains such as energy production, synthetic processes, and environmental preservation. The discovery of new catalysts has a significant influence on the sectors, mainly due to the need for rapid chemical reactions. This urgency stems from the fact that manufacturers cannot afford to wait too long to obtain products that align with the current robust market demand [1, 2].

Catalysts can be inhibited, deactivated, or destroyed by secondary processes, even if they are not consumed by the reaction itself; catalysis thus works by providing an alternative reaction pathway for the reaction product [3]. Catalysis contributes to the manufacture of most important industrial chemicals and biochemical products by accelerating chemical reactions that allow the economic formation of pure materials; thus, the use of factory catalysts can be more economical, environmentally friendly, and sustainable [2]. By increasing the reaction rate, a lower alternative pathway activation energy is generated than the catalyst-measured feed pathway; these react with one or more reactants to form intermediate products, which subsequently give rise to the final reaction product in the catalyst regeneration process [4]. Even though the catalysts are regenerated when only small amounts are used to increase the reaction rate, they can be consumed in secondary processes.

Most of all commercially produced chemical products involve catalysts at some stage in the process of their manufacture; for example, petroleum refining makes intensive use of catalysis for alkylation, catalytic cracking, naphtha reforming, and steam reforming [5]. Catalysis is applied extensively, even to fossil fuel exhaust treatment, with approximately 440 oil refineries all over the world, through catalytic converters containing platinum and rhodium, which mitigate harmful exhaust byproducts. Large-scale chemical production frequently relies on catalytic oxidation with oxygen [4; 6].

A wide array of chemical products is generated through comprehensive reduction, often facilitated by hydrogenation processes [7]. Certain reactions proceed at notably slow rates, exemplified by copper's gradual oxidation forming surface deposits in ambient conditions, while others maintain an intermediate pace, such as iron's gradual rusting upon exposure to atmospheric oxygen [8]. Changes in the speed of reactions can be influenced by various factors, including temperature, concentration, the presence of reactant catalysts, contact surface, and the nature of catalyst raw materials. Catalysis plays a crucial role not just in expediting reactions but also in the formation of bulk polymers like ethylene and propylene derivatives, as well as polyesters and polyamides [9]. Additionally, catalytic methods prove essential for synthesizing fine chemicals, encompassing both industrial techniques and intricate processes that might be unfeasible on a larger scale.

7.2 LITERATURE REVIEW

Many substances have the capacity to influence other substances, a phenomenon distinct from their propensity to engage in chemical reactions. This influence results in the decomposition of these substances, leading to the creation of novel compounds, all without becoming part of the resulting compounds' composition. The mystery of catalysts persisted until the 19th century when Swedish chemist Jean Jacob Berzelius's breakthrough in 1835 highlighted the crucial role of reaction speed, after years of dedicated research on chemical reactions, emphasizing the need for observable and result-yielding laboratory reaction rates. This previously unknown force, present in both organic and mineral substances, is now recognized as catalytic force, responsible for material decomposition known as catalysis. When a lit match flame contacts sugar, it melts without igniting. However, introducing cigarette ash or potting soil as a catalyst triggers the sugar to burn, resulting in a brilliant blue flame accompanied by crackling. Remarkably, the ash or soil remains unaltered despite its role in catalyzing the reaction, underscoring the concept of catalysts facilitating reactions in their vicinity while remaining unchanged themselves.

Amid the late 18th-century surge in chemistry's growth, the investigation of external substance effects on chemical reactions gained traction. Since 1780, significant experiments highlighted foreign substances' influence on diverse reactions, such as acids in solutions, gas-phase reactions, and solid surface absorption reactions. For example, Fusinieri pioneered the connection between surface adsorption and faster reactions aided by solid catalysts. Additionally, Faraday observed the significance of raw material surface adsorption in solid-involved reactions, while Clement and Disermes identified the vital role of nitrogen oxides in sulfur dioxide oxidation via compound formation [11]. Numerous experiments explored catalyst behavior using materials relevant to today's oil industry, including potent acids, soils, and metals like platinum.

Berzelius explored acid-induced acceleration of starch-to-sugar conversion and faster gas reactions near platinum. In 1902, Wilhelm Ostwald defined catalysts as substances altering reaction speed while remaining unchanged, clarifying their unique features. Berzelius, renowned for structuring chemical knowledge, identified foundational phenomena and underscored antecedent factors in reaction acceleration. His 1835 research pioneered the concept of catalysis, suggesting external catalysts spur chemical activity without direct involvement [12]. Berzelius's introduction of the catalytic force concept, identifying a unique chemical-capable force, was later upheld by Ostwald, though its practical utility faced critique from Liebig and others. Stavale's comments prompted a more pragmatic approach, while over 60 years, Carbertolt and Wilhelmy investigated sugarcane and ester hydrolysis. Wilhelmy's pivotal recognition of the time factor initiated foundational chemical kinetics research, preceding the exploration of catalysts' subsequent stages by analyzing reaction time's role [12, 13]. At that juncture, advanced catalytic science and technology commenced, giving rise to scientific theories about catalysts that have since been foundational in the chemical and petroleum industries, with catalytic reactions emerging as their pivotal cornerstone.

7.3 TYPES OF CATALYTIC REACTIONS

Catalysts are categorized into heterogeneous and homogeneous types, differing in the phase relationship with reactants. Despite this distinction, these catalysts share common principles: they comprise inorganic and/ or organic components, often nanoscale, with covalent bond formation or breaking being central to both types [14]. Typical instances feature solid catalysts and reactants in different phases, with adsorption occurring on catalyst surfaces in heterogeneous catalysis. Achieving high reactivity, complete product selectivity, stability, recyclability, and cost-effectiveness remains the overarching objectives in catalysis [15, 16]. Catalysts engage in a specific interaction with the surface of reactant molecules, augmenting their reactivity. An effective catalyst must possess the capability to adsorb reactant molecules with a suitable degree of strength, enabling their reaction, while also avoiding excessive adhesion of product molecules to the surface. This delicate balance ensures efficient catalytic processes without undue retention of products on the catalyst's surface.

Homogeneous catalysts operate in the same phase as reactants, fostering aligned chemical interactions, predominantly in gas or single liquid phases. They offer advantages like improved selectivity, higher yield, and adaptable optimization through ligand and metal adjustments. Despite their widespread use in commercial contexts, the challenge of economically and environmentally separating catalysts from end products limits their broader effectiveness [17]. According to van Leeuwen (2011), homogeneous catalysis provides multiple benefits due to accessible catalysts, reactivity, and precise customization for optimal results. Clark (2001) underscores the extensive industrial use of homogeneous catalysts, particularly in fine chemical processes, while highlighting the issue of catalyst separation as a significant hurdle to their sustainable implementation.

Catalysts hold immense significance in industrial processes, continually giving rise to improved and novel catalysts. This is underpinned by the rationale that heightened catalyst efficiency translates to lower requisite reaction temperatures, thereby reducing fuel consumption. In industrial settings, heterogeneous catalysis predominantly prevails due to its facilitation of product-catalyst separation. For instance, employing a solid catalyst for gas-phase reactions enables the passage of gases over the catalyst, leaving the catalyst unchanged and unreacted at the conclusion of the process.

7.4 CATALYSTS IN REFINING INDUSTRY

Catalysts hold paramount significance within the petroleum industry, a sector of utmost importance in the contemporary world. Crude oil, composed of hydrocarbon compounds, undergoes separation into distinct products through refinery processes, including distillation. Typically, imbalances in product quantities, such as surplus black oil and deficient gasoline, necessitate the conversion of excess resources into required products—a fundamental objective achieved through the utilization of catalysts.

In the refining industry, cracking is a crucial procedure that breaks down large hydrocarbon molecules into smaller, more useful components, playing a pivotal role in generating valuable compounds. This division can be achieved either under high pressures and temperatures without a catalyst or with a catalyst at milder conditions. Typically originating from naphtha or gas oil fractions obtained through crude oil's fractional distillation, these substantial hydrocarbon molecules find their source [18]. Of notable importance is catalytic cracking, a cornerstone industrial process. As emphasized, catalytic cracking is the backbone of today's petroleum refining industry, highlighting its immense significance [19, 20].

Derived from the distillation process as liquids, these fractions undergo re-vaporization prior to cracking. To facilitate this process, hydrocarbons are blended with finely powdered catalysts, forming a homogeneous mixture. This concoction is then propelled through a reaction chamber in a manner reminiscent of liquid flow, and in fluid catalytic cracking, oil is vaporized to move as a fluid through the reactor, underscoring the pivotal role of the vaporized fluid in driving the reaction [21, 22]. Referred to as fluid catalytic cracking (or fluidized catalytic cracking), this process derives its name from the liquid-like behavior exhibited by the mixture, and following the reaction, catalyst recovery takes place. The cracked mixture is subsequently separated through cooling and additional fractional distillation, making fluid catalytic cracking one of the most vital conversion processes in petroleum refineries, underscoring its significant role in the industry [23].

The conversion of products like black oil into gasoline can be effectively achieved through a thermal process known as cracking, commonly conducted under elevated pressure and temperature conditions. Cracking involves the thermal decomposition of substances, such as the heating of dense oil hydrocarbons to form gas and gasoline. This thermal degradation results in the fragmentation of heavy molecules into lighter counterparts. However, the introduction of a catalyst significantly streamlines and enhances this process, enabling greater precision and control over the transformation. Hydrocracking integrates the processes of catalytic cracking and hydrogenation, wherein the feedstock is subjected to cracking in the presence of hydrogen, resulting in the generation of more advantageous product outcomes. Fluid catalytic cracking (FCC) is a widely used method in which oil is cracked within a fluidized catalyst bed, continuously circulating between reaction and regeneration states. The complexity of defining the "active site" is exemplified by a FCC catalyst particle, with Brønsted acid sites serving as the active phase. This catalyst efficiently converts heavy crude oil fractions into valuable bulk chemicals like gasoline and propylene [7, 16].

Catalytic regeneration stands as an additional conversion process within an oil refinery, distinct from molecular fragmentation, and primarily centered on molecular structural reconfiguration. This intricate process relies on catalyst intervention and is identified as catalytic reforming. Through this transformative mechanism, gasoline characterized by low octane levels undergoes a profound conversion into gasoline boasting substantially elevated octane ratings. The efficacy of this process is frequently facilitated by the utilization of catalysts, which may encompass platinum or nickel-aluminum, thus facilitating the intricate task of structural reconstruction and ultimately enhancing the combustion characteristics of the resultant gasoline product.

Gasoline's hydrocarbons are allocated octane ratings, indicative of their engine performance. Elevated-octane variants burn more smoothly than lower-rated counterparts. Transformation processes, including catalytic reforming, convert low-octane feedstock to high-octane reformate for gasoline blending. Alkylation combines gaseous olefins with isobutane to yield liquid high-octane iso-alkanes. The oil industry enhances gasoline's combustion quality in modern engines by altering straight-chain molecules into branched-chain isomers, thus elevating octane ratings and optimizing engine performance [24, 25]. Reforming constitutes an additional procedure employed for enhancing the octane rating of hydrocarbons designated for petrol use and concurrently serves as a valuable reservoir of aromatic compounds for the chemical sector. Aromatic compounds, which are essential constituents based on benzene rings, undergo vapor-phase transit across a solid catalyst as the initial molecules. Experimental evidence underscores that the preference for specific aromatic hydrocarbons is contingent upon the lignin composition, signifying a role for both catalysts in governing selectivity [26].

7.5 CATALYSTS IN PETROCHEMICAL INDUSTRY

Chemical transformations lead to the conversion of chain molecules into ring structures while undergoing hydrogen loss. Catalytic conversion of lignin involving hydrogen encompasses reductive depolymerization and hydrodeoxygenation of lignin-derived monomers, yielding arenes, cycloalkanes, and phenols, with an emphasis on catalyst systems and reaction mechanisms. In modern times, polymers are extensively utilized across diverse applications; nevertheless, the term "plastic" inadequately encompasses the wider scope of the polymer domain. Prominent among these are polyolefins, notably polyethylene and polypropylene. Yet, recent breakthroughs in catalytic chemistry have enabled the efficient production of these crucial materials, positioning catalysts as pivotal in shaping the future polymer-centric landscape [23, 27, 28].

Catalysis is crucial in processes such as benzene treatment for nitrobenzene production involving concentrated nitric and sulfuric acids through homogeneous catalysis, with annual synthesis exceeding 5 million tons, dependent on temperature and reagent concentration. Heterogeneous catalysis occurs with ethene reacting over a silicon dioxide catalyst, offering economic benefits and environmental friendliness [29, 30]. Esters form through the acid-alcohol reaction with a concentrated sulfuric acid catalyst, displaying liquid-phase homogeneous catalysis. Reduced acid turnover (a more eco-friendly process) and increased selectivity offer economic and environmental advantages. POMs as heterogeneous catalysts in large-scale reactions replace corrosive sulfuric acid, utilize controlled nitric acid, ensure mild reactions, minimize nitrogen dioxide, allow easy isolation, enable catalyst recycling, shorten reaction times, and reduce acid waste [31].

Benzene reacts with chlorine or bromine using catalysts like aluminum chloride or iron, although iron's role is not strictly catalytic as it transforms into iron (III) chloride, acting similarly to aluminum chloride. Alkylation substitutes hydrogen in a benzene ring with alkyl groups (e.g., methyl or ethyl), utilizing aluminum chloride as a catalyst. Chloroalkanes react with benzene using aluminum chloride. Various alkyl substitutions are feasible; manganese (IV) oxide accelerates hydrogen peroxide decomposition, enhancing oxygen release in concentrated solutions. Advanced catalyst screening and characterization target issues, as recent catalyst progressions in polyether synthesis present creative product pathways [29, 32].

7.6 EFFECTS OF CATALYSTS ON REACTION RATE

Enhancing the reaction rate requires augmenting successful collisions. A catalyst offers an alternate pathway with lower activation energy, allowing reactions to occur more readily. Even with a catalyst, particles colliding with sufficient energy react similarly to when no catalyst is present in the chemical process [33, 34]. The catalyzed route prevails due to lower activation energy, impacting the frequency factor while not affecting the reactant or product energy. The low-temperature WGS reaction is vital in processes like methanol synthesis, steam reforming, and catalytic combustion. Catalyst support enhances reaction activity or holds a pivotal role. Non-noble Fe, Co, Ni, and Cu are significant catalysts for CO_2 and CO hydrogenation to hydrocarbons and alcohols. Activation energy and frequency factor vary with the catalyst. An active CO_2 hydrogenation catalyst reduces activation energy, enhancing the rate constant k(T) [35, 36].

Catalysts are classified into positive and negative catalysts, affecting reaction rates. Positive catalysts accelerate reactions by reducing activation energy, while negative catalysts inhibit reactions. This categorization highlights catalysts' role in reaction control. A study found a linear correlation between potential, work function change, and ethylene combustion rate. Enhancing catalytic site reaction rates holds potential in heterogeneous catalysis, but experimental validation is crucial [37].

7.7 CATALYSTS AND ENVIRONMENT

Major engine pollutants include NOx, CO, UBHC, and soot. Co and Ni are potent CO_2 hydrogenation catalysts, favoring methane, while Fe is active in CO formation and the reverse water gas shift reaction [36]. Three-way catalysts are essential for controlling petrol engine emissions. They utilize substrates coated with active materials and regulate the air-fuel ratio, converting pollutants into less harmful compounds, effectively reducing emissions, especially with platinum group metal catalysts. Initial use involved platinum-based oxidation catalysts for hydrocarbon and carbon monoxide reduction, followed by a platinum/rhodium catalyst for nitrogen oxide reduction and simultaneous conversion of pollutants in a single three-way catalyst [38].

Oxidation catalysts are the original type of auto catalysts, converting carbon monoxide (CO) and hydrocarbons (HC) to carbon dioxide (CO_2) and water but have little effect on nitrogen oxides (NOx). Frequently, two or more metals are used jointly, and platinum/palladium was used in some of the early oxidation catalysts, as was platinum/rhodium, which was also used under rich conditions for NOx reduction [39]. The transportation sector contributes 35% CO, 30% HC, and 25% NOx emissions. Three-way catalysts replace older systems in Europe but remain in use globally where emissions regulations are less strict, including CNG buses, motorcycles, and small engines [40].

Exhaust gas emission control employs engine modifications, fuel treatment, additives, exhaust gas recirculation (EGR), positive crankcase ventilation (PCV), and catalytic converters. The converter includes reduction and oxidation catalysts, utilizing metal oxides coated with wire mesh filter, often featuring precious metals like platinum and palladium for efficient emission oxidation. Catalytic converters, effective across vehicles like trucks, buses, cars, and motorcycles, consistently minimize noxious tailpipe emissions [41, 42].

While renewable energy aims to cut CO_2 emissions, fossil fuels will persist in the energy landscape. To combat global warming, lower CO_2 emissions and better resource use are needed. Catalytic CO_2 hydrogenation to methanol offers a carbon-neutral solution, curbing emissions and replacing fossil fuels. Advanced converters optimize catalysts with precious metals like platinum, rhodium, palladium, and gold for heightened oxidation and reduced pollutants. Modern vehicles use three-way catalytic converters, effectively handling CO, HC, and NOx emissions [34, 43].

7.8 CONCLUSION

Catalysts are vital for modern life, notably in the oil, gas, and petrochemical sectors. These industries rely on catalysts to accelerate chemical reactions efficiently, shaping petroleum production. Catalysts prevent energy-intensive temperature increases, ensuring reactions occur faster and with precision. They are essential for addressing environmental and energy challenges, promoting sustainable solutions. Novel catalyst discovery fuels industries, facilitating rapid reactions and fine chemical synthesis. Heterogeneous catalysis, predominant for its separation advantages, maintains the unchanged catalyst post-reaction. Catalytic regeneration, like reforming in refineries, adds complexity to molecular transformations, highlighting catalysts' multifaceted role.

Catalysts significantly influence reaction kinetics, classified as positive and negative catalysts based on their effect on rates. This classification underscores their essential role in controlling reactions. Furthermore, auxiliary promotion methods enhance catalytic reactions by optimizing adsorbate-catalyst interactions and reagent activity. Some methods are validated, while others remain theoretical for future exploration. The synergy of homogeneous and heterogeneous catalysis offers exciting avenues for investigation. Advancing catalytic site reactions holds promise in heterogeneous catalysis. Despite renewable energy growth, fossil fuels persist, necessitating emission reduction and resource optimization. Catalytic CO_2 hydrogenation to methanol mitigates emissions and replaces fuels. Innovative converters optimize oxidation in modern vehicles, addressing CO, HC, and NOx emissions.

ACKNOWLEDGMENT

The authors wish to express their sincere thanks and gratitude to *Universidade Federal do Pará* (UFPA) for their help and encouragement in connection with this study. We are grateful for all their assistance.

KEYWORDS

- **catalyst regeneration**
- **catalytic cracking**
- **catalytic reforming**
- **emission control**
- **fluid catalytic cracking**
- **heterogeneous catalysis**
- **homogeneous catalysis**
- **hydrocracking**
- **petroleum refining**
- **reaction kinetics**

REFERENCES

1. Harris, N. (2019). *Green Chemistry*. Scientific e-Resources. ED-tech Press.
2. Somwanshi, S. B., Somvanshi, S. B., & Kharat, P. B. (2020). Nanocatalyst: A brief review on synthesis to applications. *Journal of Physics: Conference Series, 1644*, 012046.
3. Sima, J. (2015). Catalysis of chemical processes: Particular teaching aspects. *African Journal of Chemical Education, 5*(2).
4. Chorkendorff, I., & Niemantsverdriet, W. (2007). *Concepts of Modern Catalysis and Kinetics* (2nd ed.). WILEY-VCH Verlag GmbH & Co.
5. Ananikov, V. P. (2015). Nickel: The spirited horse of transition metal catalysis. *ACS Catalysis, 5*(5), 1964–1971.

6. Suib, S. L. (2013). *New and Future Developments in Catalysis: Catalysis for Remediation and Environmental Concerns*. Newnes.
7. Al-Qahtani, K. Y., & Elkamel, A. (2010). *Planning and Integration of Refinery and Petrochemical Operations*. Wiley-VCH.
8. Young, J. S., Lee, C. W., Lee, S. Y., Na, J., Lee, U., & Hwang, Y. J. (2020). Catalyst–electrolyte interface chemistry for electrochemical CO_2 reduction. *Chemical Society Reviews, 49*(22), 6632.
9. Kalz, K. F., Kraehnert, R., Dvoyashkin, M., Dittmeyer, R., Glaser, R., Krewer, U., Reuter, K., & Grunwaldt, J. (2017). Future challenges in heterogeneous catalysis: Understanding catalysts under dynamic reaction conditions. *ChemCatChem, 9*(1), 17–19.
10. Wisniak, J. (2010). The history of catalysis. From the beginning to Nobel prizes. *Educ. quím., 21*(1), 60–69.
11. Zecchina, A., & Califano, S. (2017). *The Development of Catalysis: A History of Key Processes and Personas in Catalytic Science and Technology*. Hoboken, New Jersey: John Wiley & Sons, Inc.
12. Wiesfeld, J. J. (2019). Earth-abundant heterogeneous catalysts for the conversion of biomass into value-added chemical intermediates. [PhD thesis, Technische Universiteit Eindhoven].
13. Petit, A. (2016). Associating physics and chemistry to dissociate molecules. *Historical Studies in the Natural Sciences, 46*(3), 360–391.
14. Li, C., & Liu, Y. (2014). *Bridging Heterogeneous and Homogeneous Catalysis: Concepts, Strategies and Applications*. Wiley-VCH.
15. Ye, R., Zhao, J., Wickemeyer, B. B., Toste, F. D., & Somorjai, G. A. (2018). Foundations and strategies of the construction of hybrid catalysts for optimized performances. *Nature Catalysis, 1*, 318–325.
16. Vogt, C., & Weckhuysen, B. M. (2022). The concept of active site in heterogeneous catalysis. *Nature Reviews Chemistry, 6*, 9–111.
17. Polshettiwar, V., & Varma, R. (2010). Green chemistry by nano-catalysis. *The Royal Society of Chemistry, 12*, 743–754.
18. Da Mota, S. A. P., Da Mota, A. A. M., & Machado, N. T. (2022). Influence of fractional distillation on the yield and quality of biofuels obtained through thermal catalytic cracking of crude palm oil. *Dyna, 88*, 62–71.
19. Haridoss, S. (2017). A study on the role of catalyst used in the catalytic cracking process in petroleum refining. *International Journal of ChemTech Research, 10*(7), 79–86.
20. Nyong, B. E., & Obeten, M. E. (2021). Kinetics of the catalytic cracking of methanol with aluminium chloride under vacuum condition. *Journal of Innovation and Sustainability, 4*, 1–4.
21. Speight, J. G. (2010). *Handbook of Industrial Hydrocarbon Processes*. Elsevier Science.
22. Busia, K. (2021). *Organic Chemistry for Schools: Advanced Level and Senior High School*. Xlibris Corporation LLC.
23. Islam, M. (2023). *Pipelines: Emerging Technologies and Design Criteria*. Sustainable oil and gas development. Elsevier Science.
24. Demirbas, A., Balubaid, M. A., Basahel, A. M., Ahmad, W., & Sheikh, M. H. (2015). Octane rating of gasoline and octane booster additives. *Petroleum Science and Technology, 33*(11), 1190–1197.
25. Islam, J. S., Islam, M. R., & Islam, M. (2018). *Economics of Sustainable Energy*. Wiley.

26. Mullen, C. A., & Boateng, A. A. (2010). Catalytic pyrolysis-GC/MS of lignin from several sources. *Fuel Processing Technology, 91*(12), 1446–1458.
27. Tolinski, M. (2015). *Additives for Polyolefins: Getting the Most Out of Polypropylene, Polyethylene and TPO*. Elsevier Science.
28. Jing, Y., Dong, L., Guo, Y., Liu, X., & Wang, Y. (2018). Chemicals from lignin: A review of the catalytic conversion involving hydrogen. *ChemSusChem*.
29. Wegener, G., Brandt, M., Duda, L., Hofmann, J., Klesczewski, B., Koch, D., Kumpf, R. J., Orzesek, H., Pirkl, H. G., Six, C., et al. (2001). Trends in industrial catalysis in the polyurethane industry. *Applied Catalysis General, 303*, 335.
30. Rothenberg, G. (2017). *Catalysis: Concepts and Green Applications* (2nd ed.). Wiley-VCH.
31. Saikia, A., Puri, K., & Borah, R. (2019). Supported dual-acidic 1,3-disulfoimidazolium chlorozincate @HZSM-5 as a promising heterogeneous catalyst for the synthesis of indole derivatives. *Applied Organometallic Chemistry, 33*(3).
32. Nogueira, I. C., & Pliego, J. R. (2023). Friedel-Crafts reaction: Theoretical study of the mechanism of benzene alkylation with isopropyl chloride catalyzed by AlCl3 and Al2Cl6. *Journal of the Brazilian Chemical Society, 34*(2), 194–200.
33. Cann, P., & Hughes, P. (2020). *Chemistry* (2nd ed.). Cambridge International AS & A Level.
34. Ra, E. C., Kim, K. Y., Kim, E. H., Lee, H., An, K., & Lee, J. S. (2020). Recycling carbon dioxide through catalytic hydrogenation: Recent key developments and perspectives. *American Chemical Society, 10*, 11318–11345.
35. Grabow, L. C., Gokhale, A. A., Evans, S. T., Dumesic, J. A., & Mavrikakis, M. (2008). Mechanism of the water gas shift reaction on Pt: First principles, experiments, and microkinetic modeling. *The Journal of Physical Chemistry C, 112*(12), 4608–4617.
36. Mutschler, R., Moioli, E., Luo, W., Gallandat, N., & Zuttel, A. (2018). CO_2 hydrogenation reaction over pristine Fe, Co, Ni, Cu, and Al_2O_3 supported Ru: Comparison and determination of the activation energies. *Journal of Catalysis, 366*, 139–149.
37. Hulsey, M. V., Lim, C. W., & Yan, N. (2020). Promoting heterogeneous catalysis beyond catalyst design. *Chemical Science, 11*, 1456.
38. Twigg, M. V. (2007). Progress and future challenges in controlling automotive exhaust gas emissions. *Applied Catalysis B: Environmental, 70*, 2–15.
39. Guillén-Hurtado, N., Rico-Pérez, V., García-García, A., Lozano-Castelló, D., & Bueno-López, A. (2012). Three-way catalysts: Past, present and future. *Dyna, 79*(175), 114–121.
40. Fevre, C., Demuynck, J., & Bosteels, D. (2016). Emissions control technologies to meet current and future European vehicle emissions legislation. *Association for Emissions Control by Catalyst (AECC)*.
41. Dey, A., & Mehta, N. S. (2020). Automobile pollution control using catalysis. *Resources, Environment and Sustainability, 2*.
42. Rajesh, A. V., Sundaram, C. M., Sivaganesan, V., Nagarajan, B., & Harikishore, S. (2020). Emission reduction techniques in CI engine with catalytic converter. *Materials Today: Proceedings, 21*, 98–103.
43. Bai, S. T., Smet, G. D., Liao, Y., Sun, R., Zhou, C., Beller, M., Maes, B., & Sels, B. F. (2021). Homogeneous and heterogeneous catalysts for hydrogenation of CO_2 to methanol under mild conditions. *Chemical Society Reviews, 50*, 4259–4298.

CHAPTER 8

Shock-Assisted High Temperature Consolidation and Synthesis of Diamond Containing Powder

AKAKI PEIKRISHVILI

Ferdinand Tavadze Institute of Metallurgy and Materials Science, Tbilisi, Georgia

ABSTRACT

The WC-Co-Diamond and Cu-Diamond powder mixtures were formed into cylindrical rods or tubes using a hot shock-wave consolidation process. The aims of the investigation were to fabricate composites and to obtain a rigid coating onto the cylindrical surface of steel tubes, as well as to investigate the morphology of the remaining diamond particles behind the shock wave front. The cylindrical geometry of shock wave loading was used to consolidate and synthesize composites in hot conditions. The processing temperature was changed up to 1000°C. The intensity of loading was under 10 GPa. The investigation showed that the combination of high temperature and shock wave loading was beneficial to the consolidation of the WC-20%Co-10%Diamond composites, resulting in near-theoretical densities, high hardness values, and the formation of a transient layer between the coating and the steel tube's wall. The structure and properties of the samples obtained, as well as the formation of the transient layer, depended on the shock loading conditions and the phase content of the precursor powders. As for Cu-Diamond precursors, it was observed that depending on the temperature, there was perfection of diamond crystals, and at temperatures under 1000°C, they practically had ideal morphology.

Advanced Topics in Polymer Chemistry and Materials Science: Tamara Tatrishvili and Marc Jean Médard Abadie (Eds)

8.1 INTRODUCTION

Shortage of data concerning the study of consolidation and synthesis of refractory and ceramic alloys is caused by a number of technological difficulties associated with shock wave processing [1–5]. These difficulties are related to the impossibility of direct compacting of powders. The problem is that during the compacting of the powder blends from refractory materials, the short period of loading is not enough for the operation of the consolidation mechanism, including the plastic flow of particles, further formation of common boundaries, and the obtaining of solid solutions inside the bonding phases. The increase in the intensity of loading that is necessary to provide plastic flow for refractory and ceramic materials essentially exceeds the ultimate strength of the steel container in which the consolidated precursor is located. In the case of consolidation by weak shock waves with low intensity of loading, an additional process of sintering must be carried out. The application of high temperatures (during sintering) is not recommended because the melting processes of the steel container and the annealing processes in consolidated and plastically deformed grains begin. As a result, it is impossible to maintain the plastically deformed structure and other advantages of shock consolidation in synthesized composites. The application of low-intensity shock waves and subsequent sintering processing during the compacting of pure ceramic powders is not effective in most cases [1].

The positive role of high temperature in the processes of the consolidation of refractory and hard alloy powders is described [5–11]. The application of high temperature (up to 1000°C) in the process of consolidation/synthesis by shock waves enables the consolidation mechanism to work at essentially low intensity of loading [5, 6] through plastic flow and mutual collision of compacting particles, resulting in the melting of their surfaces and the formation of common boundaries.

The objectives of this study were: first, to demonstrate that using the hot shock wave consolidation method, the initial WC-Co-Diamond blends of powders could be consolidated with near theoretical densities and appropriate microstructures; second, that high temperature and low intensity of loading (up to 10 GPa) have an influence on the morphology of diamond particles and may provide further structural perfection.

8.2 EXPERIMENTAL AND MATERIALS

The conventional powder of WC, Co, Cu, and ACM type diamond powder made in Russia with grain sizes of 2–5 μm and 10–20 μm were applied in composites. The experimental part of the project was carried out using the Universal Automatic Experimental Robot (UAER) created in the Institute of Mining Mechanics (IMM) in 1975 and described in [12]. The equipment allows for the consolidation and synthesis of different precursors by shock waves in hot conditions in the range of temperature 20–1100°C at intensity loading up to 20GPa. Figure 8.1 illustrates the view of diamond particles; Cu-diamond blend.

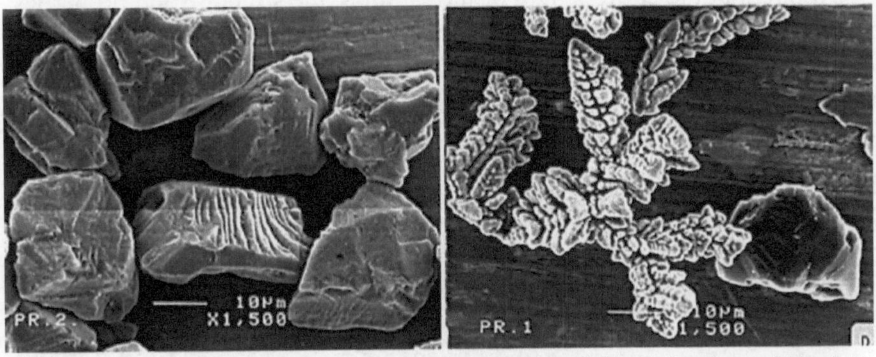

FIGURE 8.1 The view of starting materials a) diamond particles; b) Cu-diamond blend.

The method of resistance is used to heat the assembly up to the desired temperature before shock wave loading. The only ammonite (NH_4NO_3 + 20% TEN) with a detonation velocity of 3.6 km/sec and saltpeter (NH_4NO_3) as a passive additive was used as an explosive to provide the consolidation process with low-intensity shock wave loading. During the consolidation of precursors at high temperatures, the residual temperatures essentially increase as a sum of the preliminary temperature and the resulting temperature behind the shock wave front. As a result, some negative effects occur: changes in the geometry and shape of the compacts, diffusion of atoms from the steel container into the compacting powder, and the formation of undesirable phases, etc. In order to prevent the mentioned negative effects and to consolidate high-density composites with a perfect structure, the process of consolidation was fulfilled in two stages:

1. **First stage** – increasing the starting density of the compacting powders before the hot shock wave loading using different methods; a) preliminary shock wave consolidation at room temperature using cylindrical geometry of loading.
2. **Second stage** – hot shock wave compacting of the preliminary consolidated powders.

8.3 RESULTS AND DISCUSSION

The shock wave consolidation of WC-Co–10% diamond powders in hot conditions, with simultaneous sintering of the composition behind the shock wave front, results in the formation of a perfect structure in the compacted powder and a transient layer between the supporting steel surface and the obtained hard-coated material.

Throughout the entire length of the samples, the structure and properties of the transient zone completely depend on the preheating temperature and loading intensity as well. Figure 8.2 represents the fracture of the consolidated and synthesized WC-20%Co-10% diamond composite obtained at a temperature of 1000°C with a loading intensity of under 10 GPa.

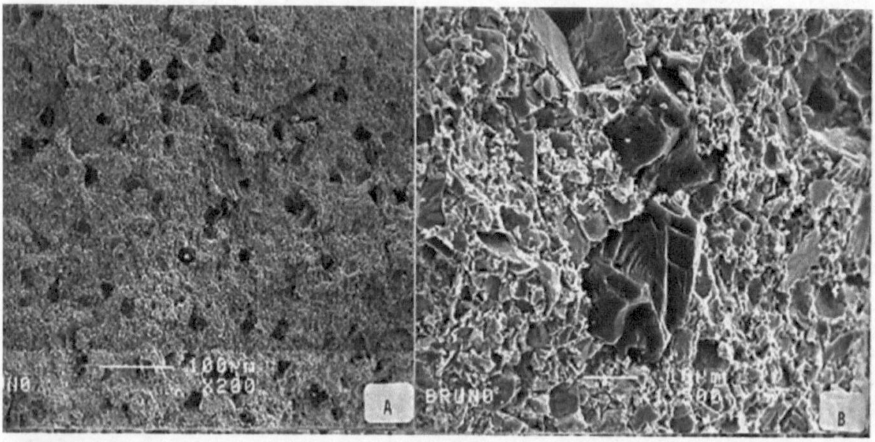

FIGURE 8.2 The fractures of the hot consolidated and synthesized WC-Co-diamond blends at temperature T= 1000°C. The intensity of loading under 10GPa.

No cracks and cavities are not observed on the structures.

As it's seen the density, as seen from the structures, the consolidated composite has high density, and no visible defects such as cracks and cavities can be observed. The diamond grains have a strong bond with the WC-Co matrix and are well combined with the structure of the alloy. There are no unconsolidated parts in the macrostructure (Figure 8.2a), and diamond particles are uniformly distributed throughout the entire volume of the obtained sample.

Figure 8.3 represents the micro and fine structures of the consolidated and synthesized WC-Co-Diamond composite obtained at a temperature of 1000°C under a loading intensity of 10 GPa. As seen from the micro and fine structure of the consolidated and synthesized composites, traces of strong plastic deformation inside the WC grains are observed. The strong traces of sliding and twinning inside the tungsten carbide grains (Figure 8.3a) result in a significant increase in microhardness from 1800 to 3000 kg/mm². After TEM investigation of individual WC grains under magnification up to X 50,000, traces of strong fragmentation and disorientation of the formed separate particles relative to each other inside the individual particles were observed (Figures 8.3b and 8.3c). The separate traces of formed dislocations as a result of plastic deformation inside the same WC grains are also observed.

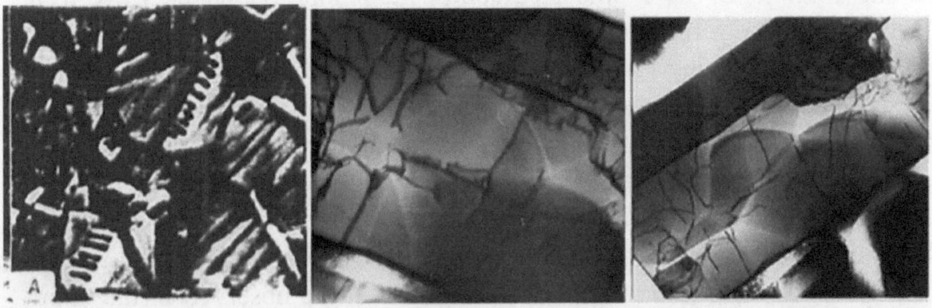

FIGURE 8.3 The structure of consolidated and synthesized WC-20%Co-10% diamond composite at temperature 1000°C under the intensity of loading 10GPa. (a) Microstructure – the strong plastic deformation inside in WC grains by sliding and twinning (magnification X2000). (b–c) fine structures of individual grains in composites showing strong fragmentation inside of WC particles with traces of formed dislocations. The inner disorientation of obtained fragments to each other may be observed too (Magnification X 42000 & 35000 correspondently).

Figure 8.4 represents the structure of transient field between the consolidated and synthesized WC-Co-Diamond composite and steel formed behind of shock wave front.

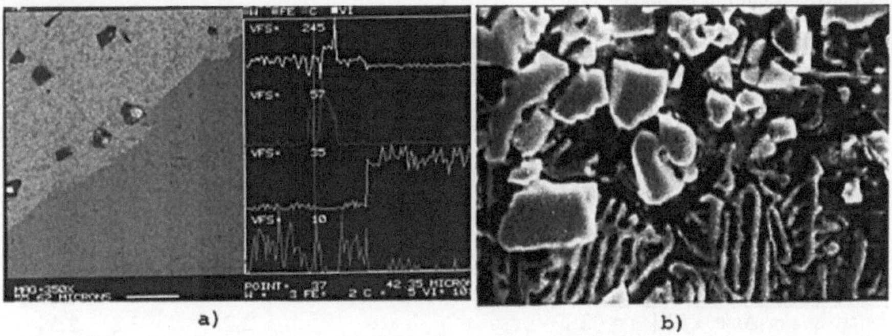

FIGURE 8.4 The microstructures of transition zone between the WC-20%Co-10%Diamond & Steel obtained at temperature T=1000°C under the intensity of loading 10GPa. (a) Microstructure of transition zone with spectral analysis, showing distribution of atoms across the border. (b) Microstructure of transient zone (magnification X3500).

The observation of the existence of the eutectic colonies demonstrates that during explosive loading at 10,000°C, melting processes occurred in the obtained samples behind the shock wave front between the steel substrate and the obtained compacts. Especially, the atoms of W (red) and Co (yellow) are deeply penetrating the steel substrate. The lower concentration of Fe and C atoms in the hard alloy may be explained by the higher resistivity of the inner hard material to compression, resulting in a lower concentration of penetrated Fe and C atoms in the synthesized composite behind the shock wave front. The observation of the transient zone at higher magnifications (Figure 8.4b) shows that the joining (transient) field forms in the process of melting and crystallization under high-intensity shock loading, which essentially facilitates the processes of diffusion under and behind the shock wave front.

The existence of the eutectic colonies in the microstructure (Figure 8.4b) confirms the aforementioned melting processes. It was established through X-ray diffraction analysis that the formation of the new phase Fe_3W_3C -carbides of high-speed steel (the "needles" of crystals can be observed in the microstructure, Figure 8.4b) took place in the transient field. Taking into account the distribution of concentration curves and the results of structural analyses, it can be said that the transient layer represents a solid solution based on Fe-Co-W atoms, reinforced by grains of WC, Fe_3W_3C, and diamond phases.

In Table 8.1 presented some strength characteristics for hot shock wave consolidated and synthesized WC-Co-Diamond composites depending on the temperature of processing at intensity of loading under 10GPa.

TABLE 8.1 Presented Some Strength Characteristics for Hot Shock Wave Consolidated and Synthesized WC-Co-Diamond Composites Depending on the Temperature of Processing at Intensity of Loading Under 10GPa

WC-20%Co-10%Diamond Characteristics	**Consolidation/Synthesis Temperature [°C]**					
	Virgin	**600**	**700**	**800**	**900**	**1000**
Hardness Hv5, kg/mm^2, loading 5 kg	–	High porosity and Cracks	High porosity and Cracks	780 Cracks observed	866	880
broadening of (200) line for WC phase	0.22	–	–	0.210	0.32	0.240
Grain size for WC phase [μ]	0.52	–	–	0.38	0.18	0.26
Micro stress [Δa/a, x10^4]	2.1	–	–	2.2	3.2	2.5

As seen from Table 8.1, the investigation showed that only at temperatures above 900°C is it possible to obtain high-density samples without visible structural defects such as porosity and cracks. The hardness value for consolidated and synthesized WC-20%Co-10%Diamond hard alloys slightly depends on the temperature, reaching 880 Hv at 1000°C, despite the fact that strong plastic deformation occurs within the WC particles (Figure 8.3). This can be explained by the matrix Co phase, which provides bonding of the hard WC and diamond particles in the composition and is in a fully annealed condition.

To investigate the effect of temperature on the morphology of diamond during shock wave loading, blends of Cu-20% diamond were consolidated at different temperatures with a loading intensity of under 10 GPa. It was found that high temperatures combined with shock loading have a positive influence on the surface morphology of consolidated diamond, and near the melting point of copper, the perfection of separate diamond grains was observed.

Figure 8.4 represents the structure of diamond grains in Cu-10%Diamond composites obtained after shock wave consolidation in hot conditions with a loading intensity of under 10 GPa.

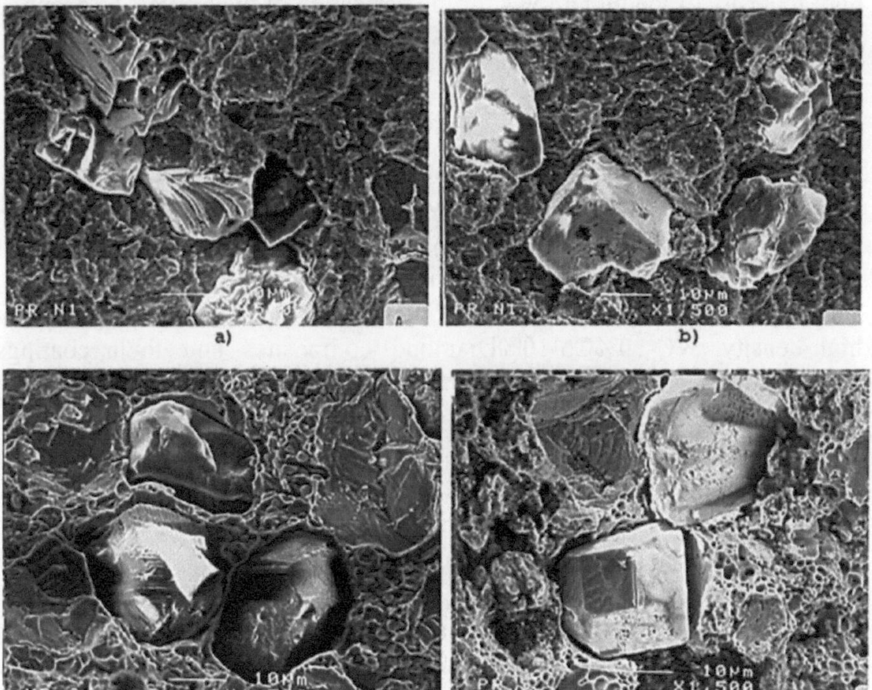

FIGURE 8.5 The microstructures Cu-10%Diamond composites obtained after high temperature shock wave consolidation with intensity of loading under 10GPa. a-b) consolidated under 4000C; b-c) consolidated at 900 & 1000°C correspondently. The traces of melting/crystallization process in copper matrix (Figure 8.5d).

As can be observed from the structures in Figure 8.5, the morphology of consolidated diamond particles is dependent on the temperature value, particularly near the melting point of copper (around 1000°C). After shock wave loading of Cu-diamond precursors with an intensity below 10 GPa, modifications and perfection of the diamond particles are observed (*see* Figures 8.5c and 8.5d).

Taking into account the low melting point of copper (1080°C), it can be suggested that during loading by shock waves with an intensity near 10 GPa, the copper matrix transforms into a liquid state. After stress relief due to the remaining temperature, melting and crystallization processes in the copper matrix also take place.

It can be assumed that at the front of the shock wave during the loading process, the diamond crystals, being in a liquid copper medium, are subjected to all-around compression. Despite the short duration of this process, the high intensity of loading allows for the perfection of the crystals to occur.

8.4 CONCLUDING REMARKS

The only high-temperature shock wave consolidation makes it possible, in one stage, to consolidate and synthesize WC-Co-Diamond composition powders near theoretical density without cracking and with a perfect structure. The temperature and intensity of loading are the main parameters, and only above 900°C and with low intensity of loading under 10 GPa can high-density WC-20%Co-10%Diamond composites and their coatings on steel surfaces be obtained, with the formation of a transient zone. The consolidation and synthesis of WC-20%Co-10%Diamond composites are accompanied by a significant improvement in structure. The strong plastic deformation in WC particles with significant fragmentation and sliding/twinning there takes place. The value of hardness of composites obtained after fabrication at 1000°C reaches to 880kg/mm^2

High-temperature treatment of diamond crystals with an intensity of loading under 10 GPa showed a positive role of temperature, and at 900–1000°C, significant improvement and perfection of its morphology were observed.

Further investigation and more attention to this topic, focusing on increasing the percentage of diamond crystals, pulse duration during the loading, and temperature values, seem to be attractive.

KEYWORDS

- **diamond crystal morphology improvement**
- **hardness of WC-20%Co-10%Diamond composites**
- **high-density composites**
- **high-temperature shock wave consolidation**
- **plastic deformation in WC particles**
- **WC-Co-Diamond composite synthesis**

REFERENCES

1. Prummer, R. (1987). *Explosivverdichtung pulvriger Substanzen*. Springer-Verlag.
2. Kiiski, A. A., et al. (1995). Explosive compaction of copper and graphite powder mixtures. In L. E. Murr, K. P. Staudhammer, & M. A. Meyers (Eds.), *Shock Wave High-Strain Rate Phenomena* (pp. 109–115). Elsevier.
3. Andreev, B. D., et al. (1991). Some properties of the compacts from AlN obtained by shock wave loading. *Powder Metallurgy (Russian), 10,* ISSN 0032-4795.
4. Hoenig, C. L., & Yust, C. S. (1981). Explosive compaction of AlN, amorphous Si3N4, boron and Al2O3 ceramics. *American Ceramic Society Bulletin, 60*(11), 1175–1224.
5. Mote, J., et al. (1984). Emergent process methods for high technology ceramics. In *Emergent Process Methods for High Technology Ceramics* (pp. 695–710). Plenum Press.
6. Bhalla, A. K. (1980). Hot explosive compaction of metal powders. *Transactions of the PMAI, 7*(1).
7. Peikrishvili, A., & Chikhradze, N. (1986). Explosive working of some metals and alloys at high temperatures. In L. E. Murr, K. P. Staudhammer, & M. A. Meyers (Eds.), *Metallurgical Application of Shock Wave and High Strain Rate Phenomena* (pp. 905–915). Marcel Dekker, Inc.
8. Japaridze, L., Peikrishvili, A., et al. (1992). Importance of preheating at dynamic consolidation of some hard materials. In M. A. Meyers, L. E. Murr, & K. P. Staudhammer (Eds.), *Shock-Wave and High-Strain Rate Phenomena in Materials* (pp. 463–472). Marcel Dekker, Inc.
9. Peikrishvili, A., et al. (1995). Shock wave deformation of W-N-Fe heavy alloys at high temperatures. In L. E. Murr, K. P. Staudhammer, & M. A. Meyers (Eds.), *Metallurgical and Materials Application of Shock Wave and High Strain Rate Phenomena* (pp. 223–229). Elsevier.
10. Simonsen, Y., Horie, T., Akashi, A., & Sawaoka, A. (1992). Diamond formation in aluminum compressed with Ni-graphite under shock loading. *Journal of Materials Science, 27*(7), 1735–1740.

11. Peikrishvili, L. A., Kesckes, L., Chagelishvili, E., & Godibadze, B. (2023). Hot shock wave consolidation and syntheses of W-Mg composites. In *Tailored Functional Materials for Clean and Sustainable Development* (pp. 215–222). AAP.

CHAPTER 9

Consolidation and Synthesis of Ta-Al-B(B4C) Reactive Blends by Hot Electro Rolling Method

AKAKI B. PEIKRISHVILI, ZAQARIA MELASHVILI, ZURAB ASLAMAZISHVILI, MARGALITA DARCHIASHVILI, BAGRAT GODIBADZE, TEIMURAZ NAMICHEISHVILI, TAMAZ BATSIKADZE, and GIORGI PARUNASHVILI

Ferdinand Tavadze Institute of Metallurgy and Materials Science, Tbilisi, Republic of Georgia

ABSTRACT

The Ta-Al-B (B4C) blend of powder was formed into plates using combined self-propagating high-temperature synthesis and hot electrical rolling processes. The aims of the investigation were to synthesize and fabricate different TaxAly-TaBx (B4C) composites with near theoretical density in the form of sheets having a perfect structure and high strength properties. The preliminary investigation showed that the combination of the high-temperature electrical rolling method with the SHS processes of the Ta-Al-B reactive blends of powders at the beginning of rolling was beneficial to the consolidation/synthesis of the TaxAly-TaBx composites, resulting in near theoretical densities, high hardness values, and the formation of different intermetallic compounds of TaxAly. The structure and properties of the samples obtained depended on the loading conditions and the phase content of the precursor powders. The processing temperature at the beginning of rolling was changed up to 1000°C. The intensity of loading was under 60 MPa. The aforementioned features, as well as other aspects of the structure/property relationship of the consolidated and synthesized plates of TaxAly-TaBx

Advanced Topics in Polymer Chemistry and Materials Science: Tamara Tatrishvili and Marc Jean Médard Abadie (Eds)

(B4C) composites as a function of the loading conditions (experimental set-up, intensity of loading, preheating temperature, pre-density), will also be presented and discussed.

9.1 INTRODUCTION

The development of nuclear technologies and equipment requires new approaches and the development of innovative materials characterized by a good ability to absorb neutron radiation, greater stability, and relatively low cost.

Rolling refractory and brittle materials, in contrast to plastic ones, is a very complicated problem. Such alloys, in particular those obtained by the methods of powder metallurgy, fail under the action of static loading due to their low plasticity. However, it has been known [1–3] that pre-heating reduces the tendency to crack even under dynamic loading, since with an increase in the initial temperature of some materials, such as heavy alloys and cemented carbide alloys, their hardness decreases and plasticity increases.

The application of high temperatures during the rolling of refractory alloys enables not only the prevention of cracking in the obtained billets but also the production of bimetal compositions from different pairs. During the rolling of powder materials, with the increasing temperature, the plastic flow of consolidated particles under static loading increases, and at high temperatures, due to the mutual collision of surfaces, the formation of joint boundaries occurs. Proceeding from these considerations, pre-heating before rolling is highly recommended for brittle and PM materials and is promising for any other materials, since it changes their physical-mechanical properties.

The innovative technology of self-propagating high-temperature synthesis (SHS)-electric rolling is the only one among existing technologies that ensures the continuity of SHS samples and the process of hot deformation (rolling) while obtaining products with the required longitudinal dimensions. The operating principle of this method is as follows: a filled container with a charge is brought to the rolls of a special rolling mill, and a small grip is made, ensuring reliable electronic contact between the rolls and the container. Through this contact, an electric current is supplied to the deformation zone, and the Joule heat released in the initial section of the specified deformation zone heats the charge and initiates the synthesis process. The combustion front of the charge moves along the container, and at the moment when a certain combustion zone of the charge (an area of planned size) is created in front of the area where the material enters the deformation zone, the rolls are turned on, and electric rolling begins.

From the mentioned area, the charge in a hot, viscous-plastic state is continuously supplied to the deformation site. A necessary condition for the implementation of this process is the equality of the speeds of movement of the synthesis and rolling fronts, as well as the compensation of heat losses from the contact of the container with the rolls, due to the conduction of a heating current, which ensures the existence of a hot viscous-plastic mass in the deformation zone. The duration of the SHS process is determined by the length of the container loaded with the charge.

The objectives of this study were to demonstrate that using two non-standard powder metallurgy technologies, SHS and hot electric rolling technology, it is possible to obtain high-density sheets based on tantalum and aluminum intermetallic compounds.

9.2 EXPERIMENTAL

Based on the above and taking into account the data from the Ta-Al phase diagram, we can conclude that the selection of the phase composition and the use of various passive alloying elements to regulate the synthesis temperature are important technological parameters in their subsequent static or dynamic consolidation-synthesis processes.

The SHS-electric rolling process is innovative, has no analogues, and was created and developed at the Ferdinand Tavadze Institute of Metallurgy and Materials Science [4].

When studying the SHS-electric rolling process, it should be taken into account that the process of obtaining the SHS charge—composition, packaging, i.e., the density of the briquette, the temperature of the isothermal combustion process, and the speed of movement of the combustion front—are directly related to the main technological parameters of electric rolling. Therefore, a theoretical study of these two completely independent processes was carried out, taking into account the joint implementation of these processes.

It is important to note that the Ta-Al system belongs to a reactive pair, and in this composition, self-propagating high-temperature synthesis occurs at a high temperature of approximately 2500°C. Depending on the aluminum content, the intensity of exothermic reactions changes, and accordingly, the synthesis temperature also changes.

Figure 9.1 represents the set-up of rolling processing.

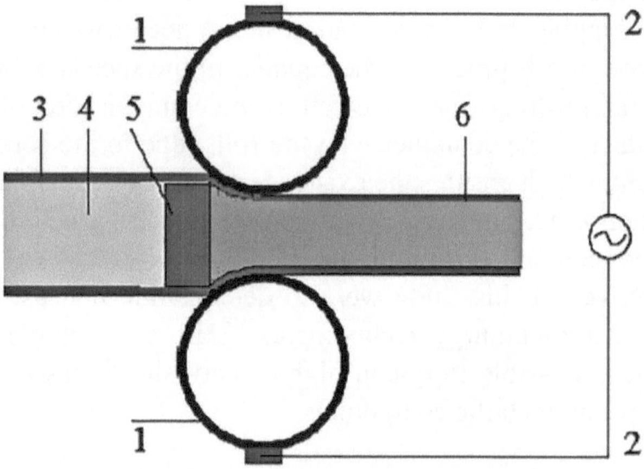

FIGURE 9.1 SHS-electric rolling set-up (1. Rollers; 2. Electrical contacts of the power supply; 3. Container; 4. Reactive powder blend; 5. Combustion front, 6. Rolled product).

The process of obtaining high-quality products from SHS electric rolling is influenced by the following parameters: dispersity, drying, activation, and degree of mixing of the solution, which determine the volumetric homogeneity of the solution. Dry mixing of powders is carried out in a ball mixer for 24 hours at a speed of 600 turns/min.

Cold briquetting forces determine the density of the briquette. The relative density of the briquette is selected in the range of 0.5–0.7, with pressing forces of 5–20 tons. The thickness of the briquette is selected within the range of 16–18 mm, and the weight of one briquette is from 130 to 150 g. The location of the briquettes in the container determines the degree of binding of the briquettes in the product.

Container dimensions determine the degree of roll deformation and gripping conditions during electric rolling. For example, to obtain the required material thickness of 6–7 mm and a width of 50–55 mm, to achieve a relative deformation of 60%, the cross-sectional dimensions of the container are 18–20 x 46–48 mm. The dimensions of the briquettes are determined by the dimensions of the mold and are 45 x 45 mm. Briquettes are placed in a container in one row. Each container contains 4 briquettes (total weight – 0.5–0.6 kg). Pre-densification of briquettes to form a sample occurs under a force of 200 tons.

TABLE 9.1 The Technological Parameters of Synthesis and Electrical Rolling

Content	SHS Reaction Velocity mm/sec	Pre-Densification Force at Room Temperature, Kg	Velocity of Roll's Turning Turn/min	Heating Electrical Current, A	Compression %	Average Specific Pressure Kg/mm^2
Ta-31%Al	6	1500	0,45	2100	60	6,8
Ta-8%Al-11% B4C	3	1500	0,225	4300	60	12,1
Ta-8%Al-10% B	10	1500	0,76	2500	60	7,4
Ta-24%Al-6% B	5	1700	0,375	2200	60	11,0
TaB2-9%Al	5	7500	0,375	2300	60	10,5

9.3 RESULTS

Figure 9.2 represents the view of rolled strips from of Ta-Al-B composition.

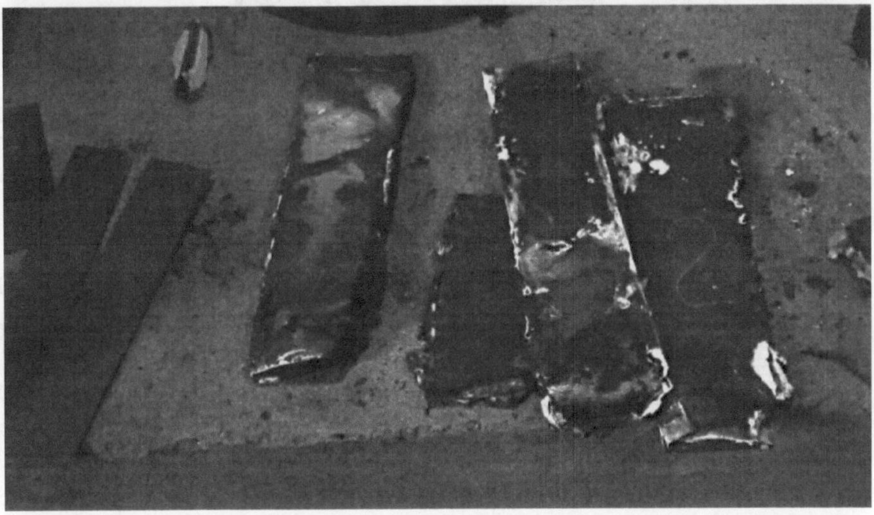

FIGURE 9.2 The billets of Ta-19%Al-6%B & Ta-8%Al-10%B composition obtained after rolling at high temperature according to the Figure 9.11.

The obtained billets were cut, polished and prepared for further structural investigation.

Figure 9.3 represents the fractures of different hot rolled samples obtained from Ta-Al-B reactive pre-densified blends.

FIGURE 9.3 The View of fractures after hot rolling Ta-19%Al-6%B and Ta-8%Al-10%B composites. (a) View of fractures; (b) fracture after 15 sec delay before rolling; (c) fracture without delay parallelly with SHS reaction.

Figure 9.4 represents the macrostructure of hot rolled Ta-19%Al-6%B pre-densified blends with correspondent spectral and quantitative composition analysis.

As it's seen from view of cross section of hot rolled Ta-19%Al-6%B sample at low magnification (x30), it's free from any defects of processing and no visible cracks on the surface of obtained samples not observed.

Figure 9.5 shows the microstructures of hot rolled and synthesized Ta-Al-B composites depending on its phase content. Rolled samples of the composition Ta-24%Al-6%B (Figures 9.5a and 9.5b) consist of two phases. Black particles are distributed on a white background. Both phases are predominantly oriented in the rolling direction, although they occur in an unevenly distributed deformed form. The white phase is in the form of independent particles. Rolled samples of the composition Ta-8%Al-10%B (Figures 9.5c and 9.5d) are also two-phase. Black particles are unevenly distributed on a white background. In contrast to the Ta-24%Al-6%B composition, an insignificant orientation in the rolling direction

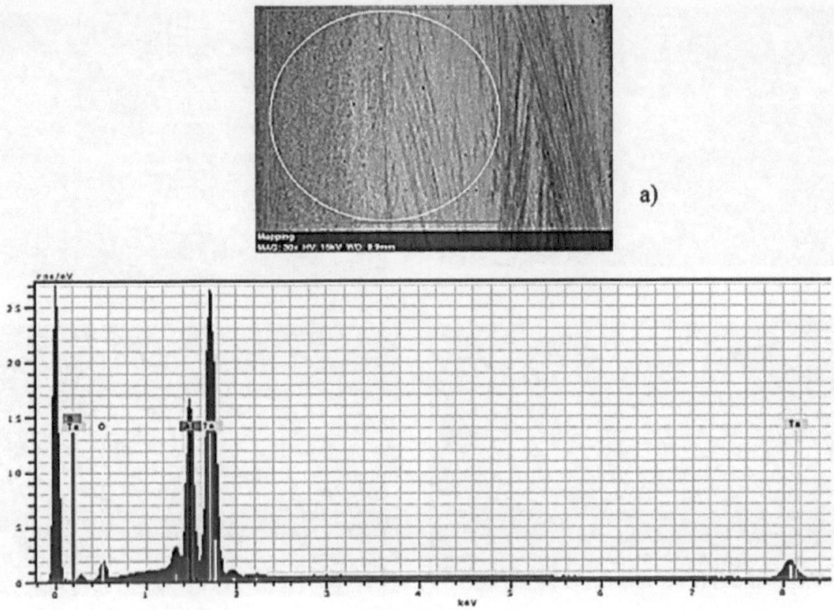

a)

b)

Spectrum: Point

Element	AN	Series	norm. C [wt.%]	Atom. C [at.%]
Tantalum	73	L-series	71.97	22.99
Aluminium	13	K-series	17.31	37.08
Oxygen	8	K-series	10.01	36.14
Boron	5	K-series	0.71	3.79
		Total:	100.00	100.00

c)

FIGURE 9.4 The macrostructure of hot rolled Ta-19%Al-6%B pre-densified blends with correspondent spectral and quantitative composition analysis. (a) Macrostructure; (b) spectral analysis; (c) quantitative composition analysis.

is observed. The white background is a collection of round particles, although in some places there are also continuous, irregularly shaped independent inclusions.

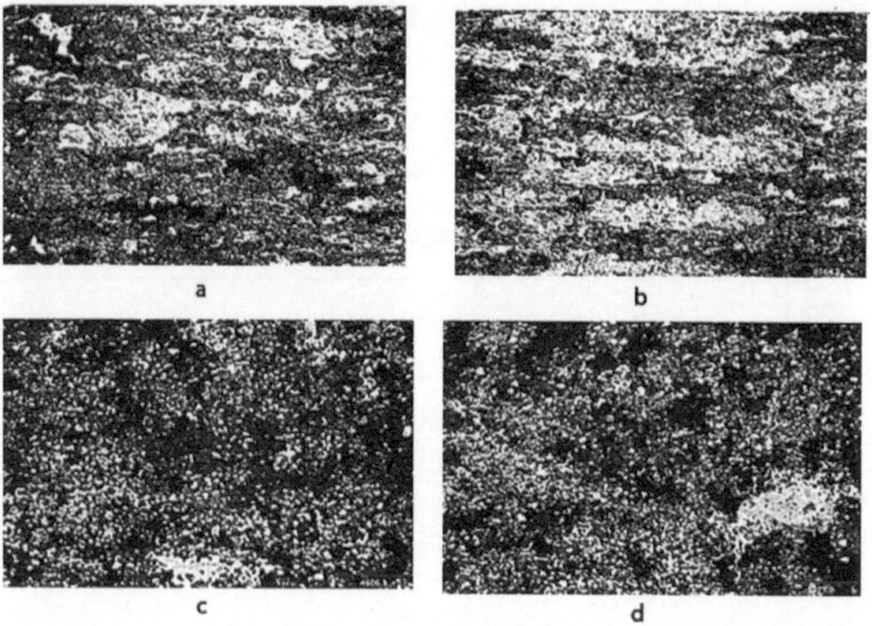

FIGURE 9.5 The microstructure of hot rolled Ta-24%Al-6%B and Ta-8%Al-10%B composites. (a–b) Ta-24%Al-6%B samples; (c–d) Ta-8%Al-10%B composites; Magnification x1000.

The microstructures of the hot-rolled sample with lower content of boron and higher content of aluminum (Figure 9.5 a–b) have a non-uniform distribution of phases, and unreacted inclusions of the aluminum phase are observed (white spots). The reduction of the amount of aluminum and the increase of boron in the composition have a positive influence on structure formation, and after rolling, it became possible to obtain high-density samples with a perfect structure and uniform distribution of phases (Figure 9.5c–d). Unfortunately, fully unreacted phases (white spots) can also be observed in the microstructures. The above-mentioned is confirmed by the X-ray diffraction pictures as well (Figure 9.6a–b), where the influence of boron content on the SHS reaction and phase formation can be clearly observed. Figure 9.6 represents the diffraction pictures of hot-rolled Ta-24%Al-6%B and Ta-8%Al-10%B composite powders.

As it's seen from diffraction pictures of hot rolled Ta-Al-B composites the composition with higher content of boron is more reactive in

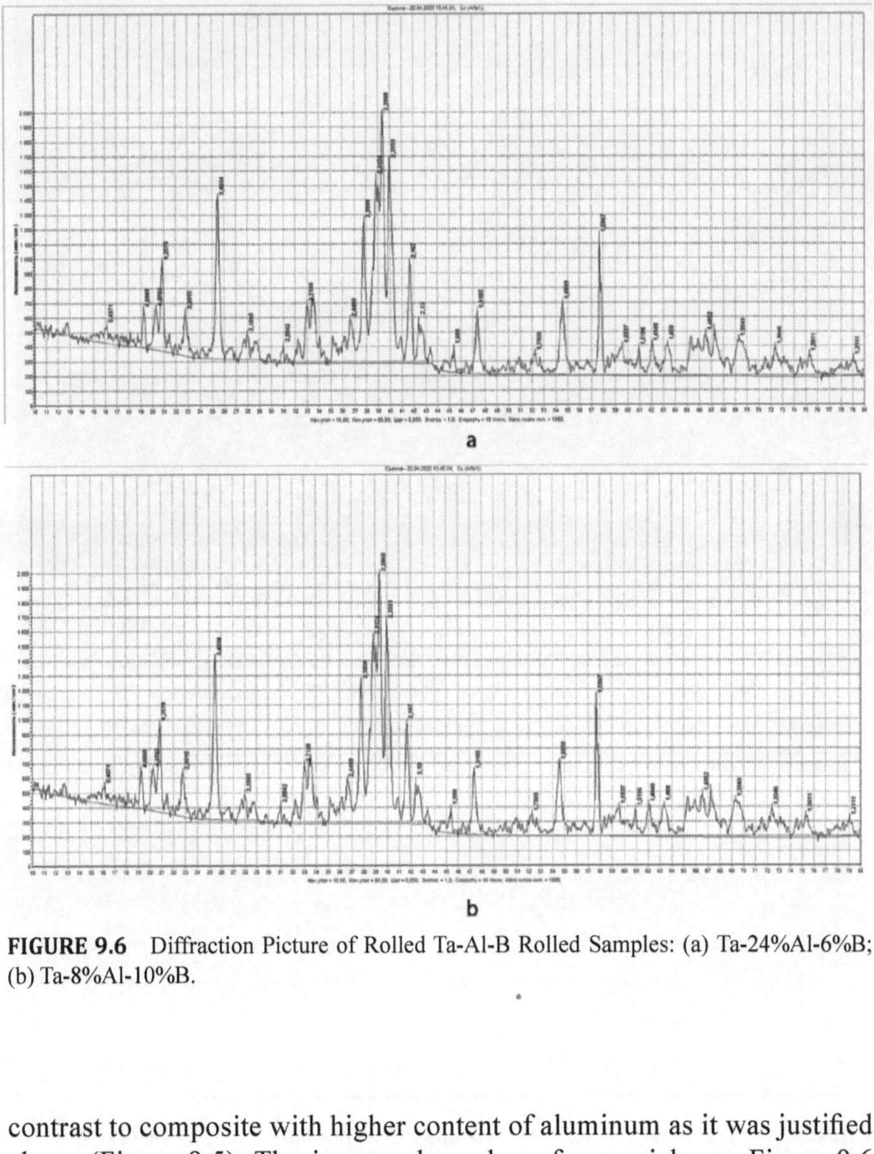

FIGURE 9.6 Diffraction Picture of Rolled Ta-Al-B Rolled Samples: (a) Ta-24%Al-6%B; (b) Ta-8%Al-10%B.

contrast to composite with higher content of aluminum as it was justified above (Figure 9.5). The increased number of new picks on Figure 9.6 b in contrast to 9.6 a confirms distribution on microstructures (Figure 9.5) where reduced unreacted phases in the forms of white spots were observed.

Figure 9.7 represents SEM microstructure of (Ta-50%Al)-10%B4C composite obtained after combined SHS & electro rolling processing.

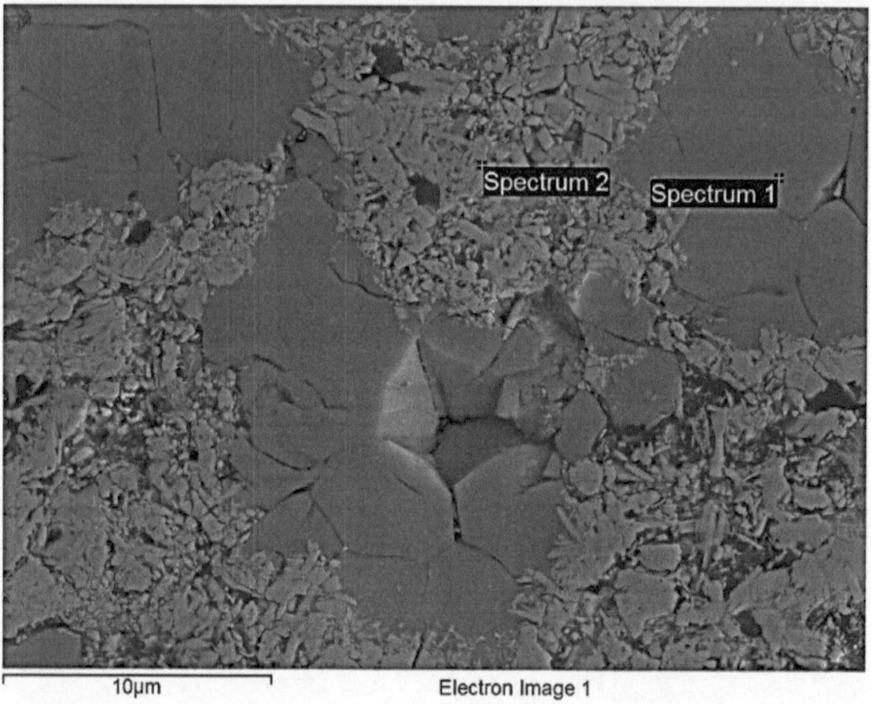

Processing option : All elements analysed

Spectrum	In stats.	B	C	O	Al	Fe	Ta
Spectrum 1	Yes	13.78	9.68	2.24	53.31		20.99
Spectrum 2	Yes	30.99	24.90	6.25	11.00	1.41	25.44
Max.		30.99	24.90	6.25	53.31	1.41	25.44
Min.		13.78	9.68	2.24	11.00	1.41	20.99

FIGURE 9.7 The microstructure and quantitative composition analysis of (Ta-50%Al)-10%B4C obtained after combined SHS and hot rolling processing.

The quantitative analysis of spectrum 1 and spectrum 2 shows that there are no unreacted phases, and in both cases, the formation of tantalum aluminates takes place. The increased amount of C and B in spectrum 2 may be the result of the formation of additional tantalum carbides and borides that, together with the formed tantalum aluminate, may be considered as a matrix phase of the obtained composition. The table 9.2 represents the distribution of hardness in hot-rolled Ta-Al-B composites.

TABLE 9.2 The Distribution of Hardness in Hot-Rolled Ta-Al-B Composites

Characteristic/composition	Ta-19%Al-6%B	Ta-8%Al-10%B
Al reach area	207kg/mm^2	875kg/mm^2
B reach area	250kg/mm^2	519kg/mm^2

Ta-Al-B (B4C) blends were preliminarily statically and dynamically densified, and by combining SHS and hot electro-rolling processes, they were fabricated in the form of sheets close to theoretical density with a definite structure and high hardness values. It was established that the initiation of the SHS reaction in compositions to fabricate high-density sheets depends on the content, and with an increase of B in powder blends, it provides full chemical reactions and the obtaining of samples with increased hardness values of up to 900 kg/mm^2.

9.4 CONCLUSION

The combination of SHS processes with hot rolling and electro-rolling makes it possible to consolidate and synthesize high-density billets in sheet form from Ta-Al-B (B_4C) composition powders. The structure and hardness of the fabricated samples are influenced by the phase content of the composition, and an increase in boron content up to 10% enables the production of samples with fully reacted and uniformly distributed phases, achieving a hardness value of up to 900 kg/mm^2.

ACKNOWLEDGMENTS

This work was supported by Shota Rustaveli National Science Foundation of Georgia –SRNSFG (Grant number STEM 1113).

KEYWORDS

- **hardness values**
- **high-density billets**
- **hot electro-rolling**
- **phase distribution**
- **self-propagating high-temperature synthesis (SHS)**
- **Ta-Al-B reactive blends**

REFERENCES

1. Mindeli, E., Licheli, G., Mgeladze, Z., Peikrishvili, A., & Chagelishvili, E. (1978). Pre-heating effect upon the explosive hardening of hard alloys. *FGV, 4*, 192.
2. Mote, J., et al. (1984). Emergent process methods for high technology ceramics. In *Emergent Process Methods for High Technology Ceramics* (pp. 695–710). Plenum Press.
3. Peikrishvili, A., et al. (1995). Possibility of obtaining combined samples from tungsten base alloys by high temperature shock wave treatment. In L. E. Murr, K. P. Staudhammer, & M. A. Meyers (Eds.), *Metallurgical and Materials Applications of Shock Wave High-Strain-Rate Phenomena* (pp. 99–107). Elsevier.
4. Namicheishvili, T., Tutberidze, A., Melashvili, Z., Tavadze, G., Aslamazashvili, Z., Oniashvili, G., & Zakharov, G. (2016). A method for producing products from inorganic materials from powdered crucibles (Final report, p. 6541). IMMS.

CHAPTER 10

Oxidation of Antimony (III) Sulfide with Elemental Sulfur in Lithium Alkaline Medium

KAKHA RUKHAIA,[1] MAIA RUSIA,[2] and MANUCHAR CHIQOVANI[3]

[1]*Doctor of Chemistry, Ministry of Environment Protection and Agriculture, Georgia*

[2]*Department of General, Inorganic, and Metalorganic Chemistry, Faculty of Exact and Natural Sciences, Iv. Javakhishvili Tbilisi State University, Georgia*

[3]*Department of the Faculty of Exact and Natural Sciences, Akaki Tsereteli State University, Georgia*

ABSTRACT

As a result of the research, it was determined that the transformation products of the LiOH-Sb_2S_3-S-H_2O system are crystalline hydrate salts containing metaantimonate and tetrathioantimonate anions, with a molar ratio of 3:5. Their molecular formulas, respectively, can be expressed as follows: $LiSbO_3 \cdot 2.5H_2O$ and $Li_3SbS_4 \cdot 9.5H_2O$.

10.1 INTRODUCTION

The synthesis of mixed oxothioarsenates (V) of alkali metals with the composition of M3AsO_4-xSx, where M is an alkali metal and x varies in the range of 1–3, is well studied [1]. As for antimony(V) products, as can be seen from literature data [2], they are not formed under such conditions.

Advanced Topics in Polymer Chemistry and Materials Science: Tamara Tatrishvili and Marc Jean Médard Abadie (Eds)

A significant difference is also observed in the preparation of arsenic (V) and antimony (V) tetrathioacid salts. Thus, for example, sodium tetrathioantimonate (V) is obtained in practice as follows [3, 4]:

$$8NaOH + Sb_2S_3 + 6S \rightarrow 2Na_3SbS_4 + Na_2SO_4 + 4H_2O \qquad (10.1)$$

At the same time, sodium dithioarsenate (V) is the main product of the system $NaOH–As_2S_3–S–H_2O$ [1]:

$$6NaOH + As_2S_3 + 2S \rightarrow 2Na_3AsO_2S_2 + H_2S + 2H_2O \qquad (10.2)$$

Similarly to the synthesis of sodium thioarsenate [1],

$$6NaOH + As_2O_3 + 2S \rightarrow 2Na_3AsO_3S + 3H_2O \qquad (10.3)$$

In the previous work we tried to obtain the same metal thioantimonate (V), but all our attempts to isolate a chemically pure target product using the following reaction:

$$6NaOH + Sb_2O_3 + 2S \rightarrow 2Na_3SbO_3S + 3H_2O \qquad (10.4)$$

were unsuccessful. As a result, we obtained sodium metaantimonate (V) as a solid phase and tetrathioantimonate (V) of the same metal in solution. Taking into account the experimental data obtained, the corresponding redox reaction can be expressed by the following equation:

$$6NaOH + Sb_2O_3 + 4S + 19.5H_2O \rightarrow 3NaSbO_3 \cdot 3.5H_2O + Na_3SbS_4 \cdot 9H_2O \qquad (10.5)$$

Previously, sodium tetrathioantimonate (V) was obtained by reaction (Eq. 10.1) [1]. Without taking into account the solid phase of the reaction mass, we assumed that it was a homogeneous mixture of sulfide of antimony (III) and residual sulfur. However, in the course of the work, it was found [6] that the color of this mixture differed significantly from the color of sulfur and, moreover, from the color of tri-sulfur antimony (Sb_2S_3). This fact, as well as the above-mentioned difference in the chemical behavior of antimony (III) chalcogenides, prompted us to test reaction (Eq. 10.1), although it has long been recommended [4, 7] as the main method of obtaining tetrathioantimonates (V) of alkali metals by hydrochemical means.

10.2 EXPERIMENTAL METHODS AND MATERIALS

To better understand and appreciate the present work, we will look at specific examples of the experiment.

Example 1. 5.60 g $LiOH{\cdot}H_2O$, 60 ml H_2O were added to a 250 ml *Erlenmeyer flask with screw cap* and, under constant stirring (magnetic stirring), 10 g of antimony(III) sulfide and 1.9 g of finely dispersed sulfur were added. We fitted the flask with a reflux condenser and a calcium chloride tube and carried out the reaction under constant stirring and heating. After 5 hours, the reaction zone acquired a stable gray color. We cooled the contents of the flask, filtered them, and after washing the resulting precipitate several times, air-dried it until a constant mass was obtained. We got 4.50 g of lithium metaantimonate crystal hydrate, which is 92.2% of the theoretical. Found, %: Sb 54.68; H_2O 20.39. $LiSbO_3{\cdot}2.5H_2O$. Calculated, %: Sb 54.90; H_2O 20.29.

To the mother liquor two volumes of 96% ethyl alcohol were added to remove $LiSbO_3{\cdot}2.5H_2O$. After filtering the small amount of precipitate obtained, the sulfur-containing substance was crystallized from the solution (by concentration), filtered, washed with ethyl alcohol, and dried in a vacuum desiccator on phosphorus pentoxide and paraffin to a constant weight.14.63 g of lithium tetrathioantimonate crystallohydrate was obtained, which is 90% of the theoretical amount. Found, %: Sb 27.44; H_2O 38.82. $Li_3SbS_4{\cdot}9.5H_2O$. Calculated, %: Sb 27.56; H_2O 38.70.

Example 2. From 8.4 g $LiOH{\cdot}H_2O$, 80 ml water, 15 g Sb_2S_3 and 2.8 g sulfur, – 6.85 g $LiSbO_3{\cdot}2.5H_2O$ (93.4% of theoretical) and 22.41 g Li_3SbS_4 $\cdot 9.5H_2O$ (91.9% of theoretical) were obtained.

Antimony was determined by the Evins method [9], sulfur [10], and crystallization water by the weighting method. In the latter case, the analyte sample was heated in a desiccator at 110–180°C until a constant mass was obtained.

10.3 RESULTS AND DISCUSSION

According to the data from earlier studies, it was established that the transformation of the NaOH–Sb2S3–S–H_2O system results in the formation of crystalline hydrates of sodium tetrathioantimonate and sodium triocostibiate. Accordingly, the oxidation reaction of antimony(III) sulfide by elemental sulfur in an aqueous solution of sodium alkali should be expressed by the equation.

$$18NaOH + 4Sb_2S_3 + 8S + 46.5H_2O \rightarrow 5Na_3SbS_4{\cdot}9H_2O + 3NaSbO_3{\cdot}3.5H_2O \quad (10.6)$$

Thus, the behavior of the second chalcogenide (sulfide) of sulfur (III) in the NaOH-Sb_2S_3-S-H_2O system has been established. The reaction products were found to be crystal hydrates of sodium methaantimonate and sodium tetrathioantimonate with a molar ratio of 3:5 [8].

Obviously, after the above-mentioned studies, the replacement of caustic sodium in the reaction system with lithium alkali and the study of interaction products $LiOH–Sb_2S_3–S–H_2O$ of the obtained substances was on the agenda.

In order to determine the similarity or difference between the processes taking place in the above-mentioned systems ($NaOH–Sb_2S_3–S–H_2O$ and $LiOH–Sb_2S_3–S–H_2O$), the reactants, similarly to the first one, were taken in the following molar ratios: $LiOH:Sb_2S_3:S = 9:2:4$.

After 5 hours of constant stirring (magnetic stirrer) and heating, the contents of the flask separated into solid (light gray powdery substance) and liquid (golden solid solution) phases. When maintaining the optimal conditions of the process, the yield of lithium metaantimonate (V) exceeds 90%. The non-reactive precipitation of the latter from the reaction area is explained by its low solubility in the mother solution. The filtrate can be freed from lithium metaantimonate (V) by re-precipitating it with ethyl alcohol and filtering the mother solution. The isolation of lithium tetrathioantimonate (V) in the individual state by careful evaporation of the filtrate on a water bath is then not a problem.

The current reaction should be presented as follows:

$$18LiOH+4Sb_2S_3+8S+46H_2O \rightarrow 3LiSbO_3\cdot 2.5H_2O+5Li_3SbS_4\cdot 9.5H_2O \quad (10.7)$$

Thus, as a result of the studies it was found that the transformation products of the system of $LiOH–Sb_2S_3–S–H_2O$, as well as the $NaOH–Sb_2S_3–S–H_2O$ system, are crystalline hydrate salts containing tetrathioantimonate and metaantimonate anions with a molar ratio of 5:3. Accordingly, molecular formulas of these salts can be expressed as follows: $Li_3SbS_4\cdot 9.5H_2O$ and $LiSbO_3\cdot 2.5H_2O$.

10.4 CONCLUSION

Thus, as a result of the studies it was found that the transformation products of the system of $LiOH–Sb_2S_3–S–H_2O$, are crystalline hydrate salts containing tetrathioantimonate and metaantimonate anions with a molar ratio of 5:3. Accordingly, molecular formulas of these salts can be expressed as follows: $LiSbO_3\cdot 2.5H_2O$ and $Li_3SbS_4\cdot 9.5H_2O$. Accordingly, it can be said that a new method for the simultaneous synthesis of lithium (V) tetrathioantimonate and lithium (V) metaantimonate crystalline hydrates has been developed.

KEYWORDS

- **alkaline**
- **lithium alkali**
- **lithium metaantimonate (V)**
- **lithium tetrathioantimonate (V)**
- **oxidation**
- **oxothioarsenates**
- **sulfide of antimony (III)**

REFERENCES

1. Brauer, G. (1985). *Handbook of Inorganic Synthesis* (Vol. 2). Mir.
2. Kizehhof, F. (1920). *Z. Anorg. Allgem. Chem.*, *112*, 67.
3. Schlippe, C. F., & Schweiggers, J. F. (1821). *Chem. U. Physik.*, *33*, 320.
4. Klyuchnikov, N. G. (1971). *Inorganic Synthesis*. Prosveshchenie.
5. Rukhaia, K. V., Chikovani, M. I., & Rusia, M. Sh. (2020). On the issue of oxidation of antimony (III) sulfide with elemental sulfur in alkaline medium. *European Science Review, 9, 10*, 42–45.
6. Samkharadze, M. G., Rusia, M. Sh., Kipiani, E. I., & Gigauri, R. D. (2000). Synthesis and transformations of ammonates of tetrathioantimonates (V) of d-metals. *Georgian Engineering News, 2*, 117–119.
7. Karyakin, Yu. V., & Angelov, I. I. (1974). *Pure Chemicals: Guide to the Preparation of Inorganic Reagents and Preparations in the Laboratory* (4th ed.). Chemystre.
8. Rukhaia, K., Rusia, M., Samkharadze, M., Kikalishvili, M., & Gigauri, R. (2007). Oxidation products of antimony (III) oxide by elemental sulfur in lithium alkaline medium. *Chemical Journal of Georgia, 7*(1), 19–20.
9. Freidlina, R. H. (1945). *Synthetic Methods in the Field of Organometallic Arsenic Compounds*. Izd. of the Academy of Sciences, USSR.
10. Charlo, G. (1965). *Methods of Analytical Chemistry*. Khimya.

PART III
Composites

CHAPTER 11

Composites Based on Secondary Thermoplastic Polymers and Some Minerals

LANA SHAMANAURI

R. Dvali Institute of Machine Mechanics, Tbilisi, Georgia

ABSTRACT

The polymer composites on the basis of secondary polyethylene and the minerals spread (andesite from Bakuriani, Sachkhere quartz sand and Okami slag) have been obtained. There are studied some physical-mechanical properties, thermal stability and materials hydrophobicity. It is shown that the ultimate strength and thermal stability of the composites extremely depend on the type and concentration of the filler. For composites containing binary filler (quartz sand + slag) discovered the synergistic effect- anomaly increasing of the ultimate strength at definite ratio of the fillers.

11.1 INTRODUCTION

Environmental ecological protection and the utilization of industrial waste are currently very important and relevant problems. From the scientific and technical literature, it is known that the development of composites based on secondary thermoplastic materials, in which different dispersive or natural and artificial fiber fillers are used, can spare about 40% of primary ores. In secondary polymer composites, industrial technological waste such as trimmings, injection molding heads, technical tare, films, bottles, and so on are used. The content of such waste varies in the range of 10% to 60%.

Advanced Topics in Polymer Chemistry and Materials Science: Tamara Tatrishvili and Marc Jean Médard Abadie (Eds)

Polyethylene is now one of the most widely used polymers, which is due to its many positive properties and low cost [1, 2]. In our work, high-pressure polyethylene industrial waste is used as a binder.

11.2 EXPERIMENTAL

11.2.1 RESEARCH OBJECTS AND METHODS OF OBTAINING

There were fine dispersive powders used, obtained as a result of grinding different polyethylene bags of domestic origin. Most of them are made from high-pressure polyethylene (with low density). Three types of minerals—Bakuriani andesite, Sachkhere quartz sand, and Okami slag—were used as composite fillers with a wide range of concentrations. Below, a short description of these fillers is presented.

Andesite (word basis -American mountains Ands) volcanic origination dark red color dense, but sometimes is porous material. This mineral is wide spread material in the AdJara-Trialeti mountain (Borjomi-Bakuriani, Tsikhisjvari) kazbegi region (mkinvartsvery, Kabarjina), in the sources of the rivers Liakhvi, Ksani, and Aragvi, on the Javakheti plateau. Andesite is used as a building and acid proof material;

Quartz sand from Sachkhere includes the quartz particles, content of silicon oxide near70–85% and rest are iron, calcium and magnesium oxides. Besides of the sand includes 5% clay and dust particles.

Slam from Okami is red color micro-porous volcanic generated mineral with high specific surface. In Georgia, this mineral is used as a warm-insulating material. The slam belongs to the basalt-type porous variety. It is glass with aluminosilicate content (75–80%). 20% of this material is crystalline. Density is 2630 kg/m^3.

At the initial stage, the mixing of composite ingredients in the propeller mill lasts for 2–3 minutes.

As a result of mixing the polymers and different fillers, a homogeneous powder was obtained, which, after preliminary drying at 50–70°C, underwent pressing in standard press forms (cylindrical and rectangular). The samples were obtained after pressing at 8–10 MPa and a temperature of 140–150°C for 10–15 minutes.

11.2.2 METHODS OF SAMPLES TESTING

The samples were tested for strengthening under compression. Mechanical parameters were defined using the German device of type Distant. Water absorption was measured separately. Temperature stability was assessed using the Vica apparatus.

11.3 RESULTS AND DISCUSSION

First of all, it was necessary to determine the dependence of material properties on the type and concentration of fillers. With this aim, we obtained composites based on polyethylene, which included mineral powders with concentrations of 20–60 wt% (20, 30, 40, 50, and 60).

On the Figure 11.1 The dependence of composite mechanical strengthening (at compression) on the filler concentration is presented. From this figure one can see that this dependence for all samples has an extreme character, on the curves the maximums are appeared. This result is in fully agree with well-known dependence – the left side of these curves corresponds to mechanical enhances because of formation the contacts between macromolecules and filler particle surfaces, but after definite concentration in result of formation of filler particles associates (at this moment not all particles wetted by macromolecules), which is equivalent to formation of the definite structural defects, the mechanical strengthening of composites decreases step by step. This process enhances with future increasing of the filler concentrations. The difference between curves is described by the difference in filler types. This difference is expressed in the character of the particles' surface profile (partially surface smoothness). The particles with deep irregularities contribute to the penetration of the polymer segments into the micro-empties of filler particles and the formation of engagements. In this way, physical bonds are formed (formation of Van der Waals bonds), which lead to an increase in the mechanical strengthening of the composite. Besides this, there is the possibility of the formation of chemical bonds between active chemical groups on the filler surface and macromolecules, which will further strengthen the composite. The character of the noted dependence appears in the thermal stability properties – at nearly the same concentrations, thermal stability achieves stable significance, which corresponds to the limit of this parameter (Figure 11.2).

It is known that in some cases it is possible to increase the physical properties of composites by using binary fillers [3]. With this aim, we prepared the composites based on PE and binary fillers, quartz sand and slam, the concentrations of which were varied in two groups of sum concentrations: 40 wt% and 50 wt%. In accordance with the experimental data on the determination of the dependence of ultimate strengthening of composites on the

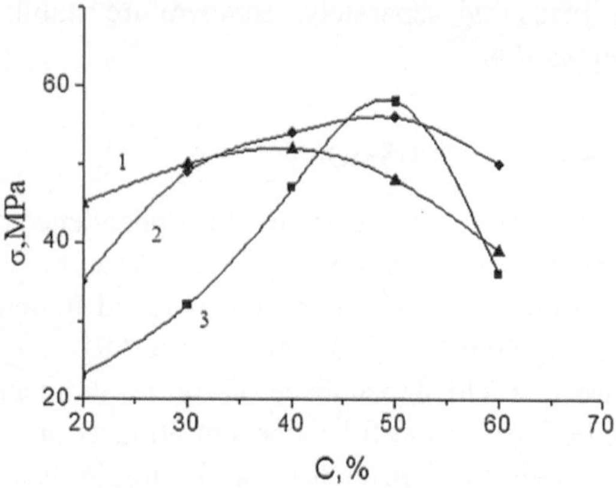

FIGURE 11.1 Dependence of the composite ultimate strength (at compression) of the PE composites containing quartz sand (1), slam(2) and andesite (3) on the filler concentration.

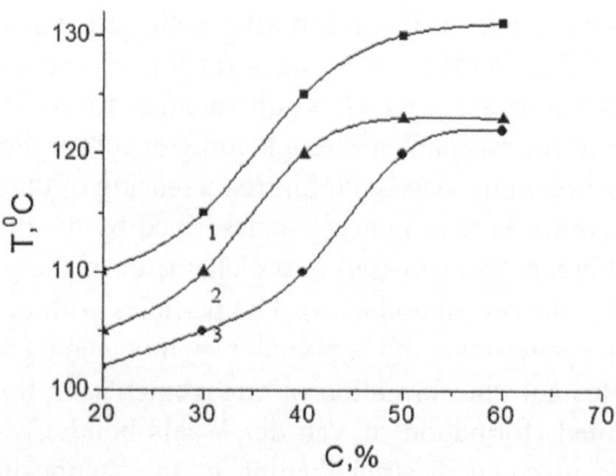

FIGURE11.2 Dependence of the composite softening (by Vica method) of the PE composites containing quartz sand (1), slam(2) and andesite (3) on the filler concentration.

different proportions of the fillers in them, these curves are characterized by maximums. In Figure 11.3, the maximum of ultimate strength appears for composites containing the fillers slam and quartz sand with a proportion of 30/70 (curve 1), when the sum of the fillers is 50 wt%. An analogous maximum is observed in curve 2 of the same dependence at a proportion of the same fillers 40/60, when the sum of the fillers is 40 wt%. The maximums on the curves correspond to the so-called synergistic effect – non-additive improvement of the material properties at definite proportions of ingredients in the binary filler [4, 5].

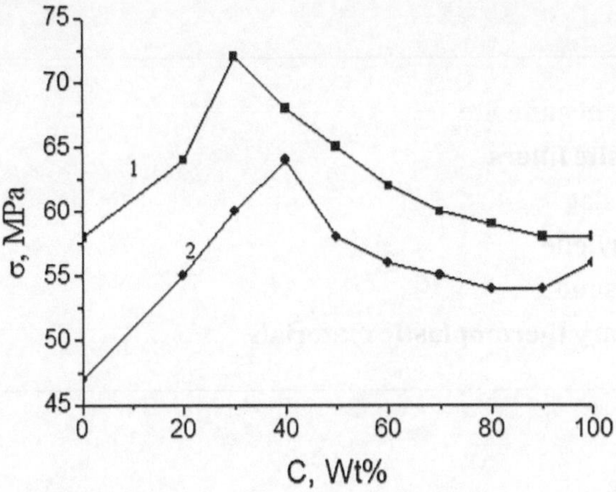

FIGURE 11.3 Dependence of the composite ultimate strength (at compression) of the PE composites with binary filler (quartz sand + slam) at the filler sum concentration 50wt% (1) and 40 wt% (2). On the *x* axis – the concentration of slam in binary filler

Comparison of numerical data of obtained dependences for composites containing binary fillers (Figure 11.3) and composites with one filler (Figure 11.1) shows that by combining the fillers, it is possible to enhance the mechanical properties of the composites. It must be noted that the experiments investigating the hydrophobic properties of the composites show that water absorption by the composites is rather low (no more than 1.5%). This fact shows that the microstructure of the composites contains a rather low amount of structural defects (partially empty spaces, cracks, etc.).

11.4 CONCLUSION

Polymer composites containing fine dispersed andesite, slag, and quartz sand, derived from industrial and domestic wastes, have been obtained and studied. Experimental results show that the physical-mechanical, thermal, and hydrophobic properties of these composites significantly depend on the type and concentration of fillers. For composites with binary fillers, an optimal filler proportion in the blend has been identified, resulting in enhanced physical properties compared to composites containing a single filler from the binary blend. Additionally, the water absorption of all investigated composites is no more than 1.5%.

KEYWORDS

- **Bakuriani andesite**
- **composite fillers**
- **Okami slag**
- **polyethylene**
- **quartz sand**
- **secondary thermoplastic materials**

REFERENCES

1. Achilias, D. S., Roupakias, C., Megalokonomos, P., & Lappas, A. (2007). *Journal of Hazardous Materials, 149*, 536.
2. Jassim, A. K. (2017). *Procedia Manufacturing, 8*, 635.
3. Asare, E., Basir, A., Tu, W., Porwal, H., Zhang, H., & Liu, Y. (2016). *Nanocomposites, 2*, 58.
4. Pedrazzoli, D., Pegoretti, A., & Kalaitzidou, K. (2015). *Journal of Applied Polymer Science, 132*, 4168.
5. Kucukdogan, N., Ozturk, S., & Sutcu, M. (2016). Archives of Material Science and Engineering, *1*, 12–18.

CHAPTER 12

Thermochromic Liquid Crystal Polymer Composites for Optical Visualization of Thermal Fields

GIA PETRIASHVILI, LALI DEVADZE, TSISANA ZURABISHVILI, NINO SEPASHVILI, ANDRO CHANISHVILI, SVETLANA TAVZARASHVILI, and MANANA ARESHIDZE

Institute of Cybernetics, Georgian Technical University, Tbilisi, Georgia

ABSTRACT

Thermochromic liquid crystal (LC) polymer films consisting of a nematic-chiral mixture with different temperature ranges were obtained using the technological process of a microencapsulation method, improved by the authors. Studies have shown that by controlling the microencapsulation process, the technological characteristics influence the thermo-optical parameters of polymer composites. Stretching of the films results in an increase in their thermosensitivity. The obtained LC polymer films are characterized by high intensity of coloring in different regions of the visible spectrum. The engineering properties determine the application of LC polymer films as rewritable thermoindicator polymer materials, which are promising for optical visualization of the thermal field distribution in various areas of thermoindication.

12.1 INTRODUCTION

The ability of cholesteric LC systems to change color with temperature changes allows them to be used for optical visualization of the thermal field distribution. Photochemically stable LC nematic-chiral mixtures, obtained

Advanced Topics in Polymer Chemistry and Materials Science: Tamara Tatrishvili and Marc Jean Médard Abadie (Eds)

by inducing LC with an optically active substance, are distinguished in such systems. Nematic-chiral mixtures have effective color-temperature parameters: high selective reflection intensity of the light wavelength (Bragg's reflection band), especially in the long wavelength region of the visible spectrum.

From a practical point of view, it is very important to obtain polymer materials based on the LC nematic-chiral mixtures. The polymer protects the integrated mixtures from external factors, preserves photochemical stability, and makes it possible to use multiple polymer materials.

The thermochromic LC system was introduced into the polymer by the method of microencapsulation, which was improved by the authors by choosing ice acetic acid as a suitable emulsifier [1, 2]. Ice acetic acid provides a stable dispersion system in polyvinyl alcohol (PVA), which contains dispersion phase-isolated compositions—aggregates of the microcapsules formed in the dispersion medium of an aqueous solution of PVA.

In the polymer composites, the initial properties of the nematic-chiral mixture are fixed in a microcapsule that preserves these parameters as much as possible and ensures the production of technologically perfect LC polymer composites with high reflection intensity. Using the technological process of the microencapsulation method, thermochromic LC polymer films [3–5] were obtained, consisting of a nematic-chiral mixture with different temperature ranges [6].

The presented article describes the technological process of the microencapsulation method for making new LC thermochromic polymer composites and discusses: a) their thermo-optical parameters based on spectral and optical-microscopic studies; b) the engineering properties of the obtained polymer films, which determine the practical application of LC polymer films as rewritable thermoindicator polymer materials.

12.2 EXPERIMENT

N and 8H thermochromic LC polymer films were obtained by the authors using the technological process of the microencapsulation method. Polymer films N were obtained by employing nematic-chiral mixtures, which contain a certified nematic substance ZLI-1939 with a cyano-biphenyl group and certified optically active substances ZLI-811 and ZLI-3786 with different ratios of components (wt%). For preparing films 8H, nematic-chiral mixtures consisting of nematic substances were employed: 4-p-n-hexyloxyphenyl-4-p'-octyloxybenzoate (HOFOOB) and 4-octyl-cyanobiphenyl (L-839) with an optically active substance tigogenin caprate (TC) with different ratios of components (wt%).

The technological process of making LC polymer films by the microencapsulation method consists of the following stages:

a) Preparation of encapsulated emulsion (layer);
b) Obtaining the polymer composites as a film;
c) Stretching the polymer film.

Obtained polymer composites consist of one encapsulated layer, containing the encapsulated composition; one protective layer, based on corresponding concentration of an aqueous solution of PVA and layer, based on a suspension of carbon black in an aqueous solution of PVA.

12.3 RESULTS AND DISCUSSION

Technologically perfect elastic and homogeneous LC polymer films are made using the technological microencapsulation method. LC polymer films 8H-1, 8H-2, and N-8, N-9, N-10 were obtained based on the nematic-chiral mixtures, the helix pitch p of which significantly depends on the temperature ($dp/dt < 0$). Such nematic-chiral thermoindicator polymer films are characterized by the shift of Bragg's reflection band towards the long wavelength region by cooling (Table 12.1). The thermo-optical parameters of the LC polymer films were measured using a fiber-optic spectrometer (Avantes-2048). Microscopic investigation of the microcapsules in the obtained polymer films was carried out using a polarizing microscope.

TABLE 12.1 Polymer Film Composition, Temperature and Wavelength Range of Selective Reflection

Polymer Film	Film Composition, wt.%	Temperature Range, °C	Wavelength Range of Selective Reflection, λ, nm
8H-1	L-839 – 76,0, HOFOOB- 9,5, TC-14,5	44,0–36,0	440–640
8H-2	L-839–82,5, HOFOOB-4,5, TC-13,0	39,5–31,0	480–680
N-8	ZLI-1939-74,5, ZLI-811 -25,5	51,0–24,0	460–650
N-9	ZLI-1939-76,0. ZLI-811 – 24,0	43,0–26,0	450–720
N-10	ZLI-1939-73,0. ZLI-3786-27,0	49,0–10,0	530–760

Microscopic studies of the films confirm that microcapsules in the polymer do not suffer disintegration upon temperature (Figure 12.1) and ultraviolet (UV) irradiation [7], which ensures maximum preservation of the properties of the composition in microcapsules. Hence, the optical parameters of LC polymer films do not change.

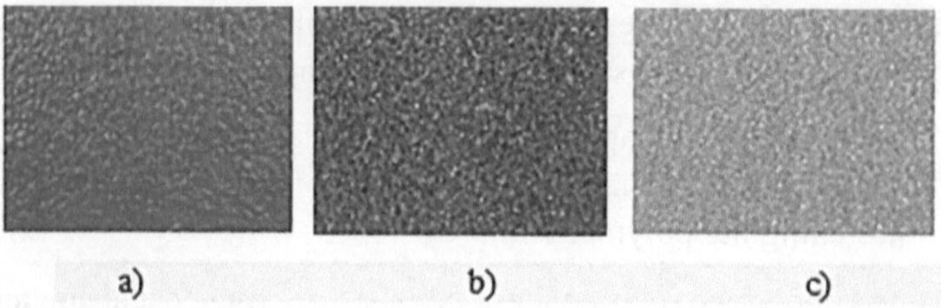

FIGURE 12.1 Microcapsules in LC polymer films: a), b), c) with temperature changes.(200x).

At all stages of the technological process, regulation of the technological characteristics, such as the sizes of microcapsules in film, the thickness of the film, and non-stretched and stretched film, influences the thermo-optical parameters of polymer composites, making it possible to obtain polymer films with high thermosensitivity.

The authors studied the influence of differing sizes of the microcapsules on the optical parameters of the thermochromic LC polymer composites and photochromic SPLC polymer composites consisting of a nematic-chiral mixture with spiropyran. Spectral and microscopic studies have shown that by controlling the stirring speed during the preparation of the encapsulated emulsion, a decrease in the size of the microcapsules increases the thermo- and photosensitivity of the polymer composites [7]. The change in thickness does not affect the intensity of selective reflection of the nematic-chiral LC polymer films, since the reflection occurs on the surface of the polymer film.

At the stage of stretching the polymer films, uniaxial deformation of the LC polymer films occurs. By stretching the non-stretched LC polymer films (Figure 12.2), a uniaxially oriented polymer is formed, the macromolecules of which, straightened by their axes, are oriented predominantly along one direction, which, together with the macromolecules, causes the ordered orientation of the microcapsules (Figure 12.3).

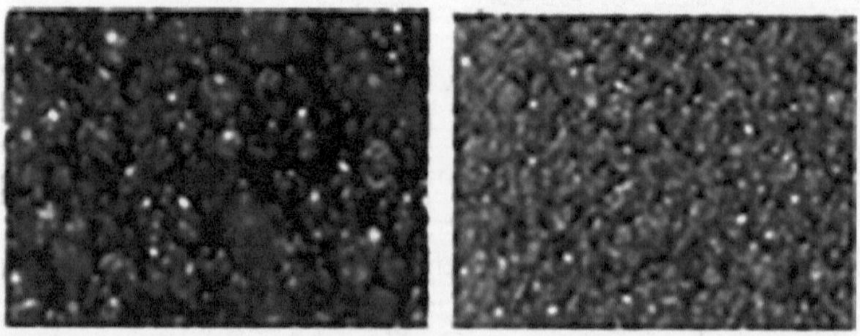

FIGURE 12.2 Non-stretched LC polymer films.

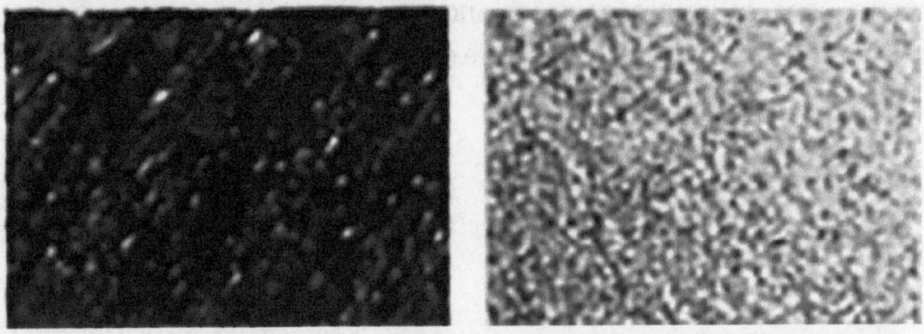

FIGURE 12.3 Stretched LC polymer films with microcapsules (200x) microcapsules (60x).

Stretching film samples leads to the formation of a planar configuration of the oriented texture of PVA macromolecules, which ensures the production of polymer films with improved optical properties in comparison with non-stretched films of the same composition. Stretching-orientation significantly increases the intensity of selective reflection of the nematic-chiral LC polymer films (Figure 12.4).

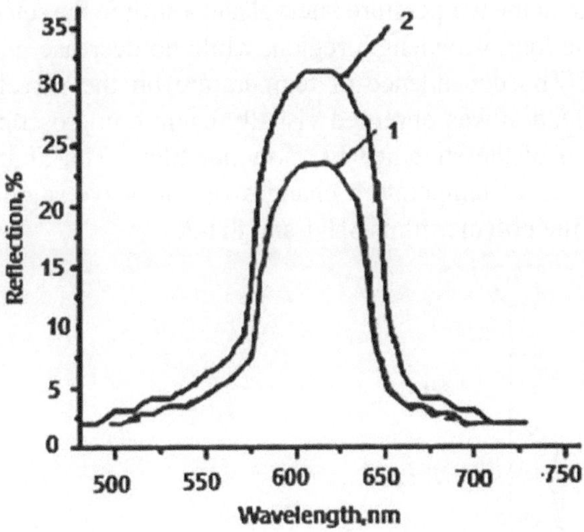

FIGURE 12.4 Intensity of selective reflection of LC polymer film: non-stretched film (1) and stretched film (2).

Thermoindicative properties of nematic-chiral stretched LC polymer films 8H-1, 8H-2, N-8, N-9, and N-10 significantly depend on temperature changes and are characterized by high intensity of coloring in the visible spectrum, especially intense green and red colors. With temperature changes, LC polymer films exhibit a full range of colors from blue to red (Figure 12.5).

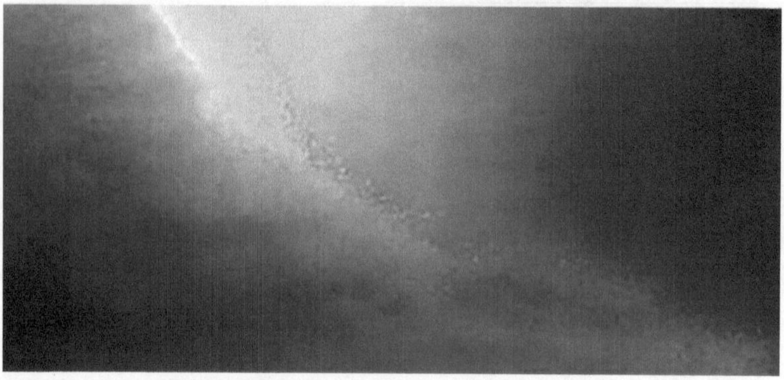

FIGURE 12.5 Visualization of temperature distribution in LC polymer film. Blue corresponds to the highest temperature, red – to the lowest temperature.

A change in the ratio of components in the films 8H-1 and 8H-2 results in both a change in the temperature interval and a shift in the selective reflection of light to the long wavelength region, while no decrease in color intensity is observed. The dependence of temperature on the wavelength of light selective reflection was observed visually using a microscope by changing the coloration of the stretched LC polymer films. The Figure 12.6 shows the dependence of temperature changes on the wavelength in the visible spectrum of the polymer films 8H-1 and 8H-2.

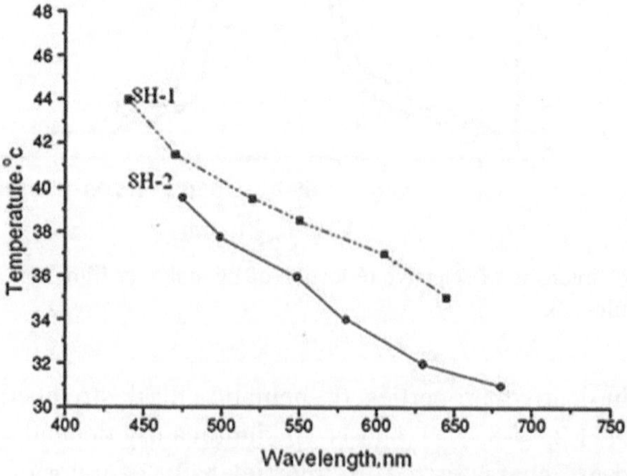

FIGURE 12.6 Color-temperature characteristics of 8H-1 and 8H-2 polymer films.

Polymer films 8H-1, 8H-2 have the high intensity of coloring, especially intense blue and red colors (Figure12.7).

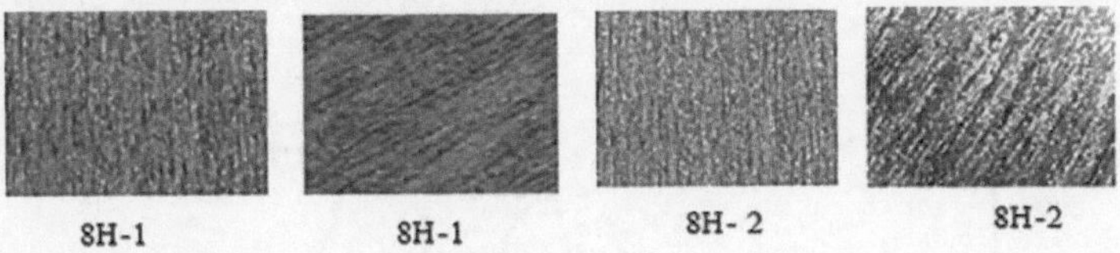

FIGURE 12.7 Coloration of the polymer films 8H-1 and 8H-2 in different regions of the visible spectrum (60x).

Polymer film N-8 is characterized by high intensity of selective reflection (Figure 12.8) and intense coloring in the blue, especially green, and red regions (Figure 12.9) of the visible spectrum.

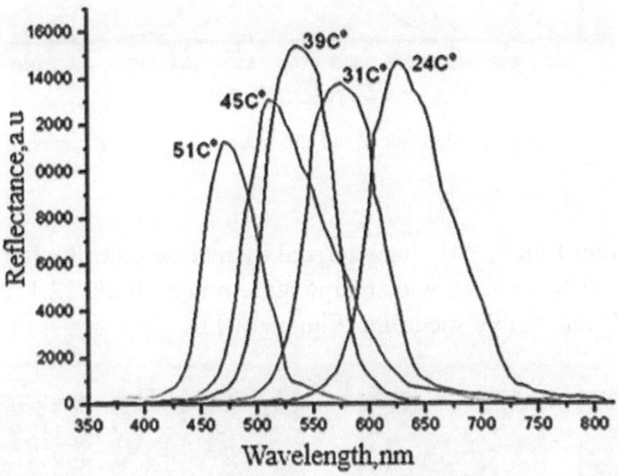

FIGURE 12.8 Temperature dependence of the selective reflection peaks of the polymer film N-8.

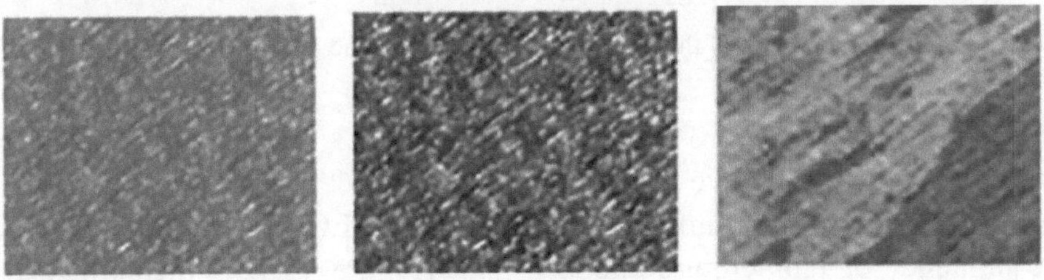

FIGURE 12.9 Coloration of the polymer films N-8 in the different regions of the visible spectrum (60x).

Polymer film N-9 has the high intensity of the selective reflection in the long wavelength region (Figure 12.10) of the visible spectrum.

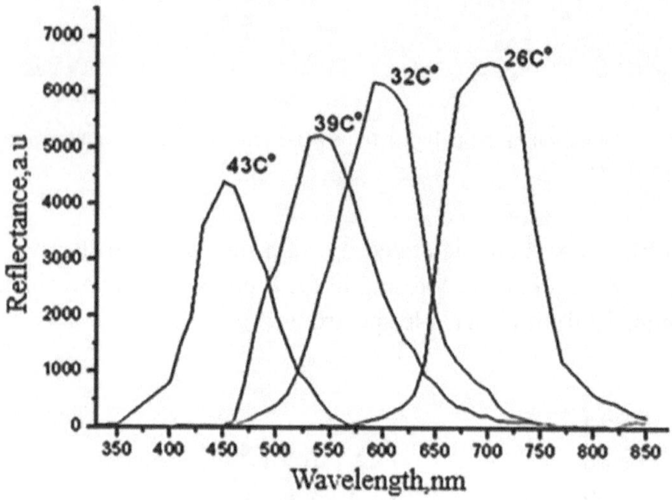

FIGURE 12.10 Temperature dependence of the selective reflection peaks of the polymer film N-9.

The polymer film N-9 is characterized by intense coloring in orange and red regions and N-10 has a wide temperature range (Table 12.1) and a large green parts of the visible spectrum (Figure 12.11).

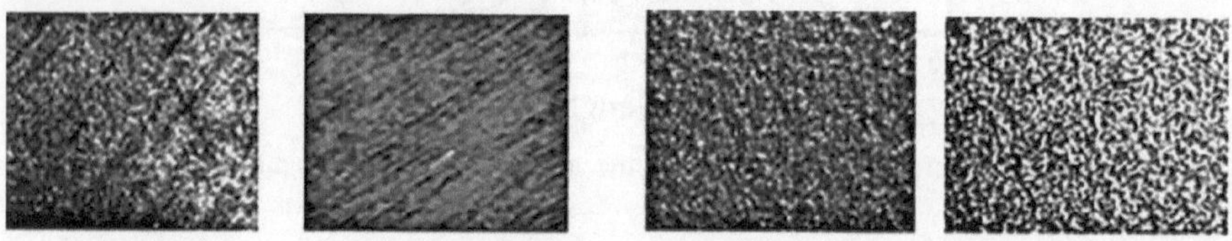

N-9 N-9 N-10 N-10

FIGURE 12.11 Coloration of the polymer films N-9 (60x) and N-10 (200x) in the different regions of the visible spectrum.

The change of selective reflection intensity in LC polymer composites is well demonstrated upon information recording in LC polymer films under the influence of temperature and UV light (Figure 12.12). Technological characteristics and thermo-optical parameters determine the engineering properties of the polymer composites, which is very important for the use of LC polymer films. The practical application of multiple uses of LC polymer films as rewritable thermoindicator polymer materials is promising for optical visualization of the thermal field distribution in various areas of thermoindication, such as:

FIGURE 12.12 Image recording in the LC polymer film induced by temperature and UV irradiation.

a) Medical thermographic diagnostics – it is possible to use the thermoindicator polymer films, the temperature range of which is in the area of healthy and pathogenic areas of the human body. With the help of such films, it is possible to visualize pathogenic areas in various diseases, including cancer (Figure 12.13).

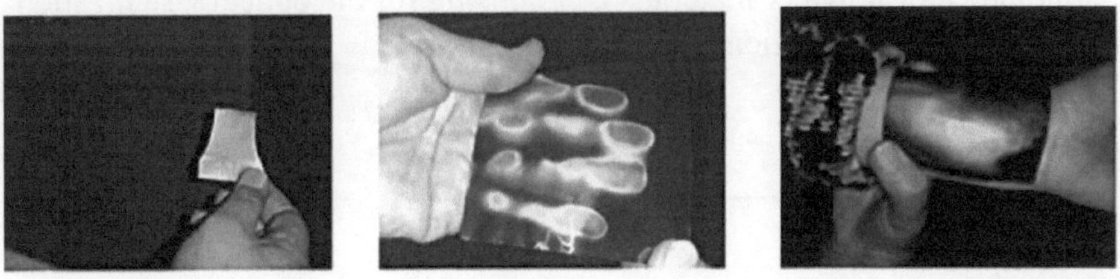

N-8 8H-2 8H-1

FIGURE 12.13 Image recording in the polymer film induced by temperature.

b) Human ecology – the received thermoindicator polymer films are an absolutely safe alternative to mercury thermometers as liquid crystal thermometry (8H-1, N-9).
c) Flaw detection – on the surface of the material to visualize the distribution of temperature fields, can use polymer films with a wide temperature range (N-8, N-10).

LC polymer films, containing a nematic-chiral mixture, the helix pitch p of which weakly depends on the temperature (dp/ dt ≈ 0), proved to be of interest as a matrix for obtaining reversible photochromic LC polymer films. Based on the LC nematic-chiral matrix doped with photochromic spiropyrans (SP), a new type of reversible photochromic SPLC polymer films was qualitatively obtained using the innovative microencapsulation method [8–10].

12.4 CONCLUSION

Thermochromic LC polymer films, consisting of nematic-chiral mixtures, were obtained using the technological process of the microencapsulation method developed by the authors. Spectral and microscopic research has shown that at all stages of the process, the regulation of technological characteristics influences the thermo-optical parameters of LC polymer composites. Stretching the films significantly increases the selective reflection intensity of LC polymer films. The obtained LC polymer films are characterized by high intensity of coloring in different regions of the visible spectrum.

It is possible to obtain technologically perfect LC polymer films of various shapes and sizes to meet practical application requirements. Technological characteristics and thermo-optical parameters determine the engineering properties of LC polymer films, and their application as rewritable thermoindicator polymer materials is promising for the optical visualization of thermal field distribution.

KEYWORDS

- **liquid crystal**
- **microencapsulation**
- **nematic-chiral mixture**
- **polymer composite**
- **polyvinyl alcohol**
- **thermochromic**
- **tigogenin caprate**

REFERENCES

1. Zurabishvili, T., Japaridze, K., Elashvili, Z., & Chelidze, G. (2002). *Thermoindicator Polymer Film*. Georgian Patent P 2683, Sakpatent Bulletin, 8.
2. Japaridze, K., Zurabishvili, T., & Petriashvili, G. (2011). *Method of Obtaining Photosensitive Liquid Crystal Polymer Film and Film Obtained by this Method.* Georgian Patent P 5232, Sakpatent Bulletin, 1.
3. Japaridze, K. G., Elashvili, Z. M., Chelidze, G. Sh., Zurabishvili, T. I., Vashakidze, E. Va., Petriashvili, G. Sh., Tavzarashvili, S. P., & Tevdorashvili, K. G. (1997). Thermochromic polymer films based on chiral nematic mixtures. *Crystallography Reports, 42*(2), 300–302.
4. Japaridze, K., Elashvili, Z., Zurabishvili, C., & Chelidze, G. (2006). Thermosensitive polymer films. *Georgia Chemical Journal, 4*(6), 461–463.
5. Japaridze, K. G., Chilaya, G. S., Elashvili, Z. M., Zurabishvili, T. I., Petriashvili, G. Sh., & Chelidze, G. Sh. (2008). Thermosensitive liquid-crystal polymer films. *Georgian Engineering News, 3*, 63–67.
6. Chilaya, G. S., Elashvili, Z. M., Ivchenko, S. P., & Vinokur, K. D. (1984). New nematic-chiral mixtures for application in thermography. *Molecular Crystals and Liquid Crystals, 106*, 67–69.
7. Devadze, L., Petriashvili, G., Chanishvili, A., Zurabishvili, T., Sepashvili, N., Chubinidze, K., Akhobadze, T., & Ponjavidze, N. (2021–2022). Technological method of increasing the photosensitivity in photochromic liquid crystal polymer films. *Nano Studies, 21/22*, 267–276.
8. Japaridze, K., Devadze, L., Maisuradze, J., Petriashvili, G., Zurabishvili, T., Mzhavanadze, I., & Sepashvili, N. (2013). A novel method of increasing the photosensitivity of spiropyran-containing systems. *Bulletin of the Georgian National Academy of Sciences, 7*(3), 57–62.
9. Petriashvili, G., De Santo, M. P., Devadze, L., Zurabishvili, T., Fonjavidze, N., Sepashvili, N., Gary, R., & Barberi, R. (2016). Rewritable optical storage with a spiropyran-doped liquid crystal polymer film. *Macromolecular Rapid Communications, 37*, 500–505.
10. Petriashvili, G., Devadze, L., Chanishvili, A., Zurabishvili, T., Sepashvili, N., Ponjavidze, N., De Santo, M. P., & Barberi, R. (2018). Spiropyran doped rewritable cholesteric liquid crystal polymer film for the generation of quick response codes. *Optical Materials Express, 8*(12), 3708–3716.

PART IV
Nanostructures

CHAPTER 13

Zeolite-Coal Nanocomposites and Soil Structure

EPRIKASHVILI LUBA, TSINTSKALADZE GIORGI, KORDZAKHIA TEIMURAZ, ZAUTASHVILI MARINE, PIRTSKHALAVA NINO, DZAGANIA MAIA, SHARASHENIDZE TINATIN, and GABUNIA VAKHTANG

Petre Melikishvili Institute of Physical and Organic Chemistry, Ivane Javakhishvili Tbilisi State University, Tbilisi, Georgia

ABSTRACT

Soil is one of the main natural resources of mankind. Restoring the productivity of degraded and contaminated soils and soil conservation is an urgent problem of modernity. Natural zeolites are unconventional mineral fertilizers capable of increasing crop yields. These minerals – crystalline hydrated aluminosilicates – are nanoporous materials with a pore size in the range of 1.5–10 nm and are of interest for nanotechnology. The main and important feature of the zeolite structure is the system of intracrystalline pores and channels, where ion exchange, occlusion, and release of small molecules can easily take place, resulting in a great ability of zeolite adsorption and desorption.

Brown coal is rich in biologically active complex humic substances, which contain humic acids and humates and act on a complex "soil-plant" system like zeolites. Humates are easily soluble salts that, after entering the nanoporous space of zeolites, form biomineral composites that have the ability to increase and accelerate physiological and biochemical processes. During their systematic use, the buffer and ion-exchange properties of the soil are improved, the phosphorus and organo-mineral balance in the soil and the plant are enhanced, as well as the mineral elements being transformed

Advanced Topics in Polymer Chemistry and Materials Science: Tamara Tatrishvili and Marc Jean Médard Abadie (Eds)

into a form available to the plant, the plant's root system is strengthened, and the vegetation period is shortened.

The goal of the work is to study the effect of organozeolite fertilizer (zeolite – brown coal), prepared on the basis of local raw materials, on the bio productivity of the agrolandschaft. Such fertilizer will help to manage brown coal waste and will contribute to the long-term enrichment of the soil with humic components using zeolite, thereby increasing fertility.

13.1 INTRODUCTION

Soil fertility is closely related to the features of its genesis, is characterized by high dynamism, and responds to changes in the factors and conditions of soil formation. Especially great influence on the level of fertility has anthropogenic intervention [1, 2]. Restoration and preservation of the productivity of degraded and polluted soils is an urgent problem of our time.

The 73rd session of the UN General Assembly officially adopted a resolution declaring 2021–2030 as the UN Decade of Ecosystem Restoration. The successful solution of this global and topical problem is closely related to agro-ecological problems of restoration and maintenance of soil fertility [1, 2].

To achieve a high yield, it is necessary to apply mineral fertilizers to the soil in a timely manner. The currently used mineral fertilizers have a significant drawback: their useful efficiency does not exceed 30–40%; the rest is washed into the soil or evaporates, worsening the environmental situation. The low coefficient of effectiveness of the fertilizer nutrients leads to an unreasonable increase in the rates of fertilizer application. High rates of application contribute to soil salinization, reduction of organic matter and humus content, accumulation of nitrates and phosphates in the soil and fruits, development of erosion, and loss of soil biota, structure, and fertility [3].

The use of environmentally friendly minerals containing nutrients necessary for plants, which at the same time improve soil properties, is of particular importance for solving the above problems. From this point of view, of great scientific and practical interest is the use of local resources — brown coal and natural zeolites as ameliorants [4, 5].

Natural zeolites are unconventional mineral fertilizers that allow for an increase in crop yields of up to 50%. Thanks to their high ion-exchange capacity and ability to retain moisture and useful components for a long time, zeolites are used as soil water-salt conditioners. Zeolite keeps moisture in the soil for a long time and slowly and continuously supplies it to the plant.

Natural zeolite improves aeration of the soil, facilitates the development of the root system and plant growth, and retains 40% of the water mass in the root zone; it acts as a reservoir of nutrients necessary for plant life and growth and enriches the soil over time. Zeolite application is especially effective for increasing the fertility of uncultivated, non-irrigated, dry soils [6–18].

Organic compounds present in brown coal play an important role in the formation of the soil biosphere. They contain 70–75% of humic substances. Humic substances are formed in nature as a result of the transformation of organic waste and are currently considered one of the promising areas of "green" chemistry, as a source of soil-renewing, economically beneficial, and environmentally friendly raw materials [19–21].

Humic preparations are easily soluble salts, which are active physiological forms of humic acids. They act at the cellular level and increase the rate of physiological and biochemical processes. Their systemic application improves the buffer and ion-exchange properties of soil, as well as the phosphorus and organic-mineral balance in soil and plants; the operation of soil microorganisms is activated, and mineral elements are transformed into a form available for plants [19–23].

When humic substances interact with zeolites, exchangeable zeolite cations are involved in the reaction. The parts of humic substances not bound to these cations move and are adsorbed on the outer surface of the zeolite. As a result of these interactions, an adsorption nanocomposite of zeolite-humic substances is formed [24–28].

13.2 EXPERIMENTAL METHODS AND MATERIALS

Two different coal ores in Georgia were used as objects of study: coal waste taken from the Tkibuli coal deposits and samples from the Vale brown coal deposits. Zeolite-clinoptilolite from the Khandaki deposit of Kaspi municipality was also used.

The preparation of the desired fraction of the samples was carried out using the following devices: Laboratory jaw crusher (RAM 35, Rantek Brand, Turkey) and Electromagnetic vibrator-shaker (BIOBASE BK-TS200, China). The processed samples were studied using chemical, X-ray diffractometric, and IR-spectroscopic research methods.

Chemical analysis was carried out on a Spectroscout XEP-04 (Germany); X-ray diffractometry analysis was carried out on a Dron-4 device (Russia). The X-ray diffractometer connected to a personal computer via the USB port multimeter AX-18B and the corresponding PC software-PC-Link, which allows the processing of experimental data in Excel format; Infrared spectrometric analysis was carried out on an Agilent Cary 630 FTIR Spectrometer in the range of 350–5000 cm^{-1} (USA).

13.3 RESULTS AND DISCUSSION

13.3.1 THE RESULTS OF CHEMICAL ANALYSIS OF SAMPLES

The results of chemical analysis of two coal samples are presented in Table 13.1.

TABLE 13.1 Chemical Analysis of Samples,%

Material Name, Quarry	Percentage Composition															
	C	O	Si	Al	H	Fe	S	Ca	N	Mg	K	Ti	Na	Cl	P	Total
Tkibuli	62	16	7.0	4.3	4.1	2.1	1.9	0.8	0.5	0.3	0.2	0.2	0.1	0.05	0.02	98.92
Vale	68	8.2	7.6	4.2	4.0	1.3	1.3	0.7	0.8	0.5	0.3	0.2	0.1	0.1	0.02	97.97

As a result of chemical analysis, besides the listed main elements, the following elements were recorded in insignificant amounts – Sr, Ba, V, Ge, Zn, Cu, Cr, Ni, Mn, Co, Rb and Zr.

The results of chemical analysis of zeolite-clinoptilolite are presented in Table 13.2.

TABLE 13.2 Chemical Analysis of Zeolite Clinoptilolite,%

Composition of Zeolite		
Formula	**Elements**	**Concentration**
SiO_2	Silicon oxide	46.91
Al_2O_3	Aluminum oxide	10.4
CaO	Calcium oxide	7.04
Fe_2O_3	Iron oxide	4.054
K_2O	Potassium oxide	1.859
MgO	Magnesium oxide	1.817
Na_2O	Sodium oxide	1.275
TiO_2	Titanium oxide	0.5647
SO_3	Sulphur oxide	0.2648
MnO	Manganese oxide	0.09538
P_2O_5	Phosphorus oxide	0.08158

Chemical analysis of zeolite shows that zeolite has a high silica content and is mainly a form of calcium.

13.3.2 THE RESULTS OF X-RAY DIFFRACTOMETRIC ANALYSIS

Coals from both deposits are of organic origin, and their structures are mostly amorphous. The study of such minerals by X-ray diffractometric analysis does not provide much information, although some nuances regarding aluminosilicates in the structure of coals can still be obtained. Figure 13.1 shows the X-ray diffractograms of the initial, heated to 100°C, and 600°C coal samples of Vale and Tkibuli.

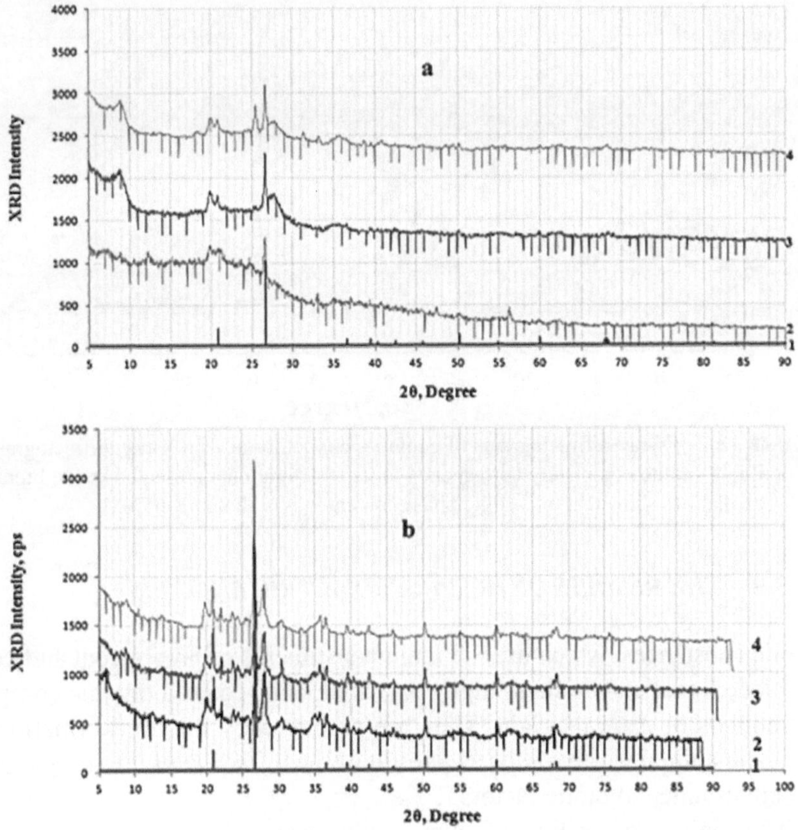

FIGURE 13.1 X-ray Diffractograms of samples. a – Brown coal from the Vale deposit with content: 1 – SiO_2, 2 – initial sample, 3 – sample heated to 100°C; 4 – sample heated to 600°C. b – Brown coal from the Tkibuli deposit with content: 1 – SiO_2, 2 – initial sample, 3 – sample heated to 100°C; 4 – sample heated to 600°C.

It is clear from the diffractograms that the structure of samples from different deposits differs from each other both qualitatively and quantitatively: in brown coal, the part of organic amorphous mass is greater than in coal, so its share in the formation of humus is higher.

The main difference between coal and brown coal is that brown coal has a lower amount of carbon and a higher amount of bituminous volatiles. Compared to coal, brown coal has a lower nitrogen content. Besides, brown coal has retained its woody plant structure [29, 30]. As a result, when applied to the soil, brown coal produces more humic substances. The zeolite-clinoptilolite structure of the Khandaki location was also studied. Figure 13.2 shows an X-ray diffractogram of clinoptilolite.

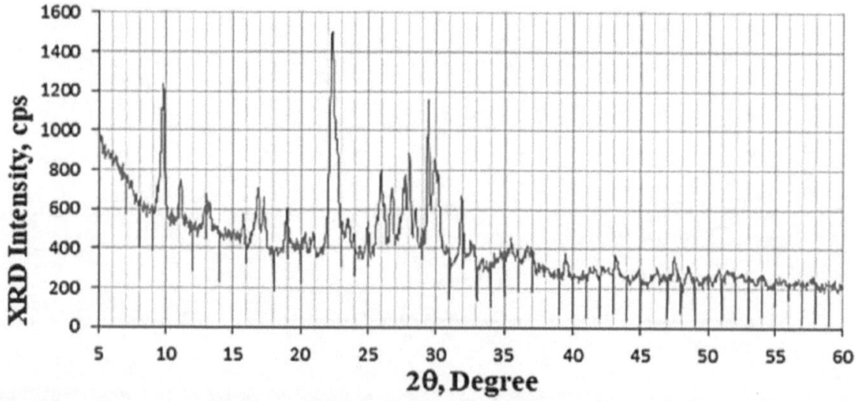

FIGURE 13.2 X-ray diffractograms of zeolite – clinoptilolite. The X-ray diffractogram of clinoptilolite zeolite completely corresponds to its structural data presented in the literature [31]. Calculations show that the zeolite phase content in it is about 60–65%.

13.3.3 THE RESULTS OF IR SPECTROMETRIC ANALYSIS

Humic substances, which are the main constituents of brown coal and coal, are biodegradation-resistant, high-molecular, dark-colored organic complex compounds of natural origin. They are produced by the transformation of plant and animal waste fossils as a result of the influence of microorganisms and environmental biotic factors [32].

The main composition of humic substances consists of humic acids, namely humic and fulvic acids. The method of FT-IR spectroscopy makes it possible to identify all the characteristic structural fragments of acids contained in humic substances and thus determine not only the composition of the corresponding substance but also their structural fragments and the differences between them [33].

Figure 13.3 presents the IR-spectra of brown coal sample from Vale and coal sample from Tkibuli in the region of 350–5000 cm^{-1}.

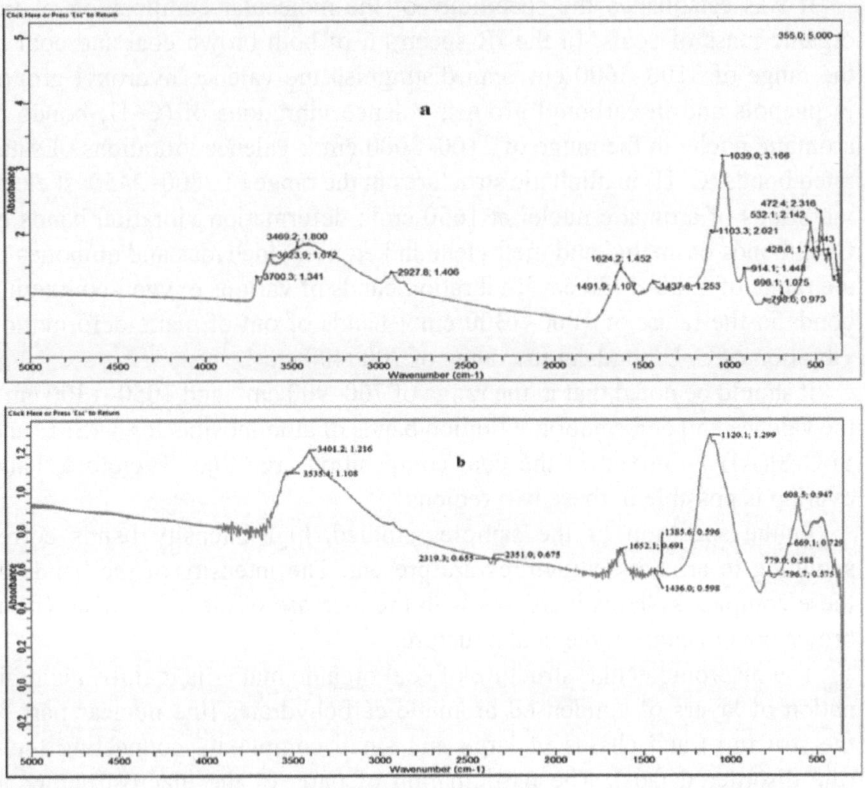

FIGURE 13.3 IR-Spectrum of the samples: a – brown coal of Vale location; b – coal of Tkibuli location.

Coal fossil is a complex composite system consisting of organic mineral micro-components and mineral inclusions.

The modern tendencies in research on coals involve the use of various physicochemical instrumental methods, among which the infrared-spectroscopic method occupies a special place. The IR-spectroscopy method has a rather high sensitivity to detect the interaction between the elements with a chemical structure and carbon-organic mass molecules [32–36].

The structural and functional composition of humic acids contained in these coals was determined by infrared spectroscopy. Methyl ($-CH_2$), carboxyl (–COOH), alcohol (–OH), thio-carbonyl (–C=S), amine ($-NH_2$), and amide ($-CONH_2$) groups were identified in these acid molecules [35, 36].

It was established the specificity of the molecular composition of the organic mass of coals. In the IR-spectrum of both brown coal and coal in the range of 3100–3600 cm^{-1} can distinguish the valence hydroxyl groups in phenols and in carbonyl groups; valence vibrations of (C–H) bonds in aromatic nuclei in the range of 3100–3000 cm^{-1}; valence vibrations of saturated bonds (C–H) in aliphatic structures in the range of 2800–2450; skeletal vibrations of aromatic nuclei at 1650 cm^{-1}; deformation vibration bands of C=O bonds of methyl and methylene in ketones, aldehydes and quinones in the range of 1350–1470 cm^{-1}; vibration bands of various oxygen-containing bonds in the range of 1000–1300 cm^{-1}; bands of out-of-plane deformation vibration of C–H bonds in the range of 700–400 cm^{-1}.

It should be noted that in the range of 700–900 cm^{-1} and 1050–1300 cm^{-1} the valence and deformation vibration bands of aluminosilicates — Si-O and Si-O-Si(Al) — mixed in the coal composition are fixed. Therefore, band overlap is possible in these two regions.

In the spectrum of the samples studied, high-intensity bands corresponding to aromatic structures are present. The intensity of the bands of these complex systems increases with the increase of metamorphism (from brown coal to coal) in the coal structure.

The macromolecular structure of coal organic matter is mainly a combination of layers of condensed aromatic carbohydrates (the nuclear part of the structure) and chains of large and small complexity connecting them (the disordered part). The participation of parts of the macrostructure in determining the overall structure depends on the degree of metamorphism of coals.

It can be said that the development of structural systems of coals, and the increase in metamorphism stage is due to a decrease in aliphatic C–H bonds (≈2900 cm^{-1}) and an increase in aromatic unsaturated C–H bonds, the vibration frequency of which is about 3040 cm^{-1} [36].

Thus, the intensity correlation of the 2920 cm^{-1} band depends on the removal of volatile substances from coals. In the obtained IR spectra, bands characterizing the polyfunctionality of humic acid molecules in the studied coals are registered. Table 13.3 shows the functional groups observed in the spectra of the studied samples.

TABLE 13.3 Vibrational Assignment of the IR-Spectrum of Studied Samples

Vibrational Assignment	Coal from Tkubuli	Brown Coal from Vale
–OH^- – intermolecular	3535, 3401	3700, 3623 3469, 3410
– CH_2 – methyl group	2945, 2855	2975, 2927 2865
O =C=O	2351	–
–C=O – valency	1652	1624
C–CH_3 – deformation	1436, 1385	1491, 1437
C–O	1120, 1100	1103, 1039
Si–O, Si–O–Si(Al) – valency		1020
C–H – Non-flat deformation rings	796, 779	914, 798
C–S – valency	669, 608	775, 696
Benzene substituted chains S– S and P=S bonds	519, 495	532, 472, 428

Thus, the infrared spectra of both samples register fluctuations of different shapes and intensities of the same bonds of different concentrations.

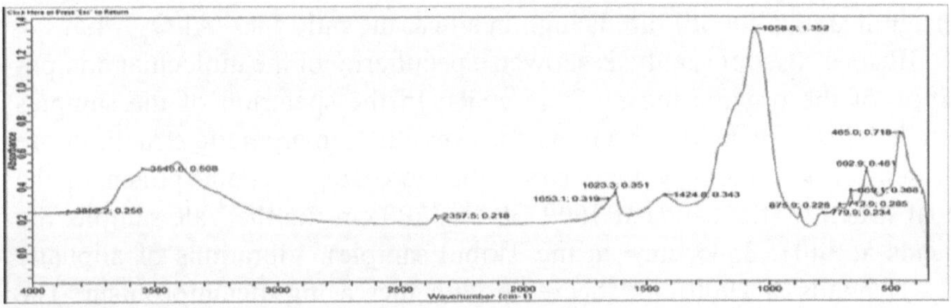

FIGURE 13.4 IR-spectrum of the zeolite – clinoptilolite.

Figure 13.4 presents the IR spectrum of zeolite-clinoptilolite from Khandaki location. The spectrum fully corresponds to the spectra of zeolites with the corresponding structure available in the literature [37, 38].

13.3.4 INFLUENCE OF NANOCOMPOSITE (ZEOLITE-BROWN COAL) PREPARED ON THE BASIS OF LOCAL RAW MATERIAL ON THE BIO PRODUCTIVITY OF THE AGROLANDSCAPE

The test of the obtained composite was carried out in the vegetative dishes. Winter wheat was used to determine changes in the fertility of soil enriched with substrates. Biometric indices of the experimental plant (germination, weight of green mass, weight of roots, average height of seed germination) were compared with the data obtained on clean soil (control). The results showed that soil enrichment with such a composite has a positive effect on the biometric indicators of grain crops. Their prolonged action and quality were also demonstrated [14–18, 26–28].

13.4 CONCLUSION

Fossil coals of two different metamorphisms (brown coal and coal) from different locations in Georgia (Tkibuli coal and Vale brown coal), as well as natural zeolite-clinoptilolite (from Kaspi, Zemo-Khandaki), were studied.

Their chemical composition, as well as their structure, is shown. The percentage content of zeolite in zeolitic rock is about 60–65%. Clinoptilolite is a form of calcium. The chemical composition of brown coal and coal differs in the amount of carbon (more in brown coal) and oxygen (more in coal); the difference between the amounts of other elements is insignificant. Clinoptilolite is high in silica, the ratio $SiO_2/Al_2O_3 = 4.5$.

IR spectroscopic analysis showed a peculiarity of the molecular composition of the organic mass of the coals. In the spectrum of the samples studied, high-intensity vibrations corresponding to aromatic structures are present. Their intensity increases with increasing metamorphism in the coal structure (the bands at 3609, 3923, 3700 cm^{-1} in the Vale sample; the bands at 3401, 3535 cm^{-1} in the Tkibul sample). Vibrations of aliphatic C—H bonds at 2900 cm^{-1} decrease with increasing metamorphism. The correlation of the intensity of this band depends on the decrease in the volume of volatile substances.

In addition, the IR spectra of coals contain bands confirming the polyfunctionality of humic acids in the studied samples. These are bands in range of: 3400–3600, 1624–1674, 1100–1020 and 914–700 cm^{-1}. Humic substances form a composite with zeolite, which is a unified system of humic substances fixed and adsorbed in the active centers and nanopores of zeolites, preserving the properties of both components. This enables us to use the nanocomposite as a humus-enriched zeolite fertilizer of prolonged action.

KEYWORDS

- **brown coal**
- **coal**
- **fulvic acids**
- **humic acids**
- **humus**
- **x-ray diffractometer**
- **zeolite-clinoptilolite**

REFERENCES

1. Burton, L., & Cooper, E. L. (2005). *Agroscience*. Thomson Press.
2. Third National Environmental Action Program of Georgia. (2017). *2017–2021, Working Version*. Tbilisi. (in Georgian)
3. Abramov, S. N. (2017). *Method for Obtaining Mineral Fertilizer*. Patent № 261462.
4. Urushadze, T. (2020). *Agrology*. Georgian Agrarian University.
5. Urushadze, T. F., & Blum, W. E. H. (2014). *Soils of Georgia*. Nova Science Publishers.
6. Andronikashvili, T., Urushadze, T., Eprikashvili, L., Gamisonia, M., & Nakaidze, E. (2008). Towards the biological activity of the natural zeolite—clinoptilolite-containing tuff. *Bulletin of the Georgian National Academy of Sciences, 2*(3), 99–108.
7. Tsitsishvili, G. V., Andronikashvili, T. G., Kirov, G. N., & Filizova. (1992). *Natural zeolites*. Ellis Horwood Limited.
8. Febles, J. A., & Núñez, G. M. S. (2010). Natural zeolite in the industry of fertilizers—agronomic results. In *Proceedings of the 8th International Conference on the Occurrence, Properties, and Utilization of Natural Zeolites* (pp. 86–87). Sofia, Bulgaria.
9. Buondonno, A., Coppola, E., de'Gennaro, B., Leone, A. P., Letizia, A., & Colella, C. (2010). Morpho-structural and chemical-physical properties of soil ped aggregate models by interaction of phillipsite-chabazitic or clinoptilolitic tuffs with different organic matter. In *Proceedings of the 8th International Conference on the Occurrence, Properties, and Utilization of Natural Zeolites* (pp. 52–53). Sofia, Bulgaria.
10. Bernardi, A. C. de C., Anchao Oliveira, P. P., de Melo Monte, M. B., & Souza-Barros, F. (2013). Brazilian sedimentary zeolite use in agriculture. *Microporous and Mesoporous Materials, 167*, 16–21.
11. Andronikashvili, T. G., & Urushadze, T. F. (2008). Use of zeolite-containing rocks in plant growing. *Agrochemistry, 12*, 63–79. (in Russian)
12. Andronikashvili, T., Urushadze, T., & Eprikashvili, L. (2009). Zeolite-containing substrates—A new way in plant production. *Annals of Agrarian Science, 7*(4), 14–45.
13. Deligianni, L., Petounis, D., & Voulgari, M. (2010). Clinoptilolite in agricultural applications. In *Proceedings of the 8th International Conference on the Occurrence, Properties, and Utilization of Natural Zeolites* (pp. 76–77). Sofia, Bulgaria.

14. Eprikashvili, L., Zautashvili, M., Kordzakhia, T., Pirtshkalava, N., Dzagania, M., Rubashvili, I., & Tsitsishvili, V. (2016). Intensification of bioproductivity of agricultural cultures by adding natural zeolites and brown coals into the soil. *Annals of Agrarian Science, 14*(2), 67–71. https://www.sciencedirect.com/science/article/pii/S1512188716300124 (accessed on 1 October 2024).
15. Eprikashvili, L., Tsitsishvili, V., Zautashvili, M., Kordzakhia, T., Dzagania, M., & Pirtskhalava, N. (2015). Influence of the soil-free substrate on the biometric parameters of bean and barley germination. *Bulletin of the Georgian National Academy of Sciences, 9*(1), 139–144. http://science.org.ge/old/moambe/9-1/Eprikashvili.pdf (accessed on 1 October 2024).
16. Eprikashvili, L., Zautashvili, M., Kordzakhia, T., Pirtskhalava, N., Dzagania, M., & Rubashvili, I. (2015). Application of composites of brown coal and natural zeolites as biological fertilizers in plant growing. In *Proceedings of the 5th International Scientific Conference "Applied Natural Sciences 2015" Perspectives in V4 Countries* (pp. 39–40). Jasná, Low Tatras, Slovak Republic.
17. Zautashvili, M. G., Eprikashvili, L. G., Kordzakhia, T. N., Pirtskhalava, N. V., Dzagania, M. A., & Rubashvili, I. V. (2013). About usage of nontraditional fertilizers based on natural composites. *Proceedings of the Georgian National Academy of Sciences, Chemical Series, 39*(1, 2), 104–106.
18. Andronikashvili, T., Zautashvili, M., Eprikashvili, L., Burkiashvili, N., & Pirtskhalava, N. (2012). Natural zeolite—One of the possibilities of transition from "chemical" to biological agronomy. *Bulletin of the Georgian National Academy of Sciences, 6*(2), 111–118. https://www.semanticscholar.org/paper/Natural-Zeolite-One-of-the-Possibilities-of-from-to-Andronikashvili-Zautashvili/0e7fb3444437b3c7adc27b8219109b1ad143a83d (accessed on 1 October 2024).
19. Zvyagintsev, D. G., Shapovalov, A. A., Putsykin, Y. G., Stepanov, A. L., & Lysak, L. V. (2008). Humic acid purification by means of soil microorganisms. *Doklady Biological Sciences, 420*, 204–206. https://link.springer.com/article/10.1134/S0012496608030186 (accessed on 1 October 2024).
20. Bratkova, S., Nikolova, K., & Chakalov, K. (2012). Potential for bioremediation of calcareous soils by rhizospheric bacteria and humic acids. *Annual of the University of Mining and Geology "St. Ivan Rilski," 55*(II), 217–222.
21. Lyons, G., & Genc, Y. (2016). Commercial humates in agriculture: Real substance or smoke and mirrors? *Agronomi, 50*(6), 1–8.
22. Leggo, P. J. (2010). Organo-zeolitic bio-fertilizers: A review of recent experiments. In *Proceedings of the 8th International Conference on the Occurrence, Properties, and Utilization of Natural Zeolites* (pp. 157–158). Sofia, Bulgaria. https://doi.org/10.4172/2168-9881.1000165.
23. Leggo, P. J. (2014). The organo-zeolitic-soil system: A comprehensive fertilizer. *International Journal of Waste Resources, 4*(3), 1, 2. https://doi.org/10.4172/2252-5211.1000156.
24. Capasso, S., Salvestrini, S., Coppola, E., & Buondonno, A. (2004). Sorption of humic acid on zeolitic tuff: A preliminary investigation. *Applied Clay Science*. https://doi.org/10.1016/j.clay.
25. Zabochnicka-Swiatek, M., & Kocela, R. (2019). Organic and mineral soil improvers intended for the cultivation of butter head lettuce. *Ekonomia i Srodowisko, 1*(68). https://ekonomiaisrodowisko.pl/journal/article/view/104 (accessed on 1 October 2024).

26. Eprikashvili, L. G., Andronikashvili, T. G., Zautashvili, M. G., Kordzakhia, T. N., Pirtskhalava, N. V., & Dzagania, M. A. (2012). Effective substrates based on natural zeolites and brown coal in plant growing. In *Proceedings of the IV International Scientific Conference "Sorbents as a Quality Factor in Life and Health"* (pp. 309–315). Belgorod.
27. Eprikashvili, L., Pirtskhalava, N., Zautashvili, M., Kordzakhia, T., Dzagania, M., & Tsintskaladze, G. (2017). Use of Georgian non-traditional agricultural resources in agriculture. *International Journal of Advanced Research (IJAR), 6*(1), 1408–1415. https://www.journalijar.com/article/21690/use-of-georgian-non-traditional-agricultural-resources-in-agriculture/ (accessed on 1 October 2024).
28. Eprikashvili, L. G., Dzagania, M. A., Kekelia, G. V., Pirtskhalava, N. B., Kordzakhia, T. N., & Zautashvili, M. G. (2020). Effect of using combined environmentally friendly methods on increasing the yield of agricultural products. *Annals of Agrarian Science, 18*(2), 195–200.
29. Gracheva, Y., Lebedev, K. S., & Platonov, V. V. (2014). Study of the chemical composition of brown coal humic substances of the "Lvovsky" quarry near Moscow. *Proceedings of Tula State University. Natural Sciences, 1*(2), 221–228.
30. Shamia, I. S., Halabi, M. N., & El-Ashgar, N. M. (2017). Humic acid determination in some compost and fertilizer samples. *IUG Journal of Natural Studies, Peer-Reviewed Journal of Islamic University-Gaza, IUGNES Special Issue*, 42–50.
31. Meier, W. M., & Baerlocher, C. (2001). Structures and structure determination. In *Molecular Sieves (Vol. 2)* (pp. 141–161). Zeolite Type Frameworks: Connectivities, Configurations and Conformations.
32. Ouatmane, A., Dorazio, V., Habidi, M., Revel, J. C., & Senesi, N. (2000). Elemental and spectroscopic characterization of humic acids fractioned by gel permeation chromatography. *Agronomie, 20*, 491–504.
33. MacCarthy, P. (1985). Infrared spectroscopy of deuterated compounds: An undergraduate experiment. *Journal of Chemical Education, 62*(7), 633.
34. Volkov, D. S., Rogova, O. B., & Proskurnin, M. A. (2021). Temperature dependences of IR spectra of humic substances of brown coal. *Agronomy, 11*(9), 1–26. https://doi.org/10.3390/agronomy11091822.
35. Prilutskaya, N. S., Korel'skaya, T. A., Popova, L. F., & Leont'eva, V. A. (2016). The study of the structural and functional composition of soil humus acids of the Euro-Arctic region by infrared spectroscopy. *Bulletin of NArFU. Series: Natural Sciences, 4*, 26–35. (In Russian).
36. Fedorova, N. I., Malysheva, V. Yu., & Ismagilov, Z. R. (2018). IR spectroscopy of coal inertinites of various stages of metamorphism. *Bulletin of the Kuzbass State Technical University, 3*, 86–92. https://doi.org/10.26730/1999-4125-2018-3-86-92.
37. Tsitsishvili, G. V., Tsintskaladze, G. P., & Charkviani, M. K. (1983). Effect of heat treatment on IR spectra of some synthetic and natural zeolites in the region of framework vibration frequencies. *Reports of the USSR Academy of Sciences, 273*(6), 1434–1439.
38. Flanigen, E. M., Khatami, H., & Szymanski, H. A. (1974). Infrared structural studies of zeolite frameworks. In E. M. Flanigen & L. B. Sand (Eds.), *Molecular Sieve Zeolites. Advances in Chemistry* (Vol. 101, pp. 460–488). American Chemical Society.

CHAPTER 14

High-Temperature Processing and Recovering of Nanoscale Carbon from Plastic Waste After Its Activation in Overheated Vapor Environment

AKAKI PEIKRISHVILI, LALI GURCHUMELIA, ARCHIL MIKELADZE, ROIN CHEDIA, MARGALITA DARCHIASHVILI, BAGRAT GODIBADZE, and SALOME GVAZAVA

Ferdinand Tavadze Institute of Metallurgy and Materials Science, Tbilisi, Georgia

ABSTRACT

The ecologically safe and economically profitable method of obtaining recovered activated carbon from plastic (PET) waste, which includes low-temperature pyrolysis of waste at the initial stage, followed by high-temperature thermo-chemical treatment in an overheated water vapor environment, will be presented. The vapor treatment at low temperatures (up to 700°C) activates the pyrolysis process and increases the primary reactions of hydrocarbon decomposition. Accordingly, the release of gases increases, the main part of which during the pyrolysis process is extracted as a liquid fraction – pyrolytic oil and the remaining amorphous coal containing different organic/inorganic impurities. After increasing the temperature (above 700°C) and simultaneously feeding vapor, the overheated water vapor increases the rates of side reactions, and full decomposition of pyrolytic remnants takes place. As a result, the loosening of the formed carbon aggregates occurs, and the formation of highly dispersed, nanoscale carbons takes place. As investigations show, the characteristics of the obtained nanoscale activated carbon

Advanced Topics in Polymer Chemistry and Materials Science: Tamara Tatrishvili and Marc Jean Médard Abadie (Eds)

structure and physical characteristics depend on treatment conditions (vapor temperature, feeding rate, treatment period, etc.), changing over a wide range and can reach up to 920 m^2/g (BET surface area) and 98.7% (purity), respectively. The mentioned, as well as other features of the structure/characteristics of recovered nanoscale carbon as a function of the processing conditions (temperature, vapor amount, time), will also be presented and discussed.

14.1 INTRODUCTION

Currently, the volume of plastic production is growing by 5–6% on average every year. The widespread use of polymers is due to their properties: durability, versatility, economy, ease of use, and low cost of production. Over the past 60 years, consumer plastic consumption has increased approximately 20-fold. Globally, annual consumption of bottled water alone has reached 500 billion units per year. At the same time, the rate of destruction of this plastic waste has increased, which has had a significant negative impact on the environment and public health. Most plastic waste is not biodegradable, is not subject to decay and corrosion, practically does not break down over time, and when burned, extremely toxic substances are released that cannot be removed from the body. Therefore, the recycling of plastic waste is a very urgent problem nowadays. In addition, the problem of plastic waste recycling becomes relevant not only from the point of view of environmental protection but also due to the fact that in the conditions of a shortage of polymer raw materials, plastic waste becomes a powerful raw material and energy resource. One of the most convenient and promising methods of plastic recycling is pyrolysis. The perspective is correct, both in terms of environmental safety (it does not pollute the environment with combustion gases) and obtaining useful products. The peculiarity of the process is the lack of oxygen for all combustion components. The production of pyrolysis products is influenced by various factors. Temperature is the dominant factor affecting the distribution of gas, liquid, and solid phase pyrolysis products and their physical and chemical properties. Additional important factors include heating rate, particle size, feedstock composition, pyrolysis time, atmospheric pressure, and catalyst presence [1–6].

The process of pyrolysis is carried out in a special container, where pre-shredded plastic waste is loaded. The container is hermetically sealed and placed in the oven, which heats up to 400–500°C. As a result, the pyrolysis process starts, and volatile gas similar to natural gas is released. During the pyrolysis process, the main part of this gas is converted into a liquid fraction (pyrolysis oil), and the non-condensable residues are used to keep the furnace burning. Solid carbon residue accumulates in the container. Thus, the pyrolytic products of plastic waste are: gaseous stream (CO and CO_2) with a product yield of –42.8%, solid waste (with benzoin, acetophenones, terephthalates, and other organic and inorganic compounds) –37.4%, and oil fraction –12.8%. Thus, by pyrolytic treatment of polymer waste, a carbonaceous solid residue (coal) containing various types of negative impurities is obtained. Therefore, it is necessary to develop additional technology for their chemical activation, mechanical, and chemical enrichment. It is evident that the development of improved technology significantly increases the cost of the pyrolysis process and the cost of the product too [7–10].

Today, existing methods of pyrolysis cannot provide high-quality, highly dispersed activated carbon. It is obtained mainly from petroleum products, which is associated with significant economic and ecological problems: a large amount of released CO_2, intensively reduced reserves of oil, and so on. Therefore, it is very important to find alternative methods for obtaining and producing activated carbon with improved properties, which will eliminate environmental problems and reduce product costs as well.

In the presented paper, an ecologically safe and economically profitable method of obtaining highly dispersed activated carbon from polymer waste is considered. The proposed two-stage method, with low-temperature pyrolysis of the waste at the first stage and high-temperature thermo-chemical treatment in an overheated water vapor environment at the second stage, contributes to the production of nanostructured activated carbon free from negative organic and inorganic impurities.

14.2 MATERIALS AND METHODS

Plastic waste – plastic bottle waste (polyethylene terephthalate – PET), which is a polymer complex ether with chemical formula $(C_{10}H_8O_4)_n$ was used experiments. The structural formula for same product is:

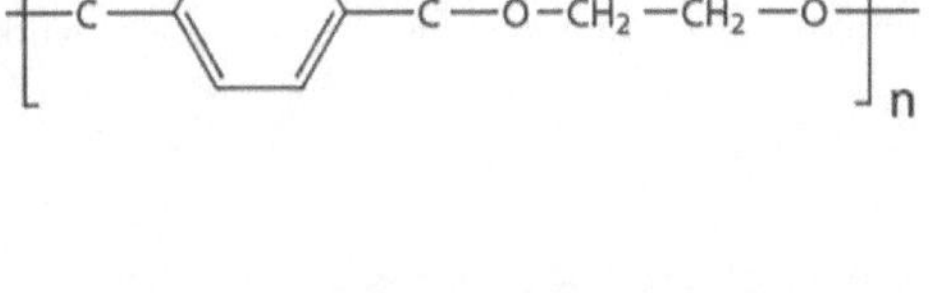

At the first stage, the bottles are washed with water to remove contents and inorganic mineral components (dust, soil, sand, etc.). After washing, the bottles are cut, air-dried, and melted in a metal pot (heated to 300–330°C). For quick solidification, the molten mass is poured into water. After cooling to room temperature and crushing, the obtained pieces are put in a drying box at 120–140°C.

The plastic waste, after drying, is loaded into a modified reactor working in a water vapor environment developed by TIMMS for recycling plastic waste (Figure 14.1). The reactor is a horizontal furnace with a length of 1200 mm and a diameter of 500 mm. The working area is made from a stainless steel tube (ϕ=130 mm & L=1200 mm). One end of the tube is closed and welded to a metal tube for supplying argon and steam during the processing. A cylindrical stainless-steel boat (ϕ=100 mm and L=630 mm), into which pre-prepared plastic waste is poured, is located in the middle of the furnace pipe. The processing of waste pyrolysis is carried out in an argon environment.

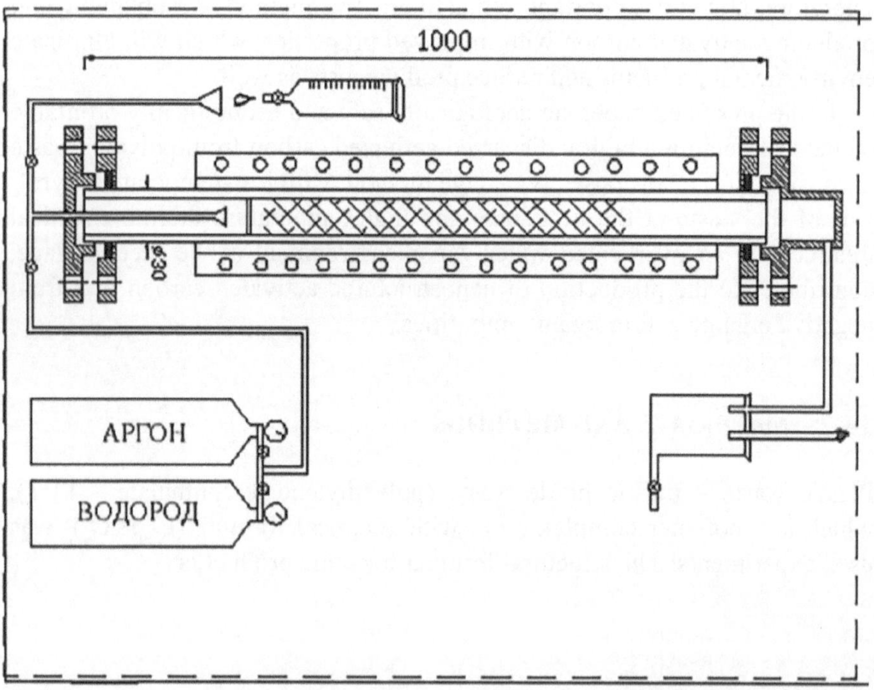

FIGURE 14.1 A modified reactor for recycling plastic waste working in water vapor environment.

14.3 RESULTS AND DISCUSSION

In order to optimize the process and determine the optimal heating rate of the reactor, the method of thermogravimetric analysis was applied (Figure 14.2). Thermogravimetric analysis of the initial raw material (pre-prepared plastic waste) was carried out for different heating rates at 10°/min and 20°/min. It was established that the temperature of destruction of plastic waste does not depend on the heating rate, and in all cases, it is equal to 450°C. The duration of the process and the yield of dry residue change according to the change in the heating rate. When the heating rate is 100°C/min, the pyrolysis process lasts for 60–65 min, while at 20°/min, the pyrolysis process is completed within 30–35 min, and an increase in the yield of dry residue is observed as well. Based on this, the heating rate of 200°C/min was selected for further experiments, which is economically beneficial (reducing process duration and energy consumption) too. Figure 14.2 represents the results of thermogravimetric analysis (TGA) of plastic waste for different heating rates.

The prepared sample is heated to 350–550°C (heating rate 200°C/min) for 35–40 minutes. At 450°C, the pyrolysis process begins, which continues up to a temperature of 550–580°C. At this time, the primary reactions of decomposition of the polymer chain are taking place. Accordingly, gases and volatile products are released. In the relatively cold part of the reactor, condensation of volatile products of pyrolysis takes place, and the separated gas leaves the reactor through a gas pipe. A dry residue of –14.2% remains in the reactor, which contains various types of organic and inorganic pyrolytic residues. After the end of pyrolysis, instead of argon, we supply 700–800 ml of water to the reaction zone (speed is 4 ml/min) and increase the temperature to 700–900°C. Activation with water vapor lasts 3–3.5 hours.

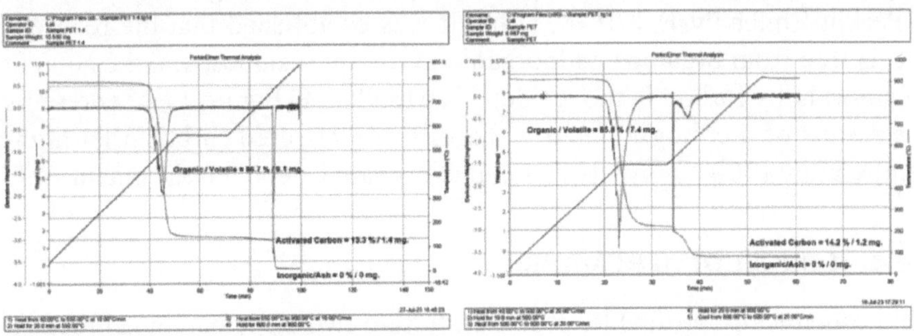

FIGURE 14.2 The results of thermogravimetric analysis (TGA) of plastic waste for different rates of heating. (a) Heating rate 100/min, the yield of pyrolysis carbon (PC) = 13.3%; (b) Heating rate 209/min, the yield of pyrolysis carbon (PC) = 14.2%.

Overheated water vapor at 700–800°C starts the activation of the received dry residue (coal). It increases the rates of side reactions (polymerization, polycondensation, cyclization) and completely destroys pyrolysis residues. At 800–900°C, the obtained carbon products are oxidized to gasification. This results in the formation of free carbon atoms in the gas, which begin to solidify in the vapor area to form highly dispersed, activated carbon "aggregates" that are jet milled/processed into high-quality, nanostructured activated carbon.

The physical and chemical characteristics of the obtained nanostructured activated carbon—dispersion, specific surface area, porosity, purity, and yield—are limited by international standards: – Dispersion was determined by ASTM B822–20 standard test method for particle size distribution of metal powders and related compounds by light scattering. A particle size analyzer—mastersizer 2000—will be used for determination.

The specific surface area and porosity were determined by the international standard – ASTM D6556 ("Standard Test Method for Carbon Black—Total and External Surface Area by Nitrogen Adsorption"). The "Gemini VII" tool (Micromeritics company) was used for measurements. Figure 14.3 represents the results of measuring the specific surface area and porosity of the obtained AC depending on processing conditions.

As it's seen from data with increasing of period of high temperature pyrolysis from 180 to 210 min, the surface characteristics and porosity changes too from 826m^2/g & 627m^2/g to 920m^2/g & 603 m^2/g correspondently. The period of treatment has positive role for surface characteristics and increases it when, at the same time the reduction of porosity in obtained product observed.

The purity and yield of the obtained activated carbon (AC) were determined by the method of thermogravimetric analysis – ASTM E1131–20 ("Standard Test Method for Compositional Analysis by Thermogravimetry"). Thermogravimetric analysis was performed using a derivatograph – Perkin Elmer Pyris 1 TGA, USA. It was established that the received AC is free from organic and inorganic pyrolytic residues, and the content of amorphous carbon in the residue is up to 98.5%. Figure 14.4 represents the results of thermogravimetric analysis of recovered carbon after high-temperature (900°C) pyrolysis in an overheated steam environment.

14.4 CONCLUDING REMARKS

An economically profitable, ecologically safe two-stage combined method of thermal treatment of polymer waste was developed. The developed method provides for low-temperature pyrolysis of waste at the initial stage

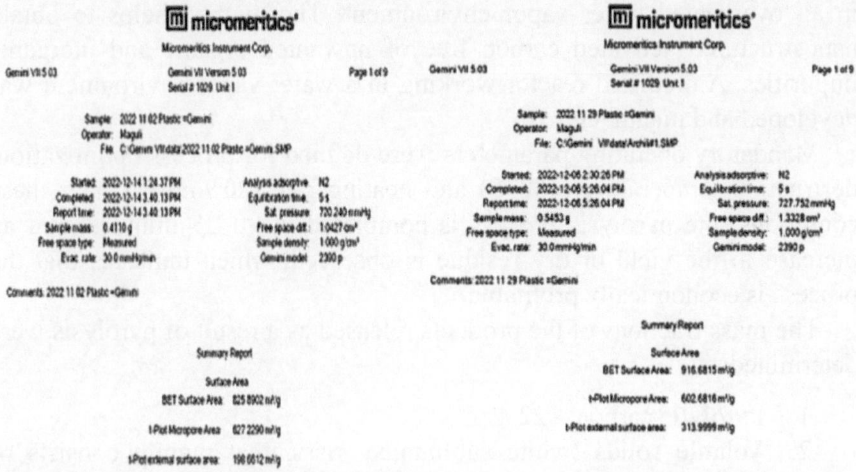

FIGURE 14.3 Results of measuring the specific surface area and porosity of the obtained AC after high temperature(900°C pyrolysis of plastics in overheated steam environment depending on processing condition. (a) Processing time 180 min, water feeding rate 4ml/min; (b) Processing time 210 min, water feeding rate 4 ml/min.

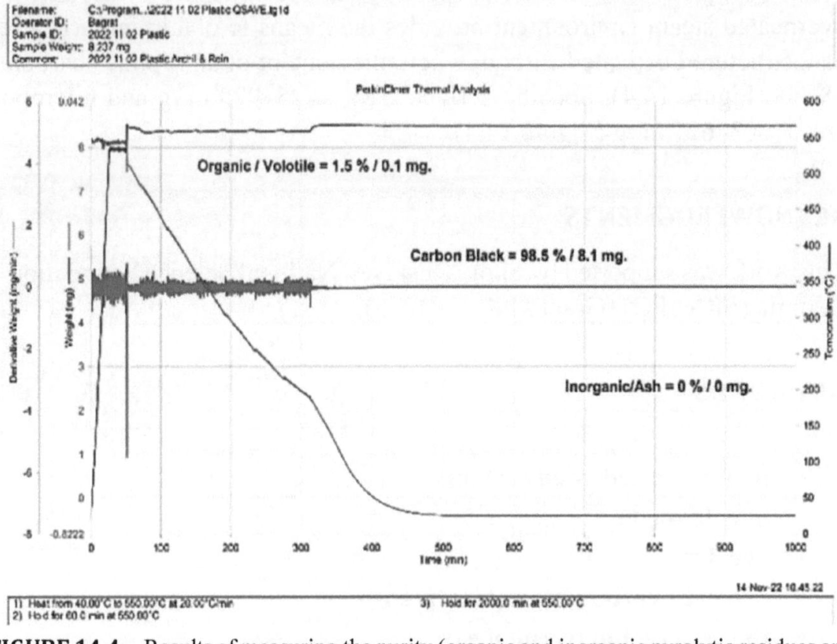

FIGURE 14.4 Results of measuring the purity (organic and inorganic pyrolytic residues and yield (98.5%) of the obtained AC.

and then, at the second stage, high-temperature thermo-chemical treatment in an overheated water vapor environment. The method helps to obtain nanostructured activated carbon free of unwanted organic and inorganic impurities. A modified reactor working in a water vapor environment was developed and manufactured.

Mandatory operating parameters were defined for process optimization: destruction temperature (450°C) and heating rate (20°/min). Under these conditions, the pyrolysis process is completed in 30–35 minutes, and an increase in the yield of dry residue is observed, which indicates that the process is economically profitable.

The mass fractions of the products released as a result of pyrolysis were determined:

1. Pyrolytic carbon – 22.4%;
2. Volatile solids (white sublimated mass, that mainly consists of benzoic and terephthalate acids) – 16.7%;
3. Gas (main products CO_2, CO, H_2O steam) – 60.9%.

It was established that pyrolytic carbon with overheated water vapor activation increases both the purity and yield of the resulting activated carbon. Thus, the presented combined method of high-temperature pyrolysis in an overheated steam environment provides the means to obtain impurity-free, nanostructured activated carbon, where the content of amorphous carbon is 98.5% (Figure 14.4); Specific surface area – 825–920 m^2/g and micropore area – 603–627 m^2/g (Figure 14.4).

ACKNOWLEDGMENTS

This work was supported by Shota Rustaveli National Science Foundation of Georgia (SRNSFG) (Grant #FR-22–4275).

KEYWORDS

- **nanoscale activated carbon**
- **plastic waste**
- **pyrolysis**
- **pyrolytic carbon vapor treatment**
- **thermogravimetric analysis**

REFERENCES

1. PET Waste Treatment Methods–Waste Types. (n.d.). Retrieved from: https://www.waste.ru/item.
2. Huang, J., Veksha, A., Chan, W. P., Giannis, A., & Lisak, G. (2022). Chemical recycling of plastic waste for sustainable material management: A prospective review on catalysts and processes. *Renewable and Sustainable Energy Reviews, 154*, 111866. https://doi.org/10.1016/j.rser.2021.111866.
3. Zhuo, C. W., & Levendis, Y. A. (2014). Upcycling waste plastics into carbon nanomaterials: A review. *Journal of Applied Polymer Science, 131*, 39931.
4. Sharifian, S., & Asasian-Kolur, N. (2022). Polyethylene terephthalate (PET) waste to carbon materials: Theory, methods, and applications. *Journal of Analytical and Applied Pyrolysis, 163*, 105496. https://doi.org/10.1016/j.jaap.2022.105496.
5. Olam, M. (2022). Production of activated carbon from waste PET chars. *International Journal of Environmental Monitoring and Analysis, 10*(2), 39–44. https://doi.org/10.11648/j.ijema.20221002.13.
6. Yuan, X., Cho, M. K., Lee, J. G., Choi, S. W., & Lee, K. B. (2020). Upcycling of waste polyethylene terephthalate plastic bottles into porous carbon for CF4 adsorption. *Environmental Pollution, 265*, 114868. https://doi.org/10.1016/j.envpol.2020.114868.
7. Kovaleva, N. I., Raevskaia, E. G., & Rozhin, A. B. (2020). Pyrolysis of plastic products: Review. *Chemical Safety Science, 4*(1), 48–79.
8. Doğan-Sağlamtimur, N., Bilgil, A., Güven, A., Ötgün, H., Yıldırım, E. D., & Arıcan, B. (2019). Producing qualified oil and carbon black from waste tyres and PET bottles in a newly designed pyrolysis reactor. *Journal of Thermal Analysis and Calorimetry, 135*, 3339–3351. https://doi.org/10.1007/s10973-018-7576-1.
9. Osman, A. I., Farrell, C., Al-Muhtaseb, A. H., & Al-Fatesh, A. S. (2020). Pyrolysis kinetic modelling of abundant plastic waste (PET) and *in-situ* emission monitoring. *Environmental Sciences Europe, 32*, Article 112.
10. Kple, M., Girods, P., Fagla, B., Anjorin, M., Ziegler-Devin, I., & Rogaume, Y. (2017). Kinetic study of low-density polyethylene using thermogravimetric analysis, Part 2: Isothermal study. *Waste and Biomass Valorization, 8*, 707–719.

CHAPTER 15

Carbon Nanotubes as Effective Remedy for Thin Oil-Water Films Removal

N. I. SALMANOVA,[1] A. B. HUSEYNOV,[1,2] and E. B. ZEYNALOV[1,2]

[1]*Institute of Geotechnological Problems of Oil, Gas, and Chemistry, Azerbaijan State University of Oil and Industry, Ministry of Science and Education, Republic of Azerbaijan*

[2]*Nagiyev Institute of Catalysis and Inorganic Chemistry, Ministry of Science and Education, Republic of Azerbaijan*

ABSTRACT

This chapter describes the process of synthesizing multi-walled carbon nanotubes (MWCNTs) as well as the results of their use as adsorbents for removing oil contaminants from the surface of water. SEM and TEM analyses were employed to characterize the synthesized MWCNTs. The sorption capacity of MWCNTs for crude oil was 3–4 ml/g, and with an increase in the specific gravity of the petroleum product, this value also increased.

15.1 INTRODUCTION

One of the priority tasks for environmental protection is the cleaning of the surface of water bodies from oil spills caused by petroleum production on the shelf, waterborne transport, and pollution by sewage from oil refineries [1]. The development of various methods of water purification, both on its surface and throughout its volume, is the subject of many studies. One of the proposed methods is the use of different variations of hydro cyclones for removing oil products from natural water reservoirs [2]. Oil-contaminated water can also be chemically purified by oxidizing it with substances such as ozone, chlorine, hydrogen peroxide, or potassium permanganate [3]. To remove oil products from natural watercourses, it is possible to use biosorbents and membrane bioreactors [4], in which a simultaneous process of biological oxidation of pollutants and membrane separation of oil emulsions takes place. Such devices make it possible to reduce the content of oil products in water from 20 mg/l to 0.5–1.2 mg/l. In addition to all of the above, various physical and chemical methods, such as flotation, coagulation, electrochemical treatment, sorption, and ultrafiltration, are widely used to remove oil products from water bodies [5].

The most common method for purifying natural water from oil products is pressure flotation [6]. When cleaning oil-contaminated water by electrocoagulation, special metal electrodes are utilized. The dissolution of these electrodes forms metal hydroxides, which reduce the stability of oil-in-water emulsions. This, under the influence of various external forces, leads to their separation [7]. The drawback of applying the electrocoagulation method for the purification of water from oil products is the high electric power consumption. The process of water purification from oil and oil products by the ultrafiltration method is carried out in membrane separators operating under extremely high pressure [8]. In a comparative study [9], the process of purifying natural waters in adsorption filters using various sorbents, as well as membrane separators with ultrafiltration modules from various manufacturers, was carried out. As a result of the study, the authors concluded that the use of activated carbon as a sorbent is the most optimal solution in terms of the ratio of quality to price of water purification.

A group of researchers proposed the use of an adsorbent based on charred moss for the purification of reservoirs polluted with oil products [10]. It was found that pre-carbonized at a temperature of 200–250°C, peat moss, which was then modified with acetic acid, has a high sorption capacity. The introduced nanocarbon adsorbents can increase the effectiveness of cleaning the water surface until the water becomes almost clean, and the residual oil content in the water is less than 0.03 g/l.

Regarding the effectiveness of certain methods, it should also be mentioned that although they demonstrate significant efficiency for the purification of the volume of water or the water surface from thick layers of oil and oil products, they have low efficiency against thin oil films. However, even thin films of oil products are a problem that needs to be addressed, as they obstruct the penetration of sunlight and air oxygen into the depths of water bodies, which leads to the death of biological organisms such as plankton and fish and causes serious damage to the environment.

Carbon nanotubes have great potential to be used as adsorbents to capture thin films of oil pollution. Carbon nanotubes have a high sorption capacity for oil products, and their low density gives them floating properties. Moreover, carbon nanotubes are reusable adsorbents due to the fact that they can withstand multiple regeneration cycles. However, carbon nanotubes have not yet been widely used: the reason for this is their relatively high price due to the costliness of the reagents used in their synthesis. In order to lower the cost of the nanocarbon adsorbents, as a raw material we preferred to use an affordable and inexpensive propane gas for the synthesis of multi-wall carbon nanotubes (MWCNT).

15.2 EXPERIMENTAL

MWCNTs were synthesized by the chemical vapor deposition (CVD) method. For the synthesis of MWCNTs using this method, an enlarged laboratory setup, as shown in Figure 15.1, was specially assembled.

FIGURE 15.1 Photo of the installation for the synthesis of MWCNT.

Synthesis of MWCNTs was carried out in a quartz reactor at 900°C and atmospheric pressure. This reactor had a diameter of 25 mm and a length of 100 cm. The precursor for the catalyst was ferrocene. The diluent gas utilized was argon. The activator gas used was hydrogen. In the gas mixture, the volume ratio of propane, which was a hydrocarbon raw material, was 7%. The synthesis process was carried out for 60 minutes.

The sorption capacity of the nanocarbon adsorbent in the process of cleaning the oil-contaminated water surface was measured using the following laboratory procedure.

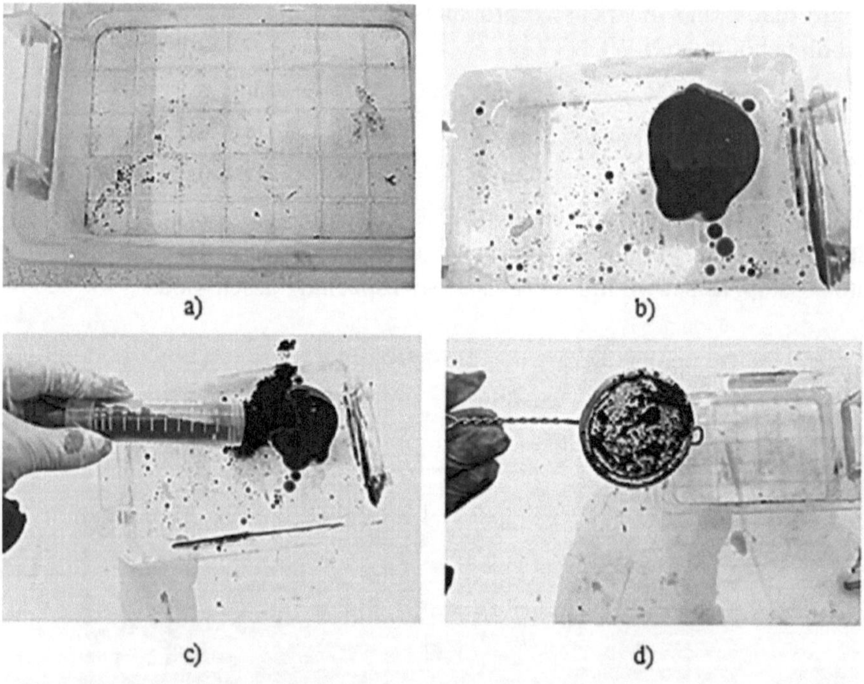

FIGURE 15.2 a – bath with clean water, b – water with oil pollution on the surface, c – with oil pollution coated with adsorbent, d – water surface, after cleaning.

A laboratory plastic bath, which had a volume of 1 liter, was filled with 0.5 L of tap water (Figure 15.2a). Using a measuring pipette, one milliliter of oil with a preliminarily determined specific gravity was placed over the water's surface (Figure 15.2b). One gram of the tested MWCNTs was placed in a plastic tube, and then the adsorbent was evenly poured from this tube onto an oil slick on the water surface (Figure 15.2c). Five minutes later, the adsorbent that captured the oil was removed from the water surface using a metal mesh (Figure 15.2d). If the oil was not completely removed from the water surface, the next portion of the adsorbent from the tube was used. The consumption of the adsorbent for the complete removal of 1 mL of oil from water is calculated by the mass of the adsorbent that remained in the tube.

15.3 RESULTS AND DISCUSSION

The synthesized MWCNTs were analyzed by SEM (Figure 15.3a) and TEM (Figure 15.3b) methods. From these microphotographs, it can be seen that although the MWCNTs synthesized from propane contain some non-tubular structures (assumedly soot or coke), there are not many of them.

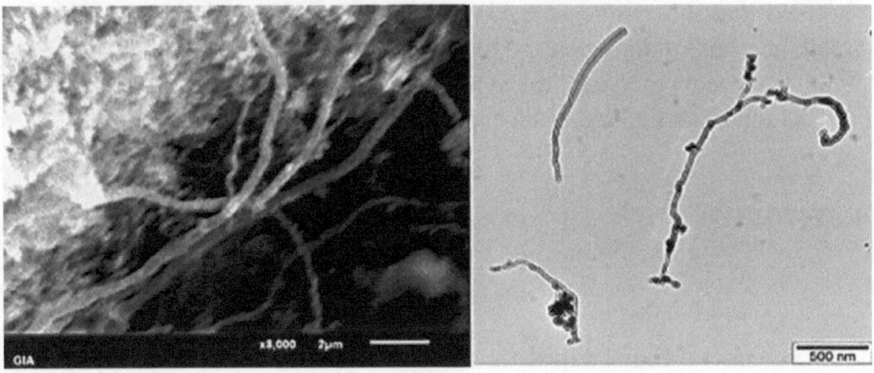

a) b)

FIGURE 15.3 SEM (a) and TEM (b) micrographs of MWCNTs synthesized from propane.

The results of measuring the sorption capacity for oil in the synthesized nanocarbon adsorbent are presented in Table 15.1.

TABLE 15.1 Sorption Capacity of a Nanocarbon Adsorbents in Relation to Oil of Different Specific Gravity

Sample №	Density of the Oil, kg/m^3	Sorption capacity, ml/g
1	0.848	3.09
2	0.877	3.37
3	0.905	3.76
4	0.932	4.05

As can be seen from Table 15.1, the numerical values of the sorption capacity for oil based on MWCNT adsorbents are in the range of 3–4 ml/g, and this value increases with increasing specific density of the oil. Through the various modifications of carbon nanotubes, there are good prospects for enhancing the absorptive capacity of MWCNTs for oil products, but this is the subject of our future studies.

15.4 CONCLUSION

In this work, the CVD method was applied for the synthesis of MWCNT. Propane was used as a raw material for the synthesis of nanotubes, and the low price and affordability of this raw material ensured the low cost of the resulting product. Based on the results of SEM and TEM analysis, it was revealed that the synthesized MWCNTs are well-structured and contain a small number of impurities. The synthesized MWCNTs were studied as floating sorbents for cleaning the water surface from oil pollution; as a result, it was found that their sorption capacity for oil products is 3–4 ml/g.

KEYWORDS

- **carbon nanotubes**
- **chemical vapor deposition**
- **environmental pollution**
- **multi-walled carbon nanotubes**
- **oil-water films**
- **pressure flotation**
- **renewable adsorbents**

REFERENCES

1. Zeynalov, E. B., & Ishchenko, N. Y. (2014). Naphthenic acids in the environment: Factors of ecological risk. *Oil and Gas Technologies, 11*, 87–90.
2. Han, T., Liu, H., Xiao, H., Chen, A., & Huang, Q. (2019). Experimental study of the effects of apex section internals and conical section length on the performance of

solid–liquid hydrocyclone. *Chemical Engineering Research and Design, 145*, 12–18. https://doi.org/10.1016/j.cherd.2019.02.040.

3. Coha, M., Farinelli, G., Tiraferri, A., Minella, M., & Vione, D. (2021). Advanced oxidation processes in the removal of organic substances from produced water: Potential, configurations, and research needs. *Chemical Engineering Journal, 414*, 128668. https://doi.org/10.1016/j.cej.2021.128668.
4. Khalifa, O., Banat, F., Srinivasakannan, C., AlMarzooqi, F., & Hasan, S. W. (2021). Ozonation-assisted electro-membrane hybrid reactor for oily wastewater treatment: A methodological approach and synergy effects. *Journal of Cleaner Production, 289*, 125764. https://doi.org/10.1016/j.jclepro.2020.125764.
5. Adelshin, A. B., Busarev, A. V., Selyugin, A. S., & Hisameeva, L. R. (2021). *Izvestiya KGASU, 2*(12), 9. E3S Web of Conferences 274, 08007. https://doi.org/10.1051/e3sconf/202127408007.
6. Khaustov, V. V., Bredikhin, V. V., & Khaustova, T. V. (2019). Treatment technology of liquid phase at industrial waste landfill. In *IOP Conference Series: Materials Science and Engineering*. https://doi.org/10.1088/1757-899X/687/6/066063.
7. Oladzad, S., Fallah, N., & Nasernejad, B. (2017). Combination of novel coalescing oil water separator and electrocoagulation technique for treatment of petroleum compound contaminated groundwater. *Water Science and Technology, 76*, 57. https://doi.org/10.2166/wst.2017.185.
8. Voigt, I., Richter, H., Weyd, M., Milew, K., Haseneder, R., Günther, C., & Prehn, V. (2019). *Chemie-Ingenieur-Technik, 91*, 1454.
9. Busarev, A., Sheshegova, I., & Khisameeva, L. (2021). Study of the processes of the purification of water from surface sources from petroleum products. *STCCE–2021, E3S Web of Conferences, 274*, 08007. https://doi.org/10.1051/e3sconf/202127408007.
10. Rotar, O. V., Iskrizhitskaya, D. V., Iskrizhitsky, A. A., & Oreshina, A. (2014). Cleanup of water surfaces from oil spills using natural sorbent materials. *Procedia Chemistry, 10*, 145–150. https://doi.org/10.1016/j.proche.2014.11.022.

CHAPTER 16

The Role of Complex Modifiers Obtained on the Basis of Carbon Nanotubes in the Hardening of High-Strength Concrete

E. B. ZEYNALOV,[1] A. B. HUSEYNOV,[1] A. A. QUVALOV,[2] M. YA. MAGERRAMOVA,[1] N. Z. AHMEDLI,[2] and DJ. N. HUSEYNOV[2]

[1]*Nagiyev Institute of Catalysis and Inorganic Chemistry, Azerbaijan*

[2]*Azerbaijan Architecture and Construction University, Azerbaijan*

ABSTRACT

To accelerate the setting of high-strength concrete, complex modifier compositions based on multiwalled carbon nanotubes (MWCNT) have been developed. The methods of incorporating multi-walled carbon nanotubes into the composition of cement mixtures, the ways of studying and eliminating problems arising in these processes have also been studied.

In the research work, a comparative test of the physical and mechanical properties of cement stone, solutions and concrete obtained by adding additives of different composition modified with carbon nanotubes was conducted. It was determined that the complex nano modifier (KNM) obtained with the application of MWCNT increased the 1-day strength of cement stone and mortar by 70–90% and 92–115%, respectively, and the strength during 10 hours of concrete hardening increased by ensures that it is higher than 100–115%. This effect is due to the formation of dispersed ettringite crystals of pomegranate, C-S-H(I), tobermorite and helium phase, which densely fill the interparticle space as a result of the acceleration of hydration processes under the influence of CNM.

Advanced Topics in Polymer Chemistry and Materials Science: Tamara Tatrishvili and Marc Jean Médard Abadie (Eds)

CNM increases the brand of concrete in terms of frost resistance and waterproofing, and even allows to reduce the consumption of cement.

16.1 INTRODUCTION

Cement is the main structural material in concrete construction. Reducing the commissioning period for the construction of residential buildings as well as industrial facilities is of great importance, as it directly affects the cost and profitability level. One of its few disadvantages is slow drying, which can be accelerated by adding chemical modifiers to the concrete mix or by heating it [1–3].

After the discovery of carbon nanotubes and the development of nanotechnology, the accelerating effect of small additions of carbon nanotubes on the hardening rate of cement-concrete structures has been found in almost all fields of science and technology. In the scientific literature, the concept that explains this effect by the formation of a large number of new crystallization centers in the studied sample is expressed [4–6]. However, there are practically no systematic studies on the investigation of this effect when using Azerbaijan-made cements.

For this purpose, research and applied works were carried out in the direction of obtaining composite materials reinforced with the addition of multiwalled carbon nanotubes (MWCNT) on the basis of cement.

Several researchers have discussed the importance of using carbon nanotubes as a modifier for cement compositions. Along with the fact that these modifiers have rare physical and mechanical properties, they have associated with the possibility of their industrial production [7, 8].

Researchers have shown that the specific surface of the particles can be increased more than 20 times by dispersing them in a liquid medium under the influence of ultrasound, taking into account the high tendency of carbon nanotubes to aggregate [9, 10].

The selection of suitable carriers and the preparation of appropriate concentrations for MWCNT in solvents used in industry and offered for this purpose is a very relevant issue today. Therefore, research is being conducted on processing MWCNT with special methods and then applying it to cement-based materials. Taking into account the above, the direction of the current research work is to improve the physical and mechanical properties of cement stone and concrete by choosing optimal carriers for MWCNT, dispersing them under the influence of ultrasound, and ensuring homogeneous distribution of the obtained complex nano modifiers in cement systems.

16.2 EXPERIMENTAL METHODS AND MATERIALS

16.2.1 *MATERIALS RESEARCH METHODS*

Cement from 3 plants was used in the research:

1. CEM I-52,5-N EN 197–1, HOLCIM.
2. CEM I-52,5-T EN 197–1, NORM.
3. CEM I-52,5-N EN 197–1, Gazakhsement.

During the preparation of the concrete mixture, 5–10 for crushed stone from the Guba stone quarry was used as a coarse filler. Grindstone has a moisture content of 0.7% and a density of 2.7 g/cm^3.

During the preparation of fine-grained concrete, river sand of the Bahramtepe deposit with a coarseness modulus of Mir=2.1, density of 2.6 g/cm^3 and stone dust from the Guba quarry with a coarseness modulus of Mir=3.4 and a density of 2.65 g/cm^3 were used as fine aggregates. The absorption of sand is 1.8%, and the absorption of stone is 1.4%.

The following modifiers were used:

1. MWCNT synthesized by propane Chemical Vapor Deposition method.
2. 40% Rheobuild 878 superplasticizer based on naphthalene sulfonate, BASF company.
3. Viscocret 20HE hyperplasticizer based on polycarboxylate, SIKA company.
4. 20% CAC superplasticizer based on polyarylsulfone sulfonate obtained by special synthesis of "Nanotechnologies in Construction Materials" SRL of Azerbaijan University of Architecture and Construction [11].

16.2.2 *RESEARCH METHODS*

The following characteristics of cement systems were determined in experiments:

1. The HORIBA nanopartisa (nanoparticle analyzer SZ-100) device in the spectral analysis laboratory of the National Academy of Sciences was used to determine the size of nanoparticles.
2. Microscopic analysis was conducted using a Jeol (JSM – 6610LV) device.

3. The effect on the strength of CNM cement stone was evaluated according to the results obtained during the determination of the compressive strength limit of 2x2x2 cm sized samples hardened under normal conditions after 1, 3, and 28 days. The determination of the strength limit was carried out on the AUTO 105/250 Cement Compression and Flexural Machine device.
4. The effect of CNM on the properties of fine-grained concrete was evaluated using 10x10x10 cm samples hardened under normal conditions.
5. Flowability of concrete is determined according to EN 206, compressive and bending strength of concrete according to GOST 10180, frost resistance according to GOST 10060, and waterproofness according to GOST 12730.5. Durability tests were performed on 1 and 28-day samples of 150x150x150 cm size stored under normal conditions. Frost resistance and waterproofing were determined on 28-day samples.

16.3 RESULTS AND DISCUSSION

The effect of different modifiers on the hardening kinetics of cement stone is given in Table 16.1.

TABLE 16.1 Kinetics of Compressive Strength of Modified Cement Stone

Composition	Additive	MWCNT	Water/cm	Density. q/sm3	Compressive Strength. MPa (day)			
					1	3	7	28
1	–	–	0.28	2.178	44.07	76.38	82.5	98.3
2	Rheobuild 878, 1%	–	0.22	2.198	49.28	77.19	89.32	96.84
3	Rheobuild 878, 1%	0.0004%	0.23	2.242	56.81	92.91	100.27	115.89
4	Vscocrete 20HE, 1%	–	0.24	2.308	57.23	91.92	93.34	99.37
5	Vscocrete 20HE, 1%	0.0004%	0.24	2.236	60.34	108.48	115.54	138.62
6	SAS-2, 1.4%	–	0.23	2.236	60.34	103.76	114.19	156.63
7	SAS-2, 1.4%	0.0004%	0.23	2.294	72.79	115.34	126.71	170.91

The analysis of the results shown in Table 16.1 indicated that the additions modified with MWCNT increase the compressive strength of cement stone by 30–65% after 1 day and by 20–75% after 28 days. Such a difference between the strength indicators of modified and non-additive samples is explained by the modifying effect of the additive on the dispersion and morphological composition of the compounds formed during cement hydration. This is what determines the higher resistance of the modified structure to collapse. Micropores are observed between the newly formed joints in unsupplemented samples magnified 1000 times in an electron microscope. According to the results of microprobe analysis, these compounds mainly consist of large crystals of highly basic hydro silicates (the size is 20–25 μm) and large plates of ettringite (Figure 16.1a, b).

The structure of the cement stone added with 0.0004% MWCNT and 1.2% SP had a denser structure and was mainly composed of low-based hydro silicates (sizes are 10–15 μm) (Figure 16.1 c, d).

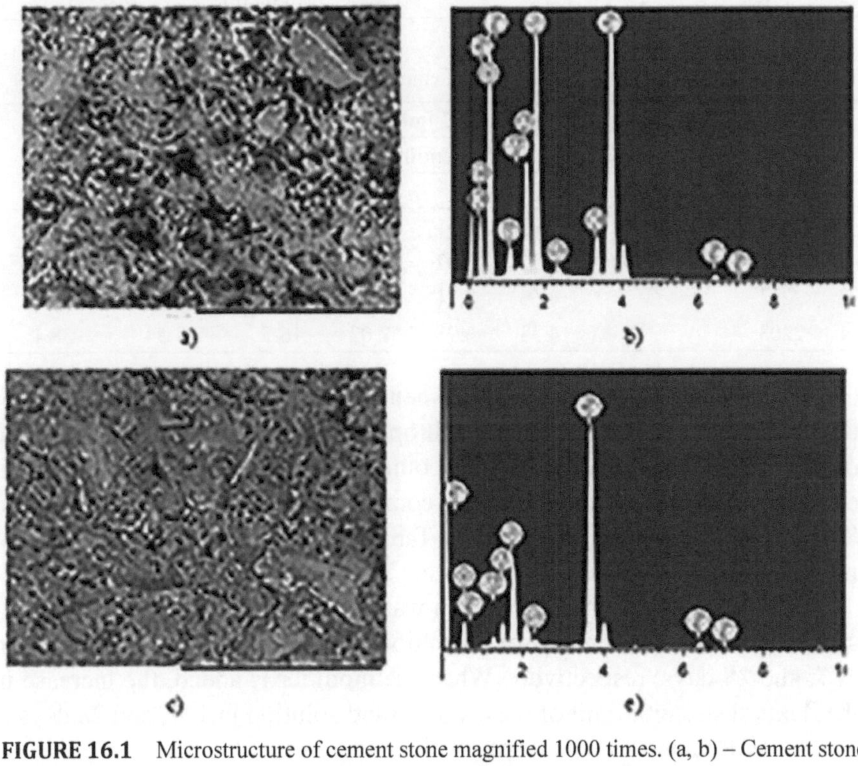

FIGURE 16.1 Microstructure of cement stone magnified 1000 times. (a, b) – Cement stone without additives; (c, d) – Cement stone with 0.0004 MWCNT and 1.0 SP added.

MWCNT and SP (Rheobuild 878), being widely applied in Azerbaijan, were used to prepare a solution to investigate their effect on the properties of fine-grained concrete. The solution was prepared using cement and filler in a 1:3 ratio. Bahramtepe river sand, CEM I 52.5-N Portland cement from the HOLCIM plant, Rheobuild 878 superplasticizer, and MWCNT synthesized from propane via the Chemical Vapor Deposition method were utilized.

The effect of the KNM and its components on the water demand of the cement-sand mixture is given in Table 16.2, and the effect on the flexural and compressive strength of the solution in 1, 7, 28 days is given in Table 16.3.

TABLE 16.2 Effect of Modifier on Water Demand of Cement-Sand Mixture

№	Cem/sand	Rheobuild 878,%	MWCNT.%	Taunit.%	Water/Cm
1	1:3	–	–	–	0.54
2	1:3	1	–	–	0.4
3	1:3	1	0.0004	–	0.37
4	1:3	1		0.0004	0.38

TABLE 16.3 Kinetics of Set Strength of Cement-Sand Mortar

№	Average Density, kq/m3	Strength Limit of Cement-Sand Solution, MPa (gün)					
		In Bending			In Compression		
		1	7	28	1	7	28
1	2051	2.93	4.11	5.01	8.6	22.33	26.75
2	2155	3.95	6.02	6.99	12.9	27.87	33.4
3	2262	4.65	7.21	8.34	18.5	36.44	40.1
4	2203	4.33	6.78	7.02	16.5	33.54	38.4

As can be seen from Table 16.2, when the KNM is added to the composition of the cement-sand mixture with optimal thickness, the water demand of the mixture decreases by 31.5% compared to the control sample, but it decreases by 7.5% compared to the composition modified with Rheobuild 878. From the results presented in Table 16.3, it is clear that the KNM increases the average density of the cement-sand solution by about 10.2%.

As it can be seen, the decrease in water demand increases the compressive strength of the cement-sand solution by 58.7%, 75.4%, and 66.5% in 1, 7, and 28 days, respectively. When the modifier is added, the increase in the flexural strength limit of the cement-sand solution in 1, 7, and 28 days is 17.7%, 19.8%, and 19.3%, respectively.

The influence of modifiers on the flowability of the concrete mixture was studied, and the kinetics of the collection of the strength of the concrete was checked in cube samples with dimensions of 15x15x15 cm made on the basis of these mixtures. The composition and properties of concrete mixtures are given in Table 16.4.

TABLE 16.4 Compositions of Concrete Mix

№	Modifier	Cement, kq	Sand, kq	Stone aggregate, kq	Crushed stone 5–10	Crushed stone 10–20	Water, l	Addition, l	Slump, sm
1	Rheobuild 878	460	530	400	345	520	173	7	16
2	Taunit	460	530	400	345	520	166	7	17
3	MWCNT	460	530	400	345	520	163	7	18

In order to determine the effect of modifiers on the kinetics of concrete strength accumulation (especially in the first periods of hardening), the strength of the samples was tested for 8, 10, 12 hours and 1, 3, 14, 28 days in accordance with GOST 10180 (Table 16.5).

TABLE 16.5 Limits of Compressive Strength of Cubic Specimens

№	Average Density, kq/m^3	Compressive Strength, MPa						
		8 hour	10 hour	12 hour	1 day	3 days	14 days	28 days
1	2380	4.8	7.4	11.1	42.7	51.8	61.4	62
2	2400	8.1	15.8	20.7	45.8	57.6	62.0	65.5
3	2430	8.3	16.0	24.1	47.0	61.8	64.2	70.6

As can be seen, adding the complex nano modifier to the composition of concrete increases the strength of concrete compared to control compositions. During the first hardening periods (10–12 hours), the strength of concrete increases significantly. After 8 hours of hardening, the strength of the modified concrete doubles compared to the control samples, but it is not allowed to remove the formwork because it cannot achieve the strength of 15 MPa adopted in the project. As concretes with modifiers exceed 15 MPa during the 10–12 hour hardening period, the opening of molds is allowed. For 8, 10, and 12 hours, the increase in strength of concrete with the modifier is 69–114% compared to control samples. The 28-day strengths of samples with complex nano-modifiers are 65.5 and 70.6 MPa, which means that B45 class high-strength concrete exceeds the design strength.

As can be seen from the studies, the complex nanomodifier modified with MWCNT synthesized from hydrocarbon gas raw materials in the 28th laboratory of the Institute of Catalysis and Inorganic Chemistry of ANAS gives better results than Taunite. This is due to better dispersion in the carrier medium as a result of the effect of ultrasound compared to MWCNT Taunite. Waterproofing of high-strength concrete samples hardened for 7 days was studied using the AGAMA-2RM device. The results of the study are given in Table 16.6.

TABLE 16.6 Waterproofing of High-Strength Concrete Modified with Complex Additives

№	Content of Additives,%			Concrete Compositions, kq/m^3			Water/cem	W
	Rheobuild 878	MWCNT	Taunit	Cement	Sand	Crushed Stone		
1	1	–	–	490	885	1000	26.9	16
2	1	0.0004	–	490	885	1000	25.5	20
3	1	–	0.0004	490	885	1000	25.9	20

As can be seen from Table 16.6, the control concrete composition reaches the water resistance grade W18 after 7 days of curing. At the same time, compositions modified with a complex additive based on various MWCNT, after 7 days of curing, have a water resistance grade exceeding W20, which is quite sufficient for use in the production of waterproof structures.

The frost resistance of concrete with the optimal content of a complex additive modified by ÇDKNB was studied using a non-destructive testing method with the Beton-FROST device on cube samples with an edge of 10 cm according to GOST 10060.3–95 at a temperature of 18 ± 2°C. The compositions of concrete and the results of their frost resistance tests are given in Table 16.7.

TABLE 16.7 Frost Resistance of High-Strength Concrete with a Complex Additive Modified by MWCNT

№	Content of Additives,%			Water/sem	Average density, kq/m^3	F
	Rheobuild 878	MWCNT	Taunit			
1	1	–	–	26.9	2490	400
2	1	0.0004	–	25.5	2580	700
3	1	0.04	–	25.9	2550	600
4	1	–	0.0004	26	2530	600
5	1	–	0.04	26.3	2500	500

The results presented in Table 16.7 show that the introduction of MWCNT in an amount of 0.0004% and 0.04% of the binder consumption leads to an increase in the frost resistance grade of high-strength concrete by 3 and 2 steps, respectively. The introduction of Taunit in an amount of 0.0004% and 0.04% of the binder consumption leads to an increase in the frost resistance grade by 2 and 1 step, respectively.

16.4 RESULTS

Thus, analyzing the results of structural changes in the cement stone modified with MWCNT and comparing them with the results of the physical, mechanical, and operational properties of high-strength concrete, it can be argued that the improvement in frost resistance is associated with an effect that promotes the formation of an additional volume of fine-crystalline new formations in the form of C-S-H(I), tobermorite 14Å, and an additional volume of gel. Fine-crystalline new formations, filling the space between the grains of large new formations in the form of portlandite and ettringite, and at the same time intertwining with them, compact the structure in the intergranular space between large grains of cement particles. This, in turn, helps to increase the density of the cement stone due to the formation of porosity in the system. When nano modifiers are added to the concrete mixture, the strength of the concrete increases by 69%, 114%, and 86% in the first periods of 8, 10, and 12 hours, respectively. As a result of using nanomodified additives, the time for opening concrete from the mold is shortened from 16 hours to 10 hours.

KEYWORDS

- **cement stone**
- **complex modifier**
- **concrete**
- **mortar**
- **multiwalled carbon nanotubes**
- **strength**

REFERENCES

1. Guvalov, A. A., & Abbasova, S. İ. (2020). Influence of polyarylsulfonsulfonate on the rheology and stability of cement suspension. *PPOR, 21*(2), 171–178.
2. Guvalov, A. A., Abbasova, S. I., & Kuznetsova, T. V. (2017). The effectiveness of modifiers in regulating the properties of concrete mixtures. *Scientific, Technical and Production Journal "Building Materials," (7)*, 49–51.
3. Guvalov, A. A. (2019). The effect of polyarylsulfonsulfonate superplasticizers on the properties of cement systems. *Information Scientific and Technical Journal Concrete Technology,* (3, 4), 24–27.
4. Ibragimov, R. A., & Kiyamova, L. I. (2015). Microstructure of cement stone modified with carbon nanotubes. *Bulletin of the Kazan Technological University, 18*(5), 71–73.

5. Pimenov, A. I., Ibragimov, R. A., & Izotov, V. S. (2015). Physico-mechanical properties of cement composites modified with nanoadditives. *Bulletin of the Kazan Technological University, 18*(1), 128–130.
6. Tolmachev, S. N., & Belichenko, E. A. (2014). Features of the influence of carbonaceous nanoparticles on the rheological properties of cement paste and technological properties of fine-grained concrete. *Nanotechnologies in Construction, 6*(5), 13–29.
7. Khuzin, A. F., Gabidullin, M. G., & Badertdinov, R. Z. (2013). Complex additives based on carbon nanotubes for high-strength accelerated hardening concrete. *Izvestia KazGASU,* (1)(23), 221–226.
8. Khuzin, A. F., Gabidullin, M. G., & Rakhimov, R. Z. (2013). Modification of cement composites with carbon nanotubes. *Bulletin of KTU, 16*(5), 115–118.
9. Gabidullin, M. G., Khuzin, A. F., & Rakhimov, R. Z. (2013). Ultrasonic treatment as an effective method for dispersing carbon nanotubes in the volume of a building composite. *Construction Materials,* (3), 57–59.
10. Samchenko, S. V., Zemskova, O. V., & Kozlova, I. V. (2017). Model and mechanism of stabilization of carbon nanotubes with a polycarboxylate-based plasticizer. *Bulletin of MGSU, 12*(7), 724–732. https://doi.org/10.22227/1997-0935.2017.7.724-732.
11. Guvalov, A. A., & Ibrahimova, M. J. (2022). Mechanism of effect of superplastic additions on cement systems. *PPOR, 23*(2), 214–223.

CHAPTER 17

Study of Electrical Properties of RGO/PDMS Nanocomposite

NATIA JALAGONIA, TINATIN KUCHUKHIDZE, NINO DARAKHVELIDZE, TAMAR ARCHUADZE, IA KURASHVILI, SOPHIO MIKABERIDZE, and LEILA KALATOZISHVILI

Ilia Vekua Sukhumi Institute of Physics and Technology, Georgia

ABSTRACT

Carbon nanostructures have attracted electronic industrial interest due to their unusual electrical properties. One of the most common applications of this material is in polymer nanocomposites to develop the most technologically promising devices. This work presents the synthesis of reduced graphene oxide (RGO)/polymer nanocomposites and discusses the effect of various factors on their electrical conductivity. RGO is a promising material that can be obtained from graphite at a relatively low cost. The electrical conductivity and surface area of RGO depend on the production method. Hence, one of the principal tasks is the synthesis of RGO, where defects will be minimized. Galvanomagnetic research methods are used for understanding the fundamental properties of electrically conductive materials. Therefore, the Hall effect plays an important role in the study of electric charge transfer. The study has shown that the electrical conductivity of RGO/polydimethylsiloxane nanocomposites increases with increasing filler concentration. The electrical conductivity is maximum at 10% RGO concentration, and was found to be 3.01 x 10^{-8} Ohm^{-1}, cm^{-1}.

17.1 INTRODUCTION

Materials based on graphene are greatly impacting the world in many facets. Graphene is one of the most recent miracle materials among them.

Advanced Topics in Polymer Chemistry and Materials Science: Tamara Tatrishvili and Marc Jean Médard Abadie (Eds)

It is particularly useful in the field of electronics because it is the thinnest, strongest, transparent, and conductive material available [1, 2].

By combining graphite with concentrated acid and an oxidizing agent, graphene oxide (GO) can be synthesized. Graphene oxide and reduced graphene oxide are valuable graphene derivatives [3]. Graphene oxide and reduced graphene oxide show different chemical and structural properties due to the differences in their chemical compositions. RGO is a two-dimensional material with a large specific surface area. It is known that RGO is a high-conductivity material; it can transfer electrons between itself and other materials. Due to these excellent properties, RGO is an ideal reinforcement phase for polymer composites [4, 5]. It is known that the homogeneous dispersion of reduced graphene oxide enhances the mechanical and electrical properties of polymer nanocomposites [6]. Therefore, at first, graphene oxide was synthesized by the intercalation method, and a stable suspension was obtained. Further reducing the graphene oxide with ascorbic acid and sonication for about 4–5 hours, aqueous RGO was obtained. RGO/PDMS polymer nanocomposite was prepared by the solution mixing method. The synthesized materials were gradually studied and identified by different methods.

17.2 EXPERIMENTAL METHODS AND MATERIALS

17.2.1 MATERIALS

The materials used in the synthesis of graphene oxide and reduced graphene oxide are mainly graphite, sulfuric acid (H_2SO_4), hydrogen peroxide (H_2O_2), and potassium permanganate ($KMnO_4$). Also included are ascorbic acid, potassium hydroxide, and PDMS granules.

17.2.2 METHODS

The following instrumental methods are used in the research process: The X-ray diffraction (XRD) analysis of the synthesized materials was carried out on a DRON-3M with diffraction molecules using CuKα radiation (λ = 1.54178 Å).

The Raman spectroscopy analysis was performed on a Witec UHTS 300, using a 532 nm laser with 1.5 mW power intensity under the following conditions: single spectra of the filler, magnification: 50x, integration time: 0.5 s; 532 nm excitation laser at room temperature.

A high-resolution scanning electron microscope (SEM) S-4800 (Hitachi) and EDS, with a resolution of 1 nm, were used for the analysis of the morphology and condition of the obtained materials.

The Hall-effect measurement equipment used was the Ecopia HMS-3000/55T.

17.2.3 SYNTHESIS OF GRAPHENE OXIDE

Synthesis of GO was carried out by modified Hummers method. 300 ml flask placed into a water bath (27–35°C), 2 g natural graphite was added into a mixture of 50 ml 98% H_2SO_4 and 6 g $KMnO_4$ and the solution was stirred during 1 h. Then temperature increased to 70°C and stirred during 1 h, after this added 200 ml H_2O and 20 mL 30% H_2O_2. Obtained GO isolated by filtration, washed with DI water and centrifuged to removal the waste graphite.

17.2.4 SYNTHESIS OF REDUCED GRAPHENE OXIDE

The obtained GO was mixed with 100 mL of water and sonicated for 30 minutes using an ultrasonic bath. Ascorbic acid (10 g) was dissolved in 100 mL of distilled water and then gradually added to the mixture at room temperature. The suspension was sonicated with the ultrasonic bath for 1 hour.

17.2.5 PREPARATION OF RGO/POLYMER NANOCOMPOSITE

The synthesis of RGO/polymer was carried out by the solution mixing method. PDMS granules were dissolved in an organic solvent, and RGO sheets with different concentrations (vol. 0.5%, 1%, 5%, 10%, 12%) were added. The mixture was homogenized at a speed of 200 rpm for 4 hours. The nanocomposite was obtained by the removal of the solvent.

17.3 RESULTS AND DISCUSSION

Many contemporary GO synthesis methods are based on the technique first described by Hummers, which involves oxidizing graphite with a H_2SO_4 containing $KMnO_4$. In the first step, GO was synthesized by a modified Hummers' method. After that, GO was reduced by ascorbic acid.

The solution mixing method is the most well-known technique to prepare polymer nanocomposites. In this work, PDMS granules and RGO sheets were dispersed in the same solvent, and the mixture was treated by ultrasonication. The nanocomposite was obtained by the removal of the solvent and was studied by XRD, SEM, Raman, and the Hall effect method. The volume percentage of RGO in the polymer matrix was 0.5%, 1%, 5%, 10%, and 12%. In XRD (Figure 17.1), we observed a peak at 200, which belongs to RGO. The Raman spectroscopy results in Figure 17.2 show a "D" peak at 1590 cm^{-1} and a "G" peak at 1350 cm^{-1}, which confirmed the lattice deformation.

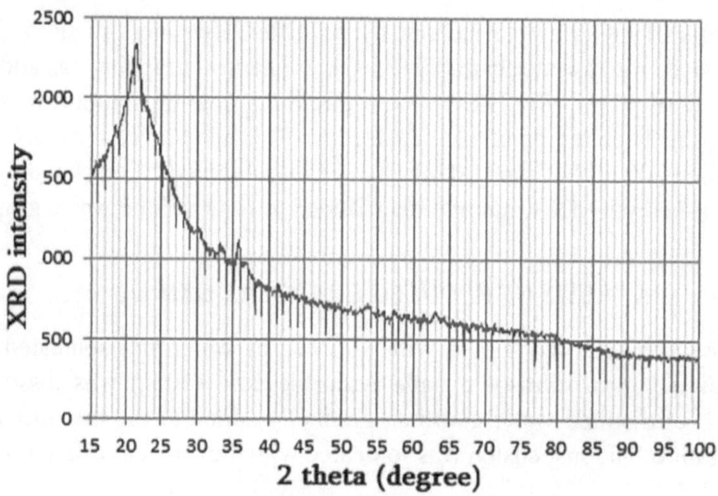

FIGURE 17.1 XRD of Reduced graphene oxide.

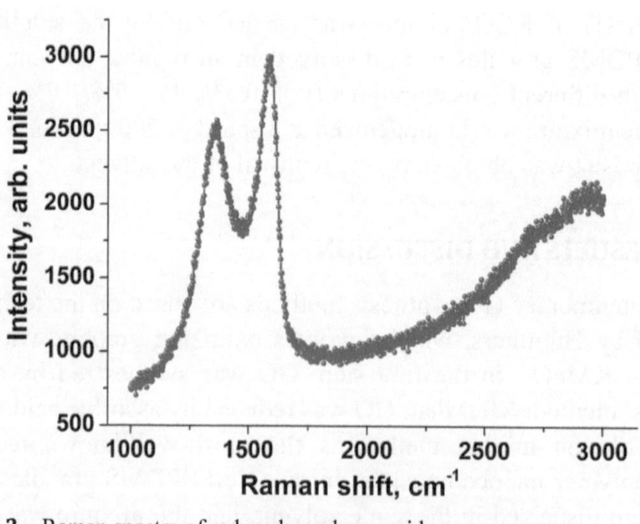

FIGURE 17.2 Raman spectra of reduced graphene oxide.

Using scanning electron microscopy, the morphology and structure of the materials were determined. The Figures 17.3 and 17.4 show the microscopic view of different samples.

FIGURE 17.3 SEM of reduced graphene oxide.

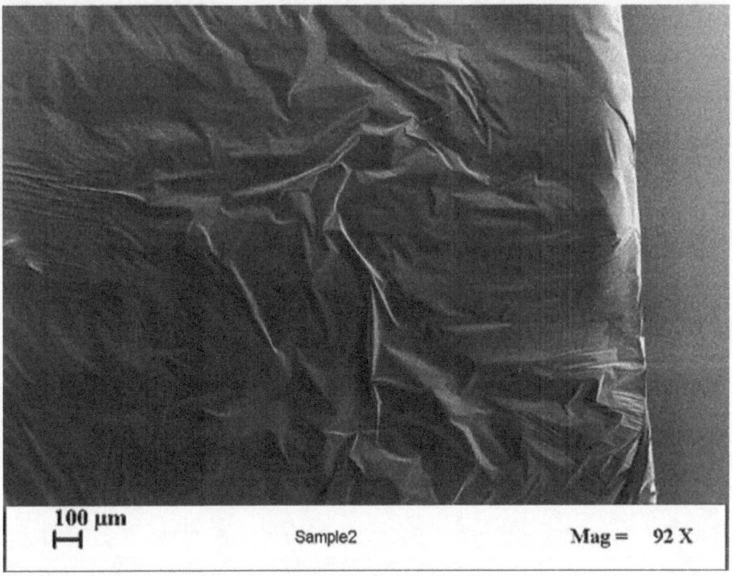

FIGURE 17.4 SEM of RGO/PDMS nanocomposite.

Because of the low density of states and disordered structure, RGO frequently exhibits the charge transfer mechanisms inherent in granular metals or disordered semiconductors over a wide temperature range [7]. We measured the electrical conductivity parameters of the samples at room temperature using the Hall-effect method. The results are given in Table 17.1.

TABLE 17.1 Electrical Properties of PDMS/GNP Nanocomposites

No.	Name	Specific resistivity ρ Ohm, cm	Electric conductivity σ Ohm^{-1}, cm^{-1}
1	PDMS	1.05 X 10^{11}	9.4 X 10^{-12}
2	PDMS/GNP 0.5%	1.6 X 10^{8}	6.25 X 10^{-9}
3	PDMS /GNP 1%	2.89 X 10^{7}	3.46 X 10^{-8}
4	PDMS/GNP 5%	1.2 X 10^{7}	8.4 X 10^{-8}
5	PDMS/GNP 10%	3.3 X 10^{7}	3.01 X 10^{-8}
6	PDMS/GNP 12%	1.1 X 10^{7}	9 X 10^{-8}

17.4 CONCLUSION

RGO is a two-dimensional material with a large specific surface area. It's known for its high conductivity, allowing it to transfer electrons between itself and other materials. Due to these excellent properties, RGO is an ideal reinforcement phase for polymer composites. It is known that the homogeneous dispersion of RGO enhances the mechanical and electrical properties of polymer nanocomposites. Research has shown that the electrical conductivity of RGO/PDMS nanocomposites increases with increasing filler concentration. The electrical conductivity is maximum at 10% graphene nanoplates concentration, and was found to be 3.01 times 10^{-8} Ohm^{-1}, cm^{-1}.

ACKNOWLEDGMENT

Shota Rustaveli National Science Foundation of Georgia (SRNSFG) (Grant number AR_18_271) supported this work.

KEYWORDS

- **carbon nanostructures**
- **conductivity**
- **graphene oxide**
- **nanocomposites**
- **polydimethylsiloxane**
- **polymer nanocomposite**
- **reduced graphene oxide**

REFERENCES

1. Huang, X., Yin, Z., Wu, S., Qi, X., He, Q., Zhang, Q., Yan, Q., Boey, F., & Zhang, H. (2011). Graphene based materials: Synthesis, characterization, properties, and applications. *Small, 7*(14), 1876–1902.
2. Dong, L. X., & Chen, Q. (2010). Properties, synthesis, and characterization of graphene. *Frontiers of Materials Science in China, 4*(1), 45–51.
3. Arthi, G., Paulchamy, B., & Lignesh, B. D. (2015). A simple approach to stepwise synthesis of graphene oxide nanomaterial. *Journal of Nanomedicine & Nanotechnology, 6*(1), 1–4.
4. Lacina, K., Kubesa, O., Horáčková, V., Moravec, Z., Kuta, J., Vanýsek, P., & Skládal, P. (2018). Graphene oxide from improved Hummers' method: Is this material suitable for reproducible electrochemical (bio)sensing? *ECS Journal of Solid State Science and Technology, 7*(10), 166–171.
5. Fernández-Merino, M. J., Guardia, L., Paredes, J. I., et al. (2010). Vitamin C is an ideal substitute for hydrazine in the reduction of graphene oxide suspensions. *The Journal of Physical Chemistry C, 114*(14), 6426–6432.
6. Du, J., & Cheng, H. M. (2012). The fabrication, properties, and uses of graphene/polymer composites. *Macromolecular Chemistry and Physics, 213*(10), 1060–1077.
7. Zhang, Y., Tan, Y. W., Stormer, H. L., & Kim, P. (2005). Experimental observation of the quantum Hall effect and Berry's phase in graphene. *Nature, 438*(7065), 201–204.

PART V
Processing

CHAPTER 18

Variants of the Thermal Decomposition of Condensed Phosphates as an Opportunity to Obtain New Cyclic or Polymeric Compounds

MARINA AVALIANI,[1] ELENA SHAPAKIDZE,[2] VAJA CHAGELISHVILI,[1] GULNARA TODRADZE,[2] and EVGENY KHICHUA[2]

[1]*R. Agladze Institute of Inorganic Chemistry and Electrochemistry, I. Javakhishvili Tbilisi State University, Georgia*

[2]*A. Tvalchrelidze Caucasian Institute of Mineral Resources, I. Javakhishvili Tbilisi State University, Tbilisi, Georgia*

ABSTRACT

A lot of important fundamental research is being carried out on condensed phosphates, so-called inorganic polymers, such as double oligomeric, cyclo-, or polymeric, and substituted phosphates. Our method of the production of condensed compounds is based on high-temperature synthesis (100–600°C) in multi-component systems M^{I}_2O-$M^{III}_2O_3$-P_2O_5-H_2O for the purpose of receiving new condensed compounds – inorganic polymers. The technology is founded on the creation of new condensed forms – double inorganic oligomers, cyclic compounds, and long-chain polymeric composites via the condensation of polyphosphoric acids, in the presence of mono- and polyvalent metals.

In this chapter, we report and discuss the outcomes of the thermal decomposition of many oligo-phosphates: double acidic diphosphates and triphosphates obtained by us for the first time during the last few years. Thus, the current data are the results of examination and analysis of the experimental records after the

Advanced Topics in Polymer Chemistry and Materials Science: Tamara Tatrishvili and Marc Jean Médard Abadie (Eds)

synthesis of new double condensed compounds of gallium, indium, scandium, and sometimes aluminum with K, Rb, Ag, and the study of the thermal behavior of some of them. We have examined the process of the phase transition of acidic oligo-phosphates and the formation of long-chain polymers and/or cyclic phosphates with a higher degree of anion condensation. The sufficient thermal stability of oligo- and polymeric phosphates has been proven.

Since, in the course of thermal conversion experiments of acidic oligo-phosphates, new cyclic compounds with a high degree of anion condensation, as well as long-chain double polymeric composites and long-chain polyphosphates were obtained, it can be concluded that this is a kind of solid-phase synthesis of new inorganic polymers. Usually, condensed double diphosphates of trivalent metals during decomposition at temperatures from 370–340°C, and between 500–750°C sometimes, form either long-chain poly-, cyclotetra-, or cyclooctaphosphates. As for the triphosphates, they are most often transformed into either cyclooctaphosphates and also normal diphosphates or long-chain polyphosphates.

18.1 INTRODUCTION

The growing interest in the synthesis of new inorganic polymers, given the potential for application in various fields of advanced technology, is essentially determined by the rapid development of various fields of chemistry over the last few decades, the attention in the search for new compounds, and the possibilities for their use in vast domains [1–7]. The essence of the extension of the explored field into condensed phosphate chemistry, in particular the chemistry of inorganic polymers of mono- and multivalent metals, is mainly due to the unique properties of these elements and their corresponding compounds [8–11]. The new inorganic polymers have found applications in microelectronics, the engineering and construction domain, in the field of high-temperature ceramics production, quantum mechanical amplifiers, lasers, and luminophores. Their use in emission materials is well known, as well as catalysts and, above all, in everyday life. Inorganic polymeric materials possess highly valuable qualities such as corrosion resistance, high-temperature resistance, increased strength, adsorption, and binding properties, serving as raw materials for the creation of phosphate glasses, as well as flame retardants for polymeric materials (polyamide, glass-filled polyamide, polycarbonate, epoxy resin). Some phosphates are effective in applying nutrients, cleaners, cement substances, ion-exchange ingredients, and catalytic agents [12–23].

In fact, many authors around the world have made a certain valuable contribution to the development of phosphate chemistry. It is impossible not to mention Corbridge, Topley, Van Weser, Griffith, d'Ivoir, Durif, Thylo, Tananaev, Chudinova, and the absolutely precious special contributions of other authors must also be mentioned [1–5, 10, 12–19, 24–29].

According to the nomenclature elaborated and proposed by famous scientists A. Durif (LEDSS, Université Joseph Fourier, Grenoble, France) and M. Th. Averbuch-Pouchot, the prior groups are cyclotriphosphates, cyclotetraphosphates, cyclopentaphosphates, cyclohexaphosphates, cyclooctaphosphates, cyclononaphosphates, cyclodecaphosphates, and cyclododecaphosphates [1, 4, 29].

In this regard, within the framework of the discovery of previously unknown compounds with catalytic and adsorption properties or other useful qualities for applications, the synthesis of double condensed phosphates, so-called inorganic polymers, with predictable properties possessing a priori adsorption and binding or other useful properties is promising. The production of condensed compounds is based on high-temperature synthesis (100–600°C) in multi-component systems with the aim of receiving new condensed compounds – inorganic polymers. The technology is founded on the creation of new condensed forms – double inorganic oligomers, cyclic compounds, and long-chain polymeric composites via the condensation of polyphosphoric acids. However, at the same time, there is another way to obtain phosphates after the condensation of already synthesized compounds: thermal decomposition of condensed forms containing protons in their structure – such as acidic diphosphates and triphosphates [26–36]. Consequently, in the presented work, we discuss the results of the examination and analysis of the experimental records: the synthesis of new inorganic polymers – double condensed phosphates of trivalent metals with some monovalent metals, and the investigation of their thermal behavior at 100–800°C. In some cases, temperatures were even raised to 1200°C. We have examined the process of decomposition of synthesized condensed double acidic diphosphates and triphosphates. The phase transition process of acidic oligophosphates and the formation of long-chain polymeric and cyclic phosphates with a higher degree of anion condensation have been studied [37–40].

18.2 EXPERIMENTAL METHODS AND MATERIALS

18.2.1 SYNTHESIS METHOD

In our previous papers, we discussed data on the synthesis of various oligophosphates during our studies in the open systems $M^I_2O–M^{III}_2O_3–P_2O_5–H_2O$

in the temperature range 130–550°C, where M^I is an alkali metal K, Rb, Ag, and M^{III} is Ga, In or Sc [30–40, 42–43, 46, 47]. Notably, H_3PO_4 at a concentration of 85%, trivalent metal oxides, and carbonates of monovalent metals were stirred thoroughly after being transferred into the carbon-glass crucible and heated at a predetermined temperature. The target objective temperature range was between 50°C and 450°C, occasionally reaching up to 650°C, with the duration of the synthesis ranging from 2 days to 28 days, depending on the type of condensed matter required. The selection of the optimal temperature and the ratio of the initial components were the main influencers on the final structural types of compounds. This method has been known for quite some time and has been approved by competent authors, as have the results we have obtained [1–5, 13–18, 29, 41, 44]. This technique allowed us to obtain numerous types of double-condensed phosphates, which are, in fact, inorganic polymers with several predetermined properties.

18.2.2 GRAVIMETRIC ANALYSIS

The basic configuration of the synthesized condensed compounds was studied by gravimetric analysis. Phosphorus was determined by the precipitation method using a molybdate mixture and a quinoline solution. The trivalent metals were determined by the hydroxyquinoline precipitation method, which was improved in our institution during many experiments for condensed compounds [8, 34, 36–37].

18.2.3 METHOD OF PAPER CHROMATOGRAPHY

Inorganic polymeric phosphates are characterized by fairly high thermal stability, which is why it is possible to classify them using the paper chromatography method. Synthesized di-, tri-, tetra-, and octahosphates, as well as polymeric compounds, were investigated by the mentioned method. In particular, crystalline samples were decomposed with an H-form cationic ion exchanger. In this case, approximately 0.1 g of the substance was contacted for 1 hour with an aqueous suspension of 2 g (KU-2, Dowex) type sulfonated cation exchanger at approximately 0°C. Later, the disintegration of the substance followed the scheme: cationite-H(solid) + M-phosphate(solid) → cationite-M(solid) + H-phosphate(liquid). The liquid mass, containing phosphoric acids, was treated with $NaHCO_3$ for neutralization and chromatographed using an acidic solvent CCl_3-COOH-CH_3COOH-CH_3OH on Filtrak FN-11 paper. The surface treatment of the obtained chromatograms was executed by spraying with a 5% solution of ammonium molybdate $(NH_4)2MoO_4$. The final step was irradiation with ultraviolet light (λ=400 nm) [33–34, 38, 40].

18.2.4 THERMO-GRAVIMETRIC ANALYSIS

DTA and TG analysis were employed for the study of thermal properties. The DTA curves were recorded. For thermogravimetric analysis (TGA), a Derivatograph Q1500-D was used with a heating rate of 10°/min in an air atmosphere and a maximum temperature of 1200°C.

18.2.5 SCANNING MICRO-SPECTROSCOPIC ANALYSIS

Scanning micro-spectroscopic analysis was accomplished using a JEOL JSM–6510LV scanning electron microscope (equipped with an energy-dispersive X–Max N20 micro-X-ray spectral analyzer from Oxford Instruments). SEM measurements were implemented using reflected BES as well as secondary SEI electrons at an accelerating voltage of 20 kV, with a working distance of 15 mm. Micrographs were taken at various magnifications. Micro-spectroscopic analysis was performed on the test point zones and their surface area.

18.2.6 ROENTGEN PHASE'S ANALYSIS

The basic structural framework of synthesized double condensed compounds is determined by X-ray diffraction method. The powder diffraction intensity data collections were performed on a DRON–3M diffractometer, with anodic Cu–K_α radiation in the range of $2\theta = 10$–60°, detector's speed 2° min^{-1}, lattice spacing d_α/n in Å, and I/I_0 – is the relative intensity (used model/standard data – by ASTM – American Society for Testing and Materials).

18.2.7 SOLID-PHASE SYNTHESIS METHOD

In addition to the foregoing, all compounds were studied and obtained by the solid-phase synthesis method, through the thermal decomposition of synthesized condensed matters from solution melts of polyphosphoric acids for the first time.

18.3 RESULTS AND DISCUSSION

18.3.1 SYSTEM CONTAINING GA AND MONOVALENT METALS

It is discovered that by crystallization from melts of polyphosphoric acids the following double condensed compounds (namely the series of new inorganic polymers) were obtained: double condensed di- and triphosphates, cyclotetraphosphates, cyclooctaphosphates, cyclooctaphosphates, at the molar ratio P_2O_5: M^I_2O: $M^{III}_2O_3$ = 15: 2.5: 1.0; 15: 5: 1.0; 15: 7.5: 1.0; 15: 10: 1.0 and 15: 3.5: 1.5; 15: 5: 1.5; 15: 6: 1.5; 15: 7.5: 1.5; 15: 8.5: 1.5; 15: 12: 1.5. The present paper examines, investigates, and analyzes the formation of several types/sequences of cyclo-, poly-, and/or diphosphates by thermal decomposition, with the aim of providing reliable statistics for the production of resistant inorganic polymers. Synthesized double acidic diphosphates are interesting because, due to the presence of protons in the structure, they are capable of further condensation to form highly condensed forms. For example, at a temperature range of 330–340°C, the acidic diphosphate of potassium-gallium forms the relevant cyclooctaphosphate:

$$2\,KGa(H_2P_2O_7)_2 \rightarrow K_2Ga_2P_8O_{24} + 4H_2O$$

The thermogavigram of the double acidic diphosphate $KGa(H_2P_2O_7)_2$ is presented in Figure 18.1 and indicates that there is a mass loss equal to 2 moles of water. The first endothermic effect of condensation is at 330–340°C and the second at 730–750°C, corresponding to the crystallization of the cyclooctaphosphate $K_2Ga2P_8O_{24}$. This compound melts by producing $KGaP_2O_7$ at 730°C, which in turn melts congruently at 930°C.

The process of the thermal decomposition of the double acidic triphosphate $KGaHP_3O_{10}$ happens at 550°C and, the process takes place in two different directions. We would like to point out that if the decomposition process proceeds slowly, not abruptly, it's more likely that the big cycles – will be formed:

$$4KGaHP_3O_{10} \rightarrow 2KGaP_2O_7 + K_2Ga_2P_8O_{24}$$

But if the temperature rise proceeds very quickly, abruptly, the decomposition reaction is as follows:

$$2KGaHP_3O_{10} \rightarrow KGaP_2O_7 + Ga(PO_3)_3 + KPO_3\,(\text{amorph}) + H_2O$$

The endothermic effect at approximately 750–775°C corresponds to the formation of cyclooctaphosphate which melts at 800–840°C (Figure 18.2).

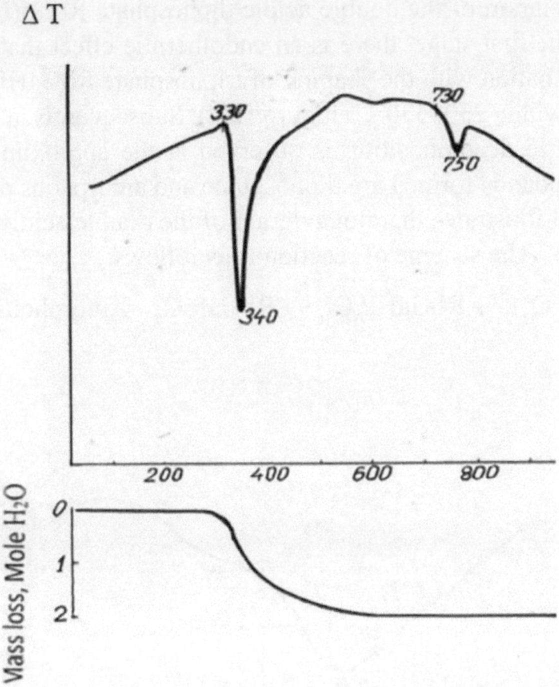

FIGURE 18.1 Thermogavigram of the double acidic diphosphate $KGa(H_2P_2O_7)_2$.

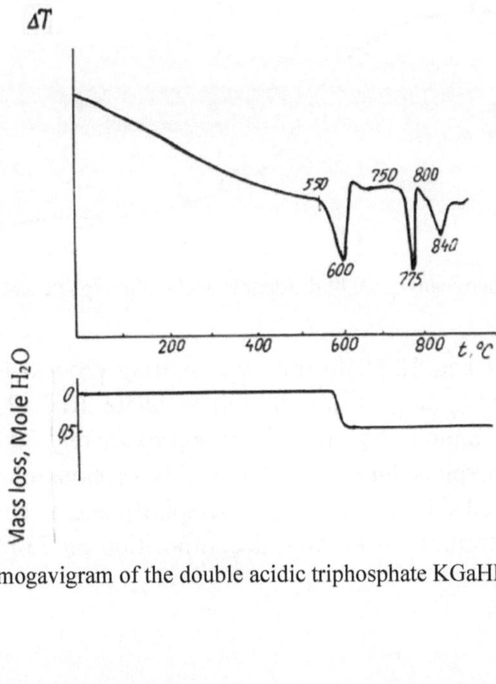

FIGURE 18.2 Thermogavigram of the double acidic triphosphate $KGaHP_3O_{10}$.

Thermogavigram of the double acidic diphosphate $RbGa(H_2P_2O_7)_2$ indicated that at the first stage, there is an endothermic effect that corresponds to the transformation with the shaping of triphosphate $RbGaHP_3O_{10}$ form II, which occurs within 230–350°C (Figure 18.2); Subsequently, a second effect corresponding to decomposition is observed at the approximately 500°C. The final compounds formed are diphosphate and amorphous phases.

Figure 18.3 illustrates thermogavigram of the double acidic diphosphate $RbGa(H_2P_2O_7)_2$. The scheme of reaction is as follows:

$$RbGa(H_2P_2O_7)_2 \rightarrow RbGaHP_3O_{10} \rightarrow RbGaP_2O_7 + \text{Amorphous phase}$$

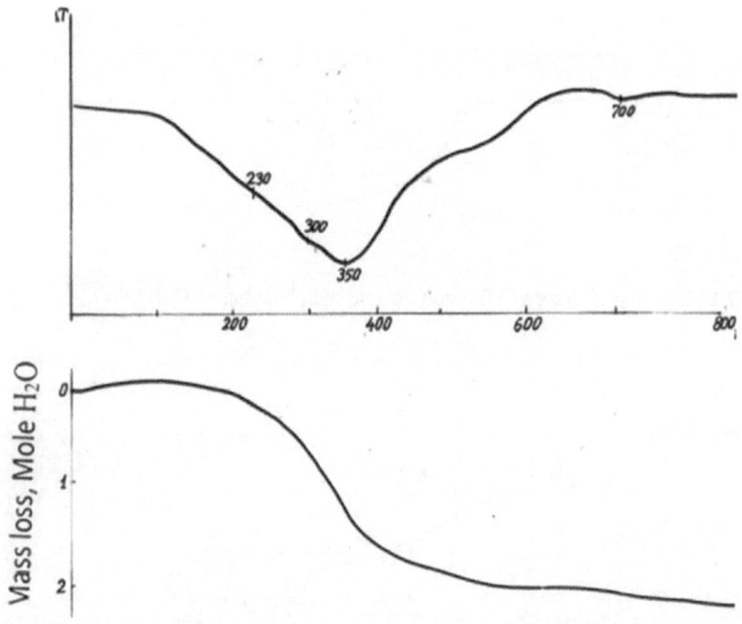

FIGURE 18.3 Thermogavigram of the double acidic diphosphate $RbGa(H_2P_2O_7)_2$

Figures 18.4 and 18.5 illustrate the melting process of cyclooctaphosphate $Rb_2Ga_2P_8O_{24}$ and of normal diphosphate $RbGaP_2O_7$ consequently. This last derivate melts congruently at approximately 800°C, and the final products are polyphosphate $Ga(PO_3)_3$ and amorphous phase. As far as the highly condensed compound – cyclooctaphosphate is concerned: it melts incongruently, forming at thermal decomposition on 730°C polyphosphate $Ga(PO_3)_3$ and melt.

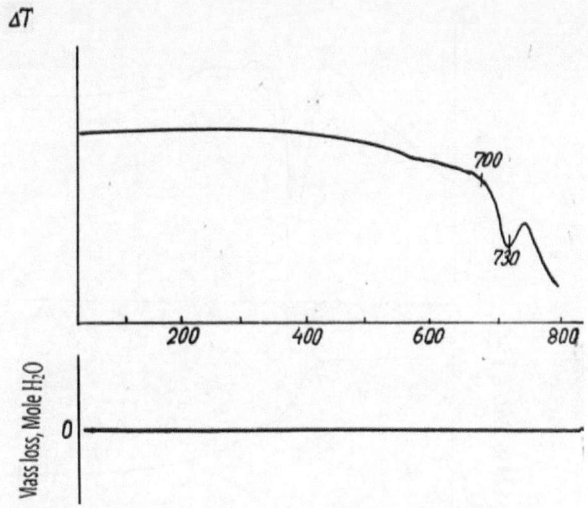

FIGURE 18.4 Thermogavigram of the double cyclooctaphosphate $Rb_2Ga_2P_8O_{24}$.

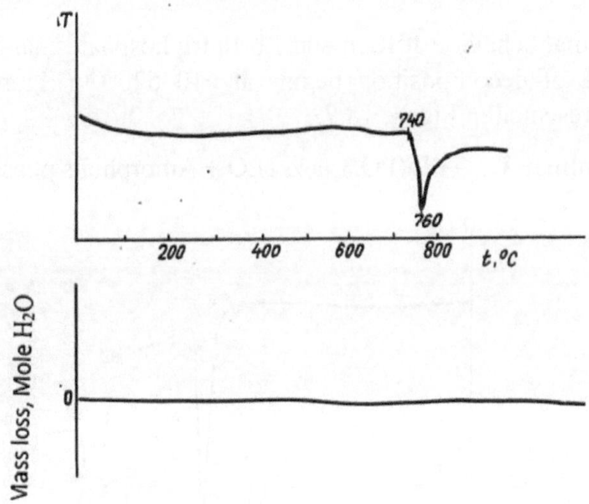

FIGURE 18.5 Thermogavigram of the double normal diphosphate $RbGaP_2O_7$.

18.3.2 SYSTEM CONTAINING K-IN AND MONOVALENT METALS

As far as the thermal decomposition of the acidic diphosphate of K-In is concerned, the thermal endothermic effect is large: 325–380°C and the loss of two moles of water is clearly visible (Figure 18.6).

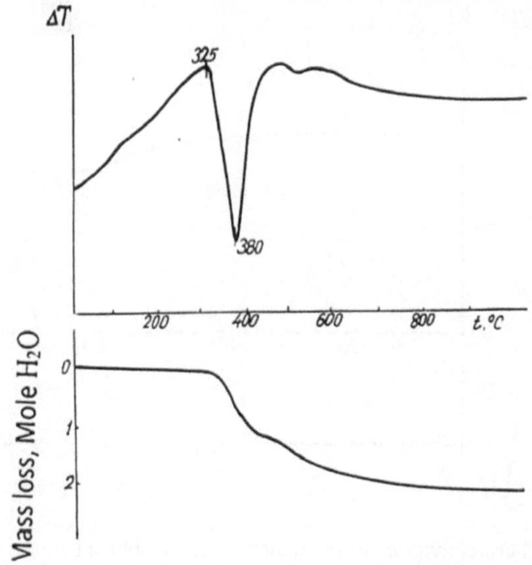

FIGURE 18.6 Thermogavigram of the double acidic diphosphate $KIn(H_2P_2O_7)_2$.

The thermal behavior of K-In and Rb-In triphosphates has been studied. The process of decomposition begins at 610–620°C (Figure 18.7), The scheme is presented in Figure 18.7.

$$RbInHP_3O_{10} \rightarrow In(PO_3)_3 + ½\ H_2O + \text{Amorphous phase}$$

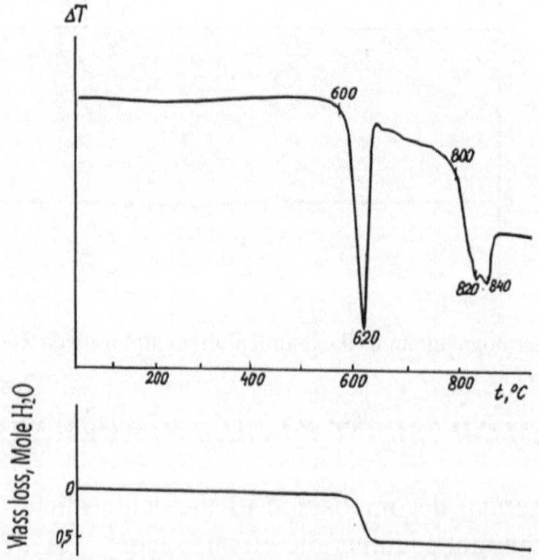

FIGURE 18.7 Thermogavigram of the double triphosphate $RbInHP_3O_{10}$.

In general, triphosphates are the most stable phases for the gallium and indium-containing systems. All triphosphates are thermally stable at temperatures around 500–600°C. Figure 18.6 shows the thermal transformation of triphosphate $RbInHP_3O_{10}$, which is stable even up to 615–620°C. It should also be noted that as the ionic radius of the alkali metal increases, the crystallization region of the acidic double diphosphates expands; whereas, in the case of the gallium-indium transition, the crystallization region of the triphosphates becomes increasingly narrow.

18.3.2 SYSTEM CONTAINING SC AND MONOVALENT METAL

During the investigation of the system with Sc and Ag, condensed phosphates were synthesized, notably the acidic double condensed triphosphate $AgScHP_3O_{10}$, obtained at 155°C with a molar ratio of Ag_2O to Sc_2O_3 equal to 5.0 to 6.5, and the acidic double diphosphate of scandium and silver $AgSc(H_2P_2O_7)2$. A thermogravimetric study of the triphosphate revealed two endothermic effects: the first effect at a temperature of 200–210°C, which corresponds to the evaporation of crystallization water, and the second effect around 480–515°C, which apparently corresponds to the detachment of chemically bound water (mass loss at the mentioned temperatures is shown in Figure 18.8). The third effect at 530–540°C is associated with the melting process of tetraphosphate $AgScP_4O_{12}$. The scheme of the reaction is as follows:

$$2AgScHP_3O_{10} \rightarrow AgScP_4O_{12} + AgScP_2O_7 + H_2O$$

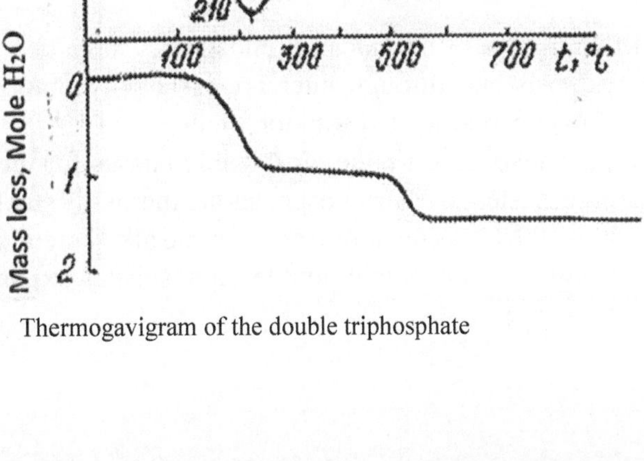

FIGURE 18.8 Thermogavigram of the double triphosphate

$$AgScHP_3O_{10}.(0,5\text{-}1,0)H_2O.$$

As previously noted, we have synthesized the acidic double diphosphate of scandium and silver, $AgSc(H_2P_2O_7)_2$, at a temperature range of 155°C to 165°C [12, 35]. This interesting compound was obtained for the first time at the molar ratio Ag_2O: Sc_2O_3=7,5; 8,0 and/or 10,0. Thermal decomposition is according to the following schema:

$$AgSc(H_2P_2O_7)_2 \rightarrow AgScP_4O_{12} + 2H_2O$$

The temperature of the final melting process of tetraphosphate $AgScP_4O_{12}$ is approximately 625–640°C. From a structural point of view, most of the obtained compounds crystallize with an arrangement that is more or less similar to Ga and In condensed phosphates; in principle, they are isomorphic in most cases. Scandium has certain resemblances to the mentioned metals, although it is difficult to speak of a clear isomorphism between them. However, at the same time, according to the Roentgen phase analysis, there are certainly some similarities.

As a final estimation, we would like to emphasize that all condensed compounds, such as oligophosphates, cyclophosphates, or polyphosphates, are extremely interesting and promising from the point of view of practical applications. This is the great merit of scientists [48–57], as evidenced by the future-oriented publications in scientific journals, as well as their inclusion in various valuable books, atlases of Infrared Spectra, etc. Some research can be found in the SAO/NASA ADS-Physics Abstract Service – Astrophysics Data System, INIS-IAEA data bulletin [12, 31, 45, 53], and in OSTI GOV U.S. data (US Department of Energy Office of Scientific and Technical Information).

18.4 CONCLUSION

All of the different types of condensed phosphates have been obtained by solid-phase synthesis and through interactions in polycomponent systems derived from polyphosphoric acid solution melts.

In general, triphosphates are the most stable phases for the gallium and indium-containing systems. All triphosphates are thermally stable at temperatures around 500–600°C. As the ionic radius of the alkali metal increases, the crystallization region of the acidic double diphosphates expands; whereas, in the case of the gallium-indium transition, the crystallization region of the triphosphates becomes increasingly narrow.

The study of the thermal behavior of double acidic diphosphates of gallium and indium with potassium and rubidium demonstrates their capacity to generate condensed phosphates with a high level of condensation—this method is one of the production routes for double cyclo-octaphosphates. The thermal decomposition of double acidic triphosphates of Ga and In with K and Rb shows their ability to form condensed diphosphates and long-chain polyphosphates as well. Obtained from molten solutions of polyphosphoric acids, the acidic triphosphate of Ag-Sc, after thermal decomposition, forms a crystalline tetraphosphate $AgScP_4O_{12}$. In other words, it has been shown that the cyclic compound can also be obtained by solid-phase condensation, and a higher degree of condensation of the anion can be achieved. They can be obtained not just from melts, but also via solid-phase synthesis, involving previously synthesized condensed compounds.

ACKNOWLEDGMENTS

I would like to express my sincere gratitude to Professor Omar Mukbaniani and Academician Ivan Tananaev for their encouragement and support of my work in the field of polymer chemistry, as well as for their important advice over the decades. I will always remember their professional guidance in the above domain and their exceptional, wonderful human qualities.

KEYWORDS

- **condensed phosphate**
- **gallium**
- **indium**
- **inorganic polymer**
- **phosphoric acid**
- **scandium**
- **synthesis**
- **thermal behavior**

REFERENCES

1. Averbuch-Pouchot, M. Th., & Durif, A. (1996). *Topics In Phosphate Chemistry.* World Scientific.
2. Tananaev, I. V. (1988). *Less-Common Element Chemistry. Rare-Earth Element Chemistry. Silicates, Germinates, Phosphates, Arsenates, Vanadates*. International Union of Pure and Applied Chemistry, pp. 45–155.
3. Zanello, P. (2012). Chains, clusters, inclusion compounds, paramagnetic labels (2nd ed.). Elsevier, pp. 100–141.
4. Durif, A. (2005). The development of cyclophosphate crystal chemistry. *Solid State Sciences, 7*, 760–766.
5. Murashova, E. V., & Chudinova, N. N. (2001). Double condensed phosphates of caesium-indium. *Neorganicheskie Materialy, 37*(12), 1521–1524.
6. Avaliani, M. A., Shapakidze, E. V., Todradze, G. V., Barnovi, N. V., Vibliani, M. B., Magradze, G. T., & Esakia, N. A. (2021). The point of view: Double condensed phosphates as analogs of inorganic polymeric compounds. In *Geopolymers–The Bilateral Materials, Reciprocals of Organic and/or Inorganic Polymers* (ICSP&AM-7), 7.
7. Avaliani, M., Chagelishvili, V., Barnovi, N., & Shapakidze, E., & Esakia, N. (2020). Advanced materials, polymers, and composites: New research on properties, techniques, and applications. *Advanced Materials, Polymers, and Composites II*, 255–276.
8. Avaliani, M., Barnovi, N., & Gvelesiani, M. (2017). Gravimetric determination of gallium, indium, and scandium by 8-oxyquinoline. *Georgian Chemical Journal*, 23–27.
9. Avaliani, M., Chagelishvili, V., Barnovi, N., Esakia, N., Gvelesiani, M., & Makhatadze, Sh. (2019). Some innovative results-oriented scientific research which lead to the development in the field of inorganic polymer's science. In *Materials Science, Composite Materials Engineering: Modeling and Technology*, 107–116.
10. Djurinsky, I. F., Tananaev, I. V., Tselebrovskaia, E. G., & Prozorovskii, A. I. (2010). Germanatophosphates of lanthanoids. *ChemInform Abstract, 22*(25). https://doi.org/10.1002/chin.199125037.
11. Avaliani, M. A., & Gvelesiani, M. K. (2006). Areas of crystallization of condensed scandium and cesium phosphates and regularities of their formation. *Proceedings of the Georgian Academy of Sciences, Chemical Series, 32*, 52–58.
12. Avaliani, M. A. (2021). Investigation and thermal behavior of double condensed phosphates of gallium, scandium, and silver. *Bulletin of the International Nuclear Information Systems INIS-IAEA, 52*, 31.
13. Marsh, T. P. (2011). Studies into the ion exchange and intercalation properties of $AlH_2P_3O_{10}\cdot 2H_2O$ (PhD thesis). University of Birmingham. Retrieved from: https://etheses.bham.ac.uk/id/eprint/1599 (accessed on 1 October 2024).
14. Mi, J. X., Borrmann, H., Huang, Y. X., & Zhao, J. T. (2003). Crystal structure of the 1M-modification of caesium gallium (III) monohydrogen triphosphate, 1M-$CsGaHP_3O_{10}$. *Für Kristallographie-News.*
15. Ben Mussa, S., Farid, M., & Trabelsi-Ayedi, M. (1995). Thermal study of double polyphosphates $M^IEr(PO_3)_4$ (M^I is H, NH_4). *Thermochimica Acta, 249*, 189–197. https://doi.org/10.1016/0040-6031(95)90692-4.
16. Zatovsky, I. V., & Strutynska, N. Y. (2008). Synthesis and characterization of phosphates in the pseudo-ternary melted systems Cs_2O–P_2O_5–$M^{II}O$ (M^{II}–alkaline earth). *Crystal Research.*

17. Averbuch-Pouchot, M. T., & Durif, A. (1991). Crystal chemistry of oligophosphates. *Annual Review of Materials*.
18. Villars, P., Cenzual, K., & Penzo, M. (2013). Inorganic substances bibliography. *Science, 3*, 3946.
19. Murashova, E. V., Chudinova, N. N., & Zakharova, B. S. (2001). Mixed-cation thallium phosphates containing di-, tri-, and pentavalent cations. *Inorganic Materials, 37*, 1170–1174.
20. Vadapalli, Chandrasekhar. (2005). Inorganic polymers: A review of synthetic strategies; problems and prospects. In *Inorganic and Organometallic Polymers* (pp. 2–26). Springer, Berlin. https://doi.org/10.1007/3-540-26215-6_1.
21. Lalena, J. N., Cleary, D. A., Carpenter, E. E., & Dean, N. F. (2008). *Inorganic Materials Synthesis and Fabrication*. John Wiley & Sons, Inc.
22. Kriven, W. M. (2015). Synthesis of $LiFePO_4$ powder by the organic–inorganic steric entrapment method. *Journal of Materials Research, 30*(15), 2133–2143.
23. Ponnusamy, S. K. (2021). Clean technologies for a sustainable environment. *IET Nanobiotechnology, 15*(2), 147–148. https://doi.org/10.1049/nbt2.12052.
24. Mukbaniani, O., Londaridze, L., Markarashvili, E., Tatrishvili, T., & Aneli, J. (2021). Triethoxysilylated styrene as a new coupling agent in wood composites. In *Proceedings of the 7th International Caucasian Symposium on Polymers and Advanced Materials–"ICSP&AM 7,"* 73.
25. Gimbert, Y., Robert, F., Averbuch, M. T., Durif, A., & Green, E. (1999). Synthesis and characterization of new binuclear Co(0) complexes with diphosphinoamine ligands: A potential approach for asymmetric Pauson–Khand reactions. *Journal of Organic Chemistry, 64*(10), 3492–3497. https://doi.org/10.1021/jo982245o.
26. Kvinikadze, N., Londaridze, L., Zedgenidze, A., Dzidziguri, D., & Mukbaniani, O. (2021). Wood polymer composites based on a new coupling agent. In *Proceedings of the 7th International Caucasian Symposium on Polymers and Advanced Materials–"ICSP&AM 7,"* 60.
27. Grunze, I., Cudinova, N. N., Palkina, K. K., Avaliani, M. A., Guzeeva, L. S., & Maksimova, S. (2009). Structure and thermal rearrangements of binary cesium–gallium phosphates. *Journal of Inorganic Materials, 23*(4), 539–544.
28. Oudahmane, A., Avignant, D., & Zambon, D. (2010). Dipotassium dialuminium cyclooctaphosphate. *Structure Reports Online, 2*(7), 149–150.
29. Durif, A. (2014). *Crystal Chemistry of Condensed Phosphates* (3rd ed.). Plenum Press.
30. Avaliani, M. A., Todradze, G., Shapakidze, U., Ukleba, M., Chikovani, N., Vibliani, M., & Magradze, G. (2020). Optimization of the gravimetric method for determination of trivalent metals using oxiquinoline. In *Proceedings of the International Conference on Analytical Chemistry "Modern Trends,"* Kyiv, Ukraine. Retrieved from: http://kcacmt.univ.kiev.ua/en (accessed on 1 October 2024).
31. Avaliani, M., Gvelesiani, M., Barnovi, N., Purtseladze, B., & Dzanashvili, D. (2016). New investigations of poly-component systems. *Proceedings of the Georgian Academy of Sciences, Chemical Series, 42*, 308–311.
32. Avaliani, M. (2016). Main types of condensed phosphates synthesized in open systems from solution-melts of phosphoric acids. *Nano Studies, 1*, 135–138.
33. Avaliani, M., Chagelishvili, V., Shapakidze, E., Gvelesiani, M., Barnovi, N., Kveselava, V., & Esakia, N. (2019). Crystallization fields of condensed scandium-silver and gallium-silver phosphates. *European Chemical Bulletin, 5*, 164–170.

34. Avaliani, M., Shapakidze, E., Barnovi, N., Dzanashvili, D., Todradze, G., Kveselava, V., & Gongadze, N. (2019). Regio-controlled synthesis of double condensed oligo-, poly-, and cyclo-phosphates, their characterization and possibilities of applications due to their solid-state properties. *Journal of Nano Studies–European Chemical Bulletin, 11*(19), 273–284.
35. Avaliani, M., Shapakidze, E., Barnovi, N., Gvelesiani, M., & Dzanashvili, D. (2017). About new inorganic polymers: Double condensed phosphates of silver and trivalent metals. *Journal of Chemical and Chemical Engineering (USA), 11*, 60–64.
36. Avaliani, M., Todradze, G., Shapakidze, E., & Kveselava, V. (2020). Synthesis of condensed phosphates of mono- and polyvalent metals: Development of optimal methods for the study of their properties and composition. *Journal of Nano Studies–European Chemical Bulletin, 20*, 71–94.
37. Avaliani, M., Chagelishvili, V., Barnovi, N., Shapakidze, E., & Esakia, N. (2021). Condensed phosphates: New inorganic polymers with a variety of applications and improvement of their gravimetric determination methods. In *Apple Academic Press, Advances in Materials, Polymers and Composites* (Chapter 18, pp. 255–276).
38. Avaliani, M. (2019). Some innovative results-oriented scientific researches which lead to the development in the field of inorganic polymer's science. In *Materials Science / Composite Materials Engineering: Modeling and Technology* (pp. 61–72). Apple Academic Press. https://doi.org/10.13140/RG.2.2.16514.32969.
39. Avaliani, M. A., Chudinova, N. N., & Tananaev, I. V. (1984). Condensed sodium-indium phosphates. *INIS-IAEA*. Retrieved from: https://inis.iaea.org/search/search.aspx?orig_q=RN:16005261 (accessed on 1 October 2024).
40. Avaliani, M., Shapakidze, E., Chagelishvili, V., & Todradze, G. (2021). Investigating the influence of the trivalent and monovalent metals ionic radius on the structure of the synthesized double condensed compounds. In *Proceedings of the IX International Conference on Chemistry and Chemical Education–Sviridov Readings*, Minsk, 109–110.
41. Tananaev, I. V., Grunze, I., & Chudinova, N. N. (1984). Prior directions and results in the domain of condensed phosphates' chemistry. *Journal of Inorganic Materials, 20*(6), 887–900.
42. Avaliani, M. A. (2011). About development in chemistry and application of condensed phosphates. In *Proceedings of the International Conference GEOHET* (Vol. 1, pp. 162–163). TSU, Tbilisi, Georgia.
43. Avaliani, M., Shapakidze, E., Chagelishvili, V., Gvelesiani, M., Barnovi, N., Vibliani, M., & Chikovani, K. (2020). Properties of the synthesized inorganic polymeric phosphate materials and the possibilities for production of new environmentally friendly and economically viable supplies from local industrial waste. *Nano Studies–European Chemical Bulletin, 20*, 63–70.
44. Tananaev, I. V. (1987). Some aspects of the chemistry of phosphates and their practical application. *Problems of Chemistry and Chemical Technology, 1*, 4–65.
45. Avaliani, M. A. (1984). Double condensed sodium-scandium phosphates. *INIS-IAEA*. Retrieved from: https://inis.iaea.org/search/search.aspx?orig_q=RN:23046460 (accessed on 1 October 2024).
46. Avaliani, M., Shapakidze, E., Todradze, G., Barnovi, N., Vibliani, M., & Magradze, G. (2021). Double condensed phosphates–As analogs of inorganic polymeric compounds. In *Proceedings of the 7th International Caucasian Symposium on Polymers & Advanced Materials* (p. 7). Tbilisi, Georgia.

47. Avaliani, M. (2022). Modern approach to the synthesis of inorganic polymers, taking into account the prospects for their application. In *Proceedings of the International Conference "Natural and Synthetic Polymers for Medical and Technical Purposes* (pp. 143–147). Minsk. https://doi.org/10.13140/RG.2.2.28245.55528.
48. Abramovich, E. A., & Selevich, A. F. (2021). Synthesis and characterization of ammonium-vanadium (III) double cyclo-phosphates $(NH_4)_2V_2P_8O_{24}$ and $(NH_4)_3V_3P_{12}O_{36}$. In *Proceedings of the IX International Conference on Chemistry and Chemical Education–Sviridov Readings* (p. 13). Minsk.
49. Weil, E. D., & Levchik, S. V. (2008). Flame retardants for plastics and textiles. *Journal of Fire Sciences, 26*(3), 243–281.
50. Jabishvili, N., Klarjeishvili, N., Karalashvili, I., & Urotadze, S. (2011). The new phosphate composition (antipyrene)–Increasing the fire resistance of various materials. In *3rd Rafael Agladze Conference RAC-3 on Applied Chemistry* (pp. 93–94). Tbilisi, Georgia.
51. Brostow, W., Cañadas, I., Fałtynowicz, H., Levinskas, R., & Garcia, J. R. (2021). Fire resistance of materials. In *Proceedings of the 7th International Caucasian Symposium on Polymers and Advanced Materials–"ICSP&AM 7"* (p. 14).
52. Selevich, A. F., Abramovich, E. A., & Ivashkevich, O. A. (2022). Synthesis of ammonium-containing double oligo-, cyclo-, and polyphosphates of trivalent metals in the ammonium polyphosphate flux. In *Proceedings of the International Conference "Natural and Synthetic Polymers for Medical and Technical Purposes"* (pp. 212–215). Minsk.
53. Palkina, K. K., Maksimova, S. I., Kuznetsov, V. G., & Chudinova, N. N. (1979). Structure of crystals of double octametaphosphate $Ga_2K_2P_8O_{24}$. *DAN [Crystallography], 245*(6), 1386–1389. https://doi.org/10.1016/j.ica.2020.116852.
54. Begum, Y. (2013). Synthesis and characterization of Mn^{3}+ condensed phosphate phases and Fe^{3}+ substituted analogues (Doctoral dissertation, University of Birmingham). eTheses Repository. Retrieved from: http://etheses.bham.ac.uk/4302/5/Begum_13_PhD[1].pdf (accessed on 1 October 2024).
55. Lyutsko, V. A., Selevich, A. F., & Kutseva, E. R. (1992). Investigation of the interaction of gallium dihydrophosphates with gaseous ammonia. *Journal of Inorganic Materials, 37*(1–6), 512–516.
56. Lesage, J., Guesdon, A., Hervieu, M., & Raveau, B. (2006). Unusual mobility of cesium via a reversible topotactic dehydration. *Chemistry of Materials, 18*(12), 2895–2900. https://doi.org/10.1021/cm060275z.
57. Melnikov, P., Guirardi, A. L., Socco, M. A. C., & Nogeira de Aguiar, E. (2008). Study of the trivalent elements polyphosphates by thermal analysis. *Journal of Thermal Analysis and Calorimetry, 94*(1), 163–167.

CHAPTER 19

Study of the Sorption Potential of Carbon Materials Derived from Polymer Waste: A Case Study of Pollutants in the *Leachate* from the Tbilisi Landfill

NATELA DZEBISASHVILI[1,2] and DAREJAN DUGHASHVILI[1,2]

[1]*Institute of Hydrometeorology, Georgian Technical University, Tbilisi, Geogia*

[2]*Raphael Agladze Institute of Inorganic Chemistry and Electrochemistry, Ivane Javakhishvili Tbilisi State University, Tbilisi, Georgia*

ABSTRACT

Today, the harmonization of the waste management process by the Georgian government with European waste management policy is one of the priority issues, which should focus on the reduction of waste generation and reuse. This, in turn, is directly related to the transition from a linear economy to circular economic development and the use of new economic management models. The introduction of circular economy principles in Georgia is unthinkable under the conditions of irrational management of one of the most difficult-to-decompose wastes – polymer waste. Our preliminary studies have shown that the share of one of the most difficult-to-decompose fractions of municipal solid waste in Georgia – plastic – is 14%, namely 126 thousand tons per year [1]. Our research includes the minimization of the amount of the most harmful and hard-to-degrade organic fractions (polymers) of municipal solid waste by obtaining sorbents and studying their sorption potential for the treatment of leachate from the Tbilisi landfill (Norio) based on some chemical indicators [2–4]. The results of our research show a high sorption

Advanced Topics in Polymer Chemistry and Materials Science: Tamara Tatrishvili and Marc Jean Médard Abadie (Eds)

potential for the main pollutants – heavy metals, biogenic substances, microbiological indicators, and organic impurities, averaging 80%.

19.1 INTRODUCTION

In November 2022, the Georgian government adopted the resolution "Vision 2030, Development Strategy of Georgia," which notes that environmental pollution from waste remains a problem [2]. This is due to the challenges of managing both municipal solid waste (MSW) and specific waste streams. Currently, the government has approved a new resolution, "National Action Plan for Waste Management (2022–2026)." Specifically, the NWAMP 2022–2026 includes 13 targets, with additions of biodegradable waste treatment and prevention measures, hazardous waste management goals and objectives, measures related to the prevention and reduction of plastics, and implementation of Extended Producer Responsibility. Georgia's commitment to sustainable development and its continuous efforts in addressing waste management challenges are validated by the aforementioned targets [3]. Regardless of the strategy adopted, one of the most pressing problems in our country remains the management of open landfills (including leachate), as well as the disposal of polymer waste without any recycling. Our preliminary studies have shown that the share of one of the most difficult-to-decompose fractions of municipal solid waste in Georgia—plastic—is 14%, namely 126 thousand tons per year [1].

The aim of our research was to find solutions to these two problems: in particular, to obtain carbon adsorbents from polymer waste and to treat leachate water from some main pollutants using the obtained sorbent [4–6]. Currently, the main part of the solid waste generated throughout the country is placed in open landfills (90%). The largest official landfill in Georgia serves the capital city, Tbilisi, and is located to the southeast of Tbilisi, east of the village of Didi Lilo (30 km from Tbilisi). The population of Didi Lilo demands the conservation of the Tbilisi MSW landfill and its relocation. The proximity of the landfill to the locals creates unbearable living conditions due to environmental pollution [1, 6, 8].

19.2 EXPERIMENTAL METHODS AND MATERIALS

19.2.1 LEACHATE WATER

Originally a reverse osmosis device was installed at the landfill to treat leachate water. The capacity of the existing reverse osmosis plant was 12

m^3/ day, which was insufficient and an urgent task was to solve this problem [10], as the leachate water cannot be treated in accordance with the legal norms for discharge into the sewerage system [Maximum Permissible Discharge (MPD)] [11]. To solve this problem, the leachate water from the sedimentation system (reservoir) is pumped to the surface of the waste cell using pumping machines; therefore, it circulates: reservoir – landfill cell. This activity moistens the waste and, consequently, activates the biodegradation process and significantly increases the degree of pollution of leachate water. In addition to all of the above, there is a high risk of environmental pollution, which is mainly due to the high probability of emergent discharge of the leachate water.

Namely, in the case of heavy precipitation, reservoirs (sedimentation systems) are the most likely to overflow and discharge contaminated water into the collector and then into the river Norikheva, which flows in the natural ravine nearby, and then into Lochini, and finally to the transboundary river Mtkvari, the main river of the capital city. From April 2021 to July 2023, we sampled leachate from the sedimentation system of water every month. Physical and chemical indicators were measured in the field (Consort 6010 Multiparameter Portable Analyzer), and then in the laboratory, the determination of Chemical Oxygen Demand (COD) (ISO 15705:2002); Biochemical Oxygen Demand (BOD_5) (ISO 5815–2:2003/2020); Total nitrogen (ISO 29441:2010); Ammonia nitrogen (ISO 5378:1978); Phenols (EPA 420.1); fats and oils (EPA 413.1); Cd, Cr, Pb, Hg (ISO 17294–2:2016); *E. coli* and Total Coliforms (ISO 9308–2:2012) was carried out.

19.2.2 OBTAIN OF SORBENT

To obtain carbon material in the framework of the study, at the first stage of the study, the polymer fraction of municipal solid waste was sorted, crushed, and the sorted secondary polymer raw material was prepared for thermogravimetric analysis in order to study the optimal conditions for obtaining sorbents. Thermogravimetric analysis of prepared samples was carried out using STA 2500 Regulus Simultaneous Thermal Analysis. The samples were heated from 25 to 550°C with a heating rate of 10°/min. Based on the results of thermogravimetric analysis, the optimal temperature for thermochemical treatment of polymer waste samples was chosen –420°C. To obtain carbon from polymer waste, a stainless-steel reactor was used, which is a sealed capsule with two stainless steel pipes – horizontal thermochemical method [5, 7]. Anaerobic conditions were achieved using a two-way tube – nitrogen flow. The reaction continued until the emission of exhaust gases ceased. In order to purify carbon material from impurities and improve its sorption properties, the material (in the mode of 24-hour exposure) was treated with an aqueous solution of HCl 1:5 [5].

19.3 RESULTS AND DISCUSSION

19.3.1 ANALYSIS OF LEACHATE WATER

In 2023, the monthly samples analysis shows that on average the leachate water have: turbidity 990 FTU, pH 7.8 (MPD 6.0–9.5); electrical conductivity 35.2 mS; TSS 120 mg/l (MPD 300 mg/l); COD is 11,300 mgO/l (MPD 600 mgO/l); Biochemical Oxygen Demand (BOD_5) >4000 mgO/l (MPD 300 mgO/l); total nitrogen 1680 mg/l (MPD 25 mg/l); phenol 0.98 mg/l l (MPD 0.25 mg/l); fats <5 mg/l (MPD 15 mg/l); Cr 1.65 mg/l (MPD 1 mg/l); Cd 0.72 mg/l (MPD 1 mg/l); Pb 1.71 mg/l (MPD 1 mg/l); Hg 0.6 mg/l (MPD 0.5 mg/l); *E-coli*<500 MPN/100 ml; Total coliforms 35,000 MPN/100ml (MPD 5000 MPN/100ml).

The conducted research shows that the main pollutants of leachate are biogenic and organic substances, some heavy metals, and microbiological indicators, which are tens of times higher than the norms (MPD) [9]. It should be noted that the content of organic compounds and heavy metals in freshwater decreases with increasing temperature and is maximum in winter, while the number of microbiological parameters is thousands of times higher in summer, despite increased dilution from abundant atmospheric precipitation.

19.3.2 LEACHATE WATER TREAT WITH OBTAIN CARBON SORBENT

1 and 2 g of acid treated and sieved, different sizes (nano and micro) carbon material were added to 50 ml of test waters and left for 1, 24 and 72 hours under static conditions. per 50 ml of test waters, 1 g of carbon sorbent, left under static conditions for 24 hours, showed the best results for all pollutants (Table 19.1).

TABLE 19.1 Sorption of the Main Pollutant of Leachate on Carbon Obtained from Polymers (2023: 1g sorbent/50ml Leachate Simple/24 Hour Holding Time)

No.	Pollutant	Unit	MPD	Primary (average), mg/l	Result	
					mg/l	% of sorption
1	NH_4^+\	mg/l	20	35, 5	12, 31	65, 32
2	BOD5	mg O/l	300	4320	1507, 68	65, 10
3	COD	mg O/l	600	11300	5164, 10	54, 30
4	phenol	mg/l	0.25	0, 98	0, 24	75, 80
5	Cr	mg/l	1	1, 65	0, 62	62, 42
6	Cd	mg/l	1	0, 72	0, 01	98, 61
7	Pb	mg/l	1	1, 71	0, 68	60, 23
8	Hg	mg/l	0.5	0, 6	0, 01	98, 33
9	Total coliforms	MPN/100ml	5000	35000	5250, 00	85, 00

Figures 19.1 and 19.2 show that after the treatment of leachate samples with the obtained carbon sorbents (micro fraction), the concentrations of the main pollutants in the leachate are less than or approaching the relevant norms.

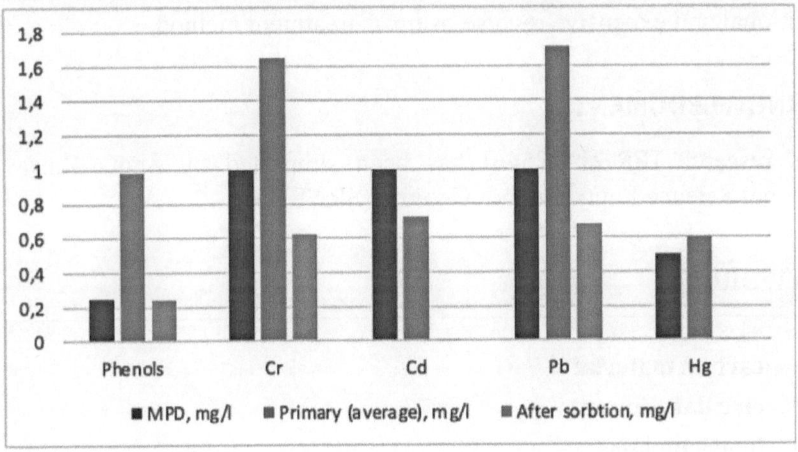

FIGURE 19.1 Content of phenols and some heavy metals before and after sorption process.

19.4 CONCLUSION

The use of the carbon material obtained from polymeric waste with a particle size of 1200–40 microns in static conditions showed that the sorption capacities are 60–99% for heavy metal ions, 50–80% for organic pollutants, and 80–95% for microbiological pollutants. According to the results of the analysis, the use of nanomaterials is ineffective when water stagnates under static conditions due to the formation of tarry solutions.

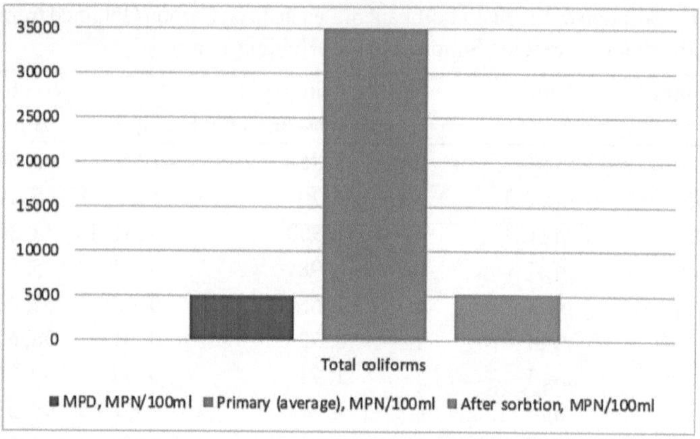

FIGURE 19.2 Content of total coliforms before and after sorption process.

The research proves that the carbon sorbent obtained from polymeric waste reduces the main impurities content of the most polluted wastewater to a much lower level than MPD. The method can be an alternative to the traditional and expensive reverse osmosis treatment method.

ACKNOWLEDGMENT

This research [FR-21–12546] has been supported by Shota Rustaveli National Science Foundation of Georgia (SRNSFG).

KEYWORDS

- **carbon material**
- **circular economy**
- **heavy metals**
- **leachate treatment**
- **microbiological indicators**
- **organic pollutants**
- **polymer waste**
- **reverse osmosis alternative**
- **sorption potential**
- **Tbilisi landfill**

REFERENCES

1. Dvalishvili, N. (2017). Impact of incineration of municipal solid waste on climate change in Georgia. In *5th World Convention on Recycling and Waste Management.* Retrieved from: https://www.longdom.org/proceedings/impact-of-incineration-of-municipal-solid-waste-on-climate-change-in-georgia-38662.html (accessed on 1 October 2024).
2. Government of Georgia. (2023). *Vision 2030: Development Strategy of Georgia.* Retrieved from: https://www.gov.ge/wp-content/uploads/2023/05/Strategy-2030-ENG-WEB2d.pdf.
3. Government of Georgia. (2022). *The National Waste Management Action Plan 2022–2026* [Ordinance].
4. Dzebisashvili, N., Tatishvili, G., Suramelashvili, E., & Dughashvili, D. (2022). Treatment of wastewater from ammonia and microbiological components using carbon materials. In *Book of Abstracts of the 4th International Scientific Conference "Natural Resources, Green Technology and Sustainable Development" (4-GREEN 2022)* (p. 116). Retrieved from: https://www.sumins.hr/wp-content/uploads/2022/09/Green2022_Book_of_Abstracts.pdf (accessed on 1 October 2024).
5. Dzebisashvili, N., Dughashvili, D., & Suramelashvili, E. (2023). Obtaining new carbon material from polypropylene waste using a horizontal type reactor and studying its sorption. *Scientific Reviewed Proceedings of the Institute of Hydrometeorology of the Georgian Technical University: Pressing Problems in Hydrometeorology and Ecology, 133*, 106–111. https://doi.org/10.36073/1512-0902-2023-133-106-111.
6. Dvalishvili, N. L. (2016). Establishment of energy potential of Norio landfill of municipal solid waste of Tbilisi. *Procedia Environmental Sciences, 35*, 377–380. https://doi.org/10.1016/j.proenv.2016.07.017.
7. Dzebisashvili, N., Tatishvili, G., & Samkharadze, Z. (2023). Advanced materials obtaining the new carbon material through anaerobic thermochemical processing of the polymer waste with the use of horizontal and vertical reactors. In *Abstracts of the 8th International Caucasian Symposium on Polymers and Advanced Materials* (p. 15). Retrieved from: https://icsp8.tsu.ge/data/file_db/icsp8/abstract_icsp8.pdf (accessed on 1 October 2024).
8. Dvalishvili, N. L., & Tabatadze, M. S. (2019). The influence of municipal solid waste of Georgia on climate changes. In S. Ghosh (Ed.), *Waste Management and Resource Efficiency* (pp. 229–235). Springer. https://doi.org/10.1007/978-981-10-7290-1_16 (accessed on 1 October 2024).
9. Government of Georgia. (2018). Resolution #431 on rules for discharging and receiving of water flowing into sewerage system and permitted limits of pollutants [Appendix 1].
10. Ministry of Environmental Protection and Agriculture of Georgia. (n.d.). Change of conditions of operation of Tbilisi solid waste landfill located in the territory of Gardabani Municipality [Scoping report]. Retrieved from: https://mepa.gov.ge/Ge/PublicInformation/11403 (accessed on 1 October 2024).
11. Government of Georgia. (2014). Technical regulations for calculating maximum permissible discharges of pollutants discharged along with wastewater into surface water bodies. Retrieved from: https://matsne.gov.ge/ka/document/view/2188404?publication=0 (accessed on 1 October 2024).

CHAPTER 20

New Type, Ecologically Friendly, Highly-Effective, Bioprotective Fire-Extinguishing Foam-Suspension

LALI GURCHUMELIA,[1] MURMAN TSARAKHOV,[1] DALI DZANASHVILI,[1] MAIA GABRICHIDZE,[2] SALOME TKEMALADZE,[1] and OLGA CHUDAKOVA[3]

[1]TSU Rafael Agladze Institute of Inorganic Chemistry and Electrochemistry, Tbilisi, Georgia

[2]University Geomedi LLC, Tbilisi, Georgia

[3]G. Tsulukidze Mining Institute, Tbilisi, Georgia

ABSTRACT

The aim of the presented investigation is the elaboration of a new type of ecologically friendly, highly effective, bioprotective fire extinguishing foam-suspensions through the use of fire-extinguishing powders made from local mineral raw materials. These foam-suspensions will have a higher fire extinguishing capacity than powder, foam, and water taken separately, and they can also regenerate damaged soil.

Foam suspensions are prepared by mechanically mixing fire-extinguishing powders (zeolite, perlite, clay, shale, and dolomite) modified with ammophos, water, and non-ionic, biodegradable surfactants. The process does not require chemical treatment of materials. Therefore, the developed technology for the production of foam-suspensions is simple and cost-effective.

The effectiveness of the foam-suspension depends on its performance properties: stability of the suspension; the type and content of the surfactants; the foam stability; and the multiplicity of foam. Performance characteristics

Advanced Topics in Polymer Chemistry and Materials Science : Tamara Tatrishvili and Marc Jean Médard Abadie (Eds)

are determined by standard laboratory methods. In order to determine fire extinguishing ability, polygon test methods are used, which consider the extinguishing of different classes of standard fires with the help of fire-extinguishing construction. Based on experimental results, it is stated that the obtained foam-suspensions are nontoxic, ecologically friendly, and characterized by higher fire-extinguishing ability compared to such powders.

At the same time, it should be noted that the use of zeolites and ammophos as combined fertilizers decreases the acidity of the soil, regulates the interchange of P, K, and N ions in the soil, and cultivates microorganisms.

Thus, we can predict that the obtained foam suspensions will not only effectively extinguish fires, but they can also regenerate the soil damaged by the fire. Therefore, they can be successfully used for the extinguishing of large-scale fires—forest fires.

20.1 INTRODUCTION

Traditionally, for extinguishing fires, CO_2, water, foam, and powders are used. However, it should be mentioned that water is characterized by high consumption, high electrical conductivity, insufficient wetting capacity, and low adhesion to the object being extinguished. Therefore, surface-active substances (surfactants) are added to water, which reduce water surface tension, increase permeability, and decrease consumption. Unfortunately, the use of foams for extinguishing large-scale fires requires significant consumption of water and surfactants. At the same time, foams are not universal, and their widespread use is ecologically dangerous. Today, the most effective fire extinguishers are powder fire extinguishers. They are the only means that can be used to extinguish practically all classes of fire. It should also be noted that mass-produced fire-extinguishing powders are finely dispersed mineral salts modified with various types of halogen, phosphorus, and nitrogen-containing hydrophobizators of organic origin. Therefore, most mass-produced powders do not meet modern requirements, primarily in terms of effective and environmentally friendly use. It should also be mentioned that fire-extinguishing powders are characterized by lower heat capacity, low permeability, and a wetting effect compared to water and foams. That's why, when extinguishing large-scale fires, they cannot wholly solve the problem of new ignition caused by the inflammation of flickering foci in open space [1–5]. From the above, it follows that for extinguishing large-scale fires, particularly forest fire extinguishing, the use of just fire extinguishing powders is less effective.

On the basis of local mineral raw materials: zeolite, perlite, clay-shale, and dolomite, our group has developed a new type of fire-extinguishing powder (patent U 2019–2013-Y), which is not inferior in efficiency to mass-produced fire-extinguishing powders. Unlike them, it is non-toxic, environmentally friendly, and characterized by high compatibility with water and foam, which allows foam suspensions to be made on its basis. [6, 7]. Also, it should be noted that the effectiveness of the foam suspension depends not only on the effectiveness of the fire-extinguishing powders but is directly related to the chemical composition of surfactants, their decomposition, and their transformation forms.

Surfactants are chemical compounds that concentrate at the phase boundary, cause a decrease in surface tension, stabilize the dispersion system, and make it impossible for particles of the dispersed phase to compact and coagulate. As a rule, these are organic compounds whose molecules contain a non-polar (hydrocarbon) part—water-insoluble hydrophobic components—and a polar part—water-soluble hydrophilic components. According to the type of hydrophilic groups, surfactants are divided into four types: anionic, cationic, amphoteric, and non-ionic. Their main quantitative characteristic is surface activity—the ability of a substance to reduce surface tension at the phase boundary. The selection criteria are: on the one hand, foaming ability; on the other hand, non-toxicity and biodegradability. Anionic and non-ionic surfactants are considered more effective. They are characterized by high surface activity, foaming ability, and biodegradability. Anionic surfactants are biodegradable (up to 80%), and non-ionic surfactants are fully biodegradable (up to 100%). Non-ionic surfactants are considered to be the least toxic compared to other types of surfactants and are safe for nature, as they completely decompose after use and do not produce toxic compounds [8, 9].

The aim of the presented investigation is the elaboration of a new type of ecologically friendly, highly effective, bioprotective fire extinguishing foam-suspensions through the use of fire-extinguishing powders (developed by us) and non-ionic, biodegradable surfactants, which will have a higher fire-extinguishing capacity than powder, foam, and water taken separately.

20.2 MATERIALS AND METHODS

Foam suspensions were prepared using fire extinguishing powders (zeolite, perlite, clay-shale, and dolomite) modified with ammophos, water, and non-ionic surfactants. Ammophos is a heterogeneous inhibitor that is well soluble in water. In cases of suspension production based on fire-extinguishing powders modified with ammophos, an increase in water inhibition is expected. The presented fire-extinguishing powders belong to hydrophobic substances. They mix easily with water. Powder particles are evenly distributed in water, and their consolidation does not occur. The additional introduction of non-ionic surfactants into suspensions causes a decrease in surface tension, stabilizes the dispersion system, prevents particles of the dispersed phase from compacting and coagulating, and causes powder flotation, which enables spraying powder together with water and foam [10, 11].

Thus, foam suspensions are prepared simply by the mechanical mixing of fire-extinguishing powders modified with ammophos, water, and non-ionic surfactants. The process eliminates the need for chemical treatment of materials, making the developed technology for producing foam suspensions simple and cost-effective.

At the same time, it should be noted that non-ionic surfactants are fully biodegradable; they completely decompose after use and do not produce toxic compounds. Thus, they are safe for nature and the soil. Zeolites and ammophos, as combined fertilizers, decrease the acidity of soil, regulate the interchange of P, K, and N ions in the soil, cultivate microorganisms, and promote their growth, which in turn are indicators of soil productivity. Thus, we can predict that the obtained foam suspensions can not only effectively extinguish fires, but they can also decrease the concentration of toxic gases in the medium, prevent the probability of their penetration into the soil, and perform a kind of regeneration of the soil damaged by the fire [12].

Based on all of the above, we can assume that the foam suspensions in our preparation will be highly effective, non-toxic, environmentally friendly, and bioprotective.

20.3 DETERMINATION OF PERFORMANCE PROPERTIES

The effectiveness of the foam suspension depends on its performance properties: stability of the suspension; the type and content of the surfactants; the foam stability; and the multiplicity of foam. Performance characteristics are determined by standard laboratory methods. The stability of the suspension determines the powder dispersity and suspension viscosity, which in turn depends on the powder volumetric concentration.

The viscosity of the suspension is determined by the capillary viscometric method with a stirring mechanism, which takes into account the time ratio of the outflow of suspension and water from the capillary tube under constant stirring conditions and is calculated by the formula:

$$\mu_s = \mu_0 \frac{\Delta_s t_s}{\Delta_0 t_0}$$

where: $\mu_0 \mu_0$ – is the viscosity of water and $\mu_0 = 0.001 \mu_0 = 0.001$ g/cm. sec; $\Delta_s, \Delta_0 \Delta_s, \Delta_0$ – the density respectively, of suspension and water, g/cm^3; $t_s, t_0 t_s, t_0$ – time of flow foam-suspension and water sec.

- powder dispersity, x (%) – granulometric composition, mass concentration of powder remains left on the sieve.
- Foam state –(τ) is the time during which 50% of the liquid phase is separated.
- Multiplicity of foam C (%), ratio of the received foam volume – V_1 to initial volume of foaming agent water solution V.

$$C = (V_1/V) \cdot 100$$

The results of an experimental study of the performance properties of fire-extinguishing foam suspensions are given in Table 20.1. An analysis of the obtained results of the performance properties of powder suspensions proved that the increase in powder volumetric concentration leads to an increase in the viscosity of the suspension and a decrease in its stability accordingly.

At the same time, it should be noted that the stability of the foam suspension is determined not only by the stability of the suspension but also by the content of the surfactants. Test results show that with the increase in the content of the surfactants, the stability of the foam suspension increases significantly.

Based on the above, considering the stability of foam suspensions, we have chosen optimal values: powder dispersity less than 250 μm; powder volumetric concentrations –20–25%; suspension viscosity – 0.015–0.018 Poise g/(cm. s); and a minimum amount of non-ionic surfactants –1% (Table 20.1).

TABLE 20.1 Performance Properties

Materials	Powder Dispersity, S (Mm)	Suspension Viscosity, P (G /Cm. Sec)	Suspension Stability, (Sec)	Foam State, T (Sec)	Multiplicity of Foam, C (%),
Powder-suspension powder (20%) + water (80%)	# 0.2–0.25	0.015	600		
Powder-suspension powder (25%) + water (75%)	# 0.2–0.25	0.018	420		
powder-suspension powder (30%) + water (70%)	# 0.2–0.25	0.025	240		
Foam-suspension powder (20%) + water (80%) + foamer 0,5%)	# 0.2–0.25	-		900	30
Foam-suspension powder (25%) + water (75%) + foamer 0,5%)	# 0.2–0.25	-		600	20
Foam-suspension powder (30%) + water (70%) + foamer 0,5%)	# 0.2–0.25	-		240	15
Foam-suspension powder (20%) + water (80%) + foamer 1.0%)	# 0.2–0.25	-		1200	40
Foam-suspension powder (25%) + water (75%) + foamer 1.0%)	# 0.2–0.25	-		900	30
Foam-suspension powder (30%) + water (70%) + foamer 1.0%)	# 0.2–0.25	-		300	20

20.4 DETERMINATION OF FIRE- EXTINGUISHING ABILITY

In order to determine fire-extinguishing ability, the polygon test methods are used, which consider the extinguishing of different class standard fires with the help of fire-extinguishing constructions and enable the determination of: minimum quantity of extinguishing materials consumed for fire focus extinguishing or minimum mass concentration of extinguish (G); extinguish time (τ); material supply intensity (I). The results of an experimental study of the fire-extinguishing ability of fire-extinguishing powders and foam suspensions are given in Table 20.2.

TABLE 20.2 Fire- Extinguishing Ability

Materials	Class of Fire	Time of Fire Extinguishing, τ (sec)	Minimum Consumption Per Unit Area, G (kg/m^2)	Minimum Mass Concentration, Cn (kg/m^3)
Zeolite + Clay shale + Perlite	A	12	1.6	2.8
Dolomite +Ammophos	B	8		1.6
Foam suspension	A	8	1.1	1.5
	B	5	–	1.3

The fire-extinguishing ability of powders and foam-suspensions based on the obtained powders are determined for the A (wood) and B (oil) class standard fires.

Test results shows how, that the fire- extinguishing ability foam-suspensions produced on the bases of powders (zeolites, perlites, clay-shales and dolomites) modified with ammophos are higher than fire -extinguishing ability of such powders.

20.5 RESULTS AND DISCUSSION

Foam suspensions are prepared simply by mechanically mixing fire-extinguishing powders from local mineral raw materials (our preparation patent U 2019 2013 Y) modified with ammophos, water, and surfactants. The process does not require chemical treatment of materials. Therefore, the developed technology for the production of foam suspensions is simple and cost-effective.

Considering the stability of suspensions, we have chosen optimal values: powder dispersity less than 250 μm; powder volumetric concentrations –20–25%; suspension viscosity –0.015–0.018 poise g/(cm•s).

We selected a type (non-ionic, biodegradable) and a minimum amount (1%) of surfactants, which decrease surface tension, stabilize the dispersion system, prevent particles of the dispersed phase from compacting and coagulating, and cause powder flotation, enabling the spraying of powder together with water and foam.

Ammophos is a heterogeneous inhibitor that is well soluble in water. In the case of suspension production based on fire-extinguishing powders modified with ammophos, water inhibition increases. Additionally, the diluting effect of burning gases and water heat capacity is also increased. Hence, ammophos significantly raises the fire-extinguishing capacity of foam suspensions. Based on the above, we can suggest that foam suspensions produced from powders (zeolites, perlites, clay shales, and dolomites) modified with ammophos are characterized by higher inhibition capacity and cooling effect than water; they have permeability (high dispersion of sprayed water) and wetting effects similar to water and foam, but unlike them, they provide both homogeneous and heterogeneous inhibition of the burning process. Thus, the foam suspensions produced will have a higher extinguishing effect than water, foams, or powders taken separately.

Fire-extinguishing powders made from local mineral raw materials such as zeolite, perlite, clay-shale, and dolomite are non-toxic and ecologically friendly. Non-ionic surfactants are fully biodegradable; they completely decompose after use and do not produce toxic compounds. Thus, they are safe for nature and soil. Zeolites and ammophos are combined fertilizers that decrease soil acidity and increase the nutrient content of the soil (P, K, and N ions). Based on the above, we can predict that the foam suspensions produced not only effectively extinguish fires but can also regenerate damaged soil.

From all of the above, one can suggest that the fire-extinguishing foam suspensions are ecologically friendly, highly effective, and bioprotective. Therefore, they can be used for extinguishing large-scale fires, such as forest fires, and do not require additional antiseptic measures.

KEYWORDS

- **anionic surfactants**
- **bioprotective**
- **ecologically friendly**
- **fire extinguishing ability**
- **fire-extinguishing foam-suspensions**
- **foam suspensions**
- **performance properties**
- **powder fire extinguishers**
- **viscometric method**

REFERENCES

1. Emme Italia. (n.d.). *CO2, Dry Powder and Foam Fire Extinguishers*. Retrieved from: https://www.emme-italia.com (accessed on 1 October 2024).
2. Nauka v Sibiri. (2000). *Fire-Extinguishing Powders*.
3. United States Patent Office. (1965). US3172852A–Dry chemical fire extinguisher composition.
4. Skybrary. (n.d.). *Fire Extinguishing Agents*. Retrieved from: https://www.skybrary.aero/index.php/Fire_Extinguishing_Agents (accessed on 1 October 2024).
5. Voitovych, T., Gusar, B. M., Kovalyshyn, V., & Koshelenko, V. V. (2018). Research on domestically produced fire-fighting foam agents for subsurface fire extinguishing. *Fire Safety*.
6. Gurchumelia, L., Tsarakhov, M., Tkemaladze, S., & Bezarashvili, G. (2017). Development of novel environmentally safe and highly efficient fire-extinguishing powders based on local mineral raw materials. *Safety and Security Engineering*, WIT Transactions on the Built Environment, 131–140. Rome, Italy.
7. Gurchumelia, L., Tsarakhov, M., Dzanashvili, D., Tkemaladze, S., Bejanov, F., & Chudakova, O. (2020). Evaluation of noble environmentally friendly and highly efficient fire-extinguishing means. *Journal of Chemistry: Education, Research and Practice, 4*(2), 26–31. Opast Publishing Group. https://opastonline.com/wp-content/uploads/2020/06/ISSN-2578-7365 (accessed on 1 October 2024).
8. Byju's. (n.d.). *Action of Surfactants, Structure, Types and Uses*. Retrieved from: https://byjus.com/jee/surfactants/ (accessed on 1 October 2024).
9. Aguirre-Ramírez, M., Silva-Jiménez, H., Banat, I. M., & Díaz De Rienzo, M. A. (n.d.). Surfactants: Physic-chemical interactions with biological macromolecules. *Biotechnology Letters*. Retrieved from: https://link.springer.com/ (accessed on 1 October 2024).
10. Gurchumelia, L., Tsarakhov, M., Bejanov, F., Tkemaladze, L., & Chudakova, O. (2017). New types, halogen less, eco-safe fire-extinguishing powders and foam-suspensions. In *International Conference "Ecology and Safety" Proceedings* (pp. 78–84). Bulgaria, Elenite.
11. Gurchumelia, L., Tsarakhov, M., Tkemaladze, S., Bejanov, F., & Tkemaladze, L. (2020). Production and performance evaluation of noble fire extinguishing foam suspension using locally available and environmentally friendly natural mineral raw materials. *Chemical Engineering Research Bulletin, 21*, 58–64. Bangladesh University of Engineering & Technology. https://doi.org/10.3329/cerb.v21i1.47372.
12. Tsitsishvili, G. V., Andronikashvili, T. G., & Kardava, M. A. (1993). *Natural Zeolites in Agriculture*. Tbilisi: Metsniereba.

CHAPTER 21

Features of Sorption Extraction of Rhenium Ions by Interactivated Hydrogels in the Lewatit CNP LF(H^+)-Poly-2-Methyl-5-Vinylpyridine Interpolymer System

T. K. JUMADILOV,[1] D. E. FISCHER,[2] and A. BAISHIBEKOV[2]

[1]*A.B. Bekturov Institute of Chemical Sciences, Almaty, Kazakhstan*

[2]*The Institute of Metallurgy and Ore Beneficiation, Almaty, Kazakhstan*

ABSTRACT

To assess the possibility of interaction between Lewatit – P2M5VP (poly-2-methyl-5-vinylpyridine) hydrogels and ammonium perrhenate, the effect of hydrogen ion concentration and electrical conductivity is considered. At 0 hours, the aqueous medium has a low pH value, indicating a high content of H^+ ions in the aqueous medium. An increase in the interaction time of the components up to 48 hours leads to a decrease in the pH value, indicating an increase in the content of H^+ ions in the rhenium solution. With the ratios 6:0, 0:6, and 5:1 in the Lewatit CNP LF (H^+)-P2M5VP intergel systems, the range of maximum and minimum pH values is observed. At ratios of 6:0 and 5:1, maximum electrical conductivity is observed in the Lewatit CNP LF (H^+)-P2M5VP system. The results obtained indicate that Lewatit CNP LF(H^+) has the maximum sorption in the interpolymer system, while P2M5VP has the opposite effect, as a result of which the interpolymer system in the presence of P2M5VP poorly sorbs rhenium, while the ion-exchange resin Lewatit CNP LF(H^+) promotes the sorption of rhenium ions. The maximum value of

Advanced Topics in Polymer Chemistry and Materials Science: Tamara Tatrishvili and Marc Jean Médard Abadie (Eds)

the rhenium content after sorption was shown in the ratios of 6:0 and 5:1 at 48 hours of remote interaction.

21.1 INTRODUCTION

Rhenium is obtained from a mineral known as molybdenum. In addition, this metal is found in technological solutions by leaching other minerals. In this regard, numerous ion-exchange sorbents for various purposes are currently being produced in the world. However, the tests carried out showed that a significant number of these sorbents are not widely used due to low efficiency and selectivity and do not have high sorption activity with respect to rare earth metals (REM) and rare metals [1]. Therefore, the task of developing methods for creating more selective and universal sorbents for the extraction of rare earth metals and rare metals is currently in demand. In this regard, the purpose of this work is the sorption extraction of rhenium from aqueous solutions and the study of the possible use of sorbents in an interpolymer system [9].

21.2 EXPERIMENTAL PART

A MARK 603 conductometer (Russia) and a Metrohm 827 pH-Lab pH meter (Switzerland) were used to measure electrical conductivity. The mass of swollen hydrogel samples for subsequent calculation of the degree of swelling (α) was determined by weighing on an electronic analytical balance, SHIMADZU AY220 (Japan).

To measure the optical density for the subsequent calculation of the rhenium concentration, a KFK-3M spectrophotometer (Unico-Sys, St. Petersburg, Russian Federation) was used.

21.2.1 MATERIALS

Industrial Ionite Lewatit CNP LF(H^+) and laboratory-synthesized P2M5VP were used for experiments. Poly-2-methyl-5-vinylpyridine was synthesized by radical polymerization of distilled monomer 2-methyl-5-vinylpyridine in the presence of benzoyl peroxide initiator and epichlorohydrin crosslinking agent in dimethylformamide medium.

21.2.2 ELECTROCHEMICAL RESEARCH

The experiments were carried out at room temperature. Studies of the intergel system were carried out in the following order: (I) each hydrogel in its dry initial state was placed in a special cell made of polypropylene mesh (the polypropylene mesh cell was permeable to low molecular weight ions but impervious to hydrogel dispersion); (II) then each hydrogel in the dry state in the polypropylene mesh cells was placed in a glass with ammonium perrhenate [6]; the electrical conductivity and pH of the solutions were measured in the presence of hydrogels, and aliquots were obtained after a certain period of time [2]. To study the processes of sorption of rhenium ions, an aqueous solution of ammonium perrhenate with a concentration equal to 100 mg/l was used.

21.2.3 METHOD OF DETERMINATION OF RHENIUM IONS

The method of determination of rhenium ions in solution is based on the formation of a colored complex compound of the organic analytical reagent thioglycolic acid with ions of REM and rare metals.

Thioglycolic acid, in the presence of $SnCl_2$, forms a C-Re complex with a 1:2 component ratio, colored pink with a maximum light absorption in the region of 320–350 nm [14]. The degree of extraction (sorption) was calculated by the formula:

$$H = C_{init} — C_{res} \times 100\%, (2).$$

where C_{init} is the initial concentration of the metal in the solution g/l, and C_{res} is the residual concentration of the metal in the solution g/l.

21.3 RESULTS AND DISCUSSION

To assess the determination of the rhenium concentration, a calibration curve was constructed (Figure 21.1).

From Figure 21.2, we can see that the maximum electrical conductivity is observed at ratios of 6:0 and 5:1 for 48 hours, which corresponds to the concentration measurements on the spectrophotometer after 48 hours. It should be assumed that high values of electrical conductivity at the maximum point indicate high concentrations of charge carriers. The results

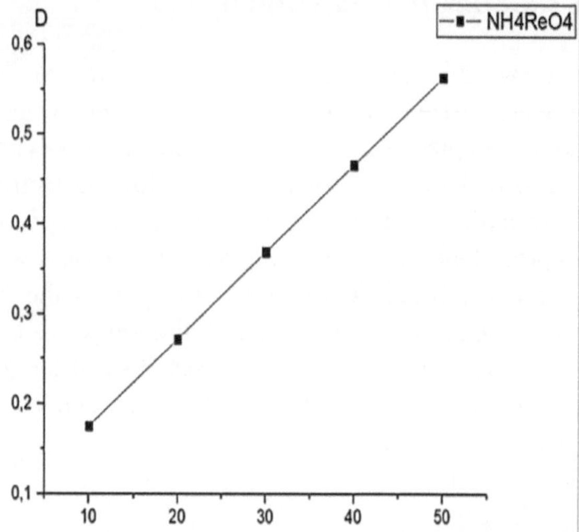

FIGURE 21.1 Calibration curve of ammonium perrhenate in an aqueous medium.

The results of measuring the electrical conductivity of an interpolymer system in an ammonium perrhenate solution are shown in Figure 21.2.

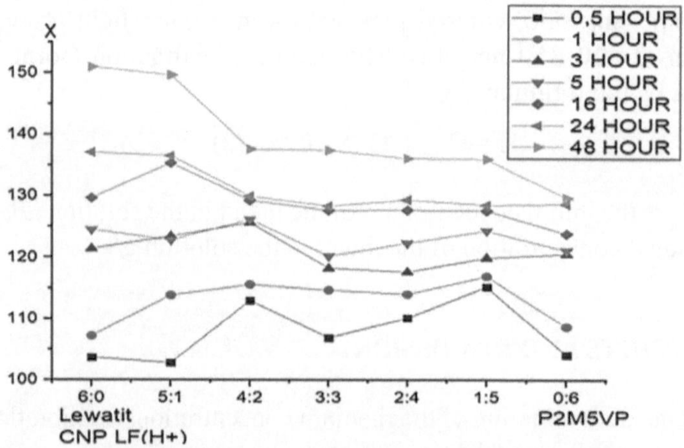

FIGURE 21.2 Dependence of the specific electrical conductivity of the Lewatit CNP LF(H+) and P2M5VP interpolymer system in the presence of an aqueous solution of ammonium perrhenate on time.

of measuring the pH of the interpolymer system in rhenium solution are shown in Figure 21.3.

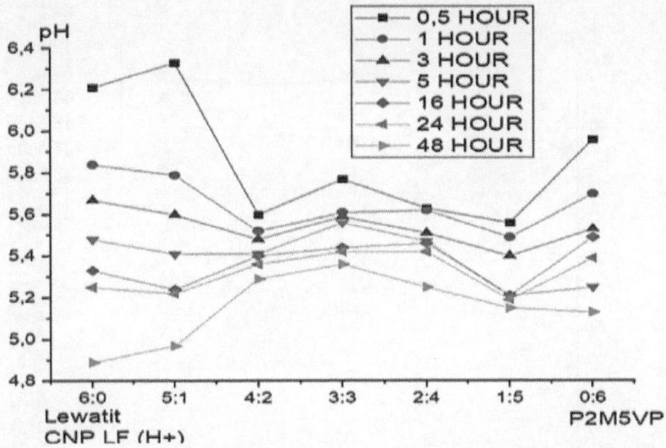

FIGURE 21.3 Dependence of the pH of the Lewatit CNP LF (H^+) and P2M5VP interpolymer system in the presence of an aqueous solution of ammonium perrhenate on time.

pH measurements also show that the minimum pH values are observed at 6:0 and 5:1 for 48 hours, where we observe high concentrations of H^+ ions. This is due to the fact that hydroxyl ions are neutralized by protons, resulting in a high concentration of positive ions [11]. Moreover, analyzing the pH data of solutions, it was found that during sorption, the acidity of solutions increased due to the replacement of H^+ ions in the Lewatit CNP LF(H^+) structure with rhenium ions, displacing the latter into the solution [5].

The presence of an intergel system in an aqueous solution of ammonium perrhenate leads to various processes affecting the electrochemical equilibrium in the solution [3]. According to the results of the residual concentration of salts in the solution, the highest degree of sorption (55.0%) was observed after 48 hours at a ratio of 6:0. This is due to the fact that at this ratio, the Lewatit CNP LF(H^+) hydrogel is in a highly ionized state [7].

According to the data on the specific electrical conductivity of solutions, a slight increase in this parameter was detected over time; the maximum values were reached after 48 hours [12]. Moreover, analyzing the pH data of solutions, it was found that during sorption, the acidity of solutions increased due to the replacement of H^+ ions in the Lewatit CNP LF(H^+) structure with rhenium ions, displacing the latter from polymers [4]. There is also a linear dependence of the concentration change at the initial moment (0.5 h of interaction), and from 1 h, clear peaks are observed at the ratios Lewatit CNP LF(H^+)-P2M5VP 6:0. Figure 21.4 shows the graph of sorption of rhenium ions by the Lewatit CNP LF(H^+)-P2M5VP interpolymer system according to the ratios.

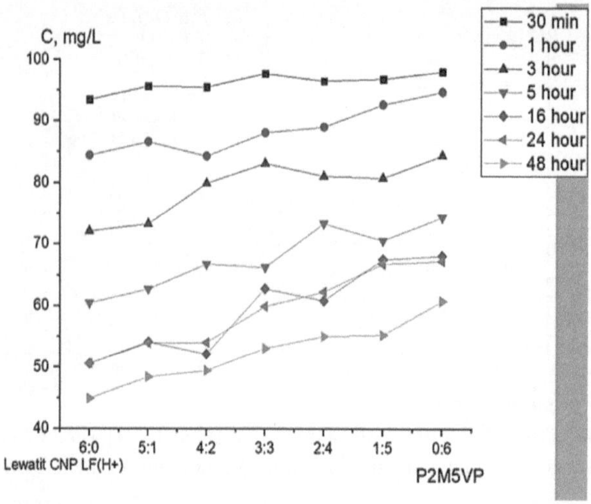

FIGURE 21.4 Graph of sorption of rhenium ions by the Lewatit CNP LF(H^+)-P2M5VP interpolymer system according to the ratios.

The results obtained show that starting from 1 hour of remote interaction, the zones of increased sorption are the ratios Lewatit CNP LF(H^+) – P2M5VP 6:0 and 5:1 [8]. The maximum sorption of rhenium occurs at 48 hours of remote interaction of ion exchangers at a ratio of 6:0, the initial concentration of rhenium in saline solution decreases from 100 mg/l to 44.93 mg/l. A significant part of the metal is sorbed from the saline solution within 48 hours.

The degree of binding of the polymer chain (relative to rhenium ions) of the Lewatit CNP LF(H^+)-P2M5VP interpolymer system is shown in Table 21.1. Figure 21.5 illustrates the graph of sorption of rhenium ions by the Lewatit CNP LF(H^+)-P2M5VP interpolymer system of various time ratios.

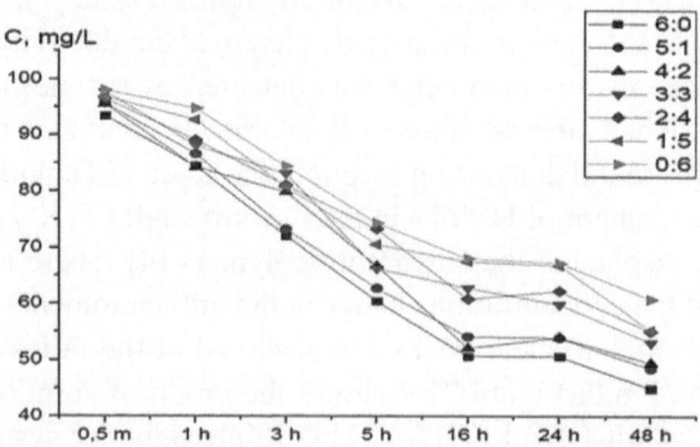

FIGURE 21.5 Graph of sorption of rhenium ions by the Lewatit CNP LF(H+)-P2M5VP interpolymer system of various time ratios.

TABLE 21.1 The Degree of Binding of the Polymer Chain with a Different Ratio of Ion Exchangers (%)

6:00	0.55	1.31	2.35	3.32	4.16	4.16	4.64
5:01	0.38	1.16	2.31	3.23	3.97	3.99	4.46
4:02	0.39	1.38	1.77	2.93	4.21	4.05	4.45
3:03	0.20	1.06	1.51	3.02	3.33	3.59	4.20
2:04	0.32	1.00	1.72	2.42	3.56	3.43	4.09
1:05	0.29	0.67	1.75	2.68	2.96	3.02	4.07
0:06	0.18	0.48	1.43	2.36	2.93	3.01	3.60

The degree of binding increases with time, and the most significant increase in the parameter is observed at the Lewatit CNP LF(H^+)-P2M5VP molar ratios of 6:0 and 5:1. The degree of binding is 0.55% and 0.38% for these ratios after 0.5 hours of remote interaction. For the other ratios, the degree of binding in this case is relatively low at 0.20% and 0.18%, respectively. After 48 hours of interaction, the degree of binding of the polymer chain increases in the ratios of 6:0 and 5:1 to 4.64% and 4.46%, respectively.

The effective dynamic volumetric capacity of the Lewatit CNP LF(H^+)-P2M5VP interpolymer system, relative to rhenium ions, shown in Table 21.2, depends on the molar ratios of ion-exchange resins and the interaction time. The sorption of rhenium ions is accompanied by a strong increase in the volumetric capacity for Lewatit CNP LF(H^+) with a ratio of 6:0 from the beginning of sorption (0.5 h)—it is 0.00029 mmol/g; while the capacity is 0.00012 mmol/g for a pair of Lewatit CNP LF(H^+): P2M5VP in the ratio of 1:5 and 0.00008 mmol/g for P2M5VP. A strong increase in the sorption parameter is observed after 48 hours; for Lewatit CNP LF(H^+) with a ratio of 6:0, it is equal to 0.00246 mmol/g, and for a pair of interpolymers in a ratio of 1:5, it is 0.00177 mmol/g; for P2M5VP—0.00151 mmol/g after 48 hours of sorption.

TABLE 21.2 Effective Dynamic Sorption Capacity

6:0	0.00029	0.00070	0.00125	0.00177	0.00221	0.00221	0.00246
5:1	0.00019	0.00058	0.00117	0.00163	0.00200	0.00201	0.00225
4:2	0.00019	0.00067	0.00086	0.00142	0.00204	0.00196	0.00215
3:3	0.00009	0.00049	0.00070	0.00141	0.00155	0.00167	0.00195
2:4	0.00014	0.00044	0.00076	0.00108	0.00158	0.00152	0.00181
1:5	0.00012	0.00029	0.00076	0.00116	0.00128	0.00131	0.00177
0:6	0.00008	0.00020	0.00060	0.00099	0.00123	0.00127	0.00151

Also, during the experiments, desorption was carried out in an aqueous solution of 2% nitric acid for 72 hours. Figure 21.6 illustrates the dependence of the relative optical density during desorption in a 2% solution of HNO_3-NaOH of ammonium perrhenate.

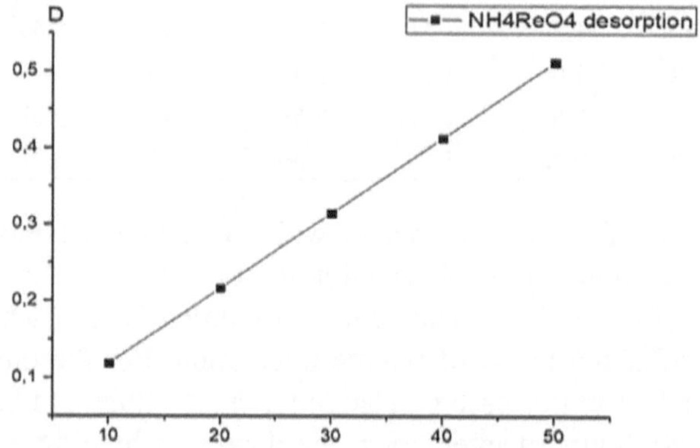

FIGURE 21.6 Dependence of the relative optical density during desorption in a 2% solution of HNO_3-NaOH of ammonium perrhenate.

A desorption schedule was drawn up after 72 hours for Lewatit CNP LF(H^+) and P2M5VP ion exchange resins. Desorption is maximal for a ratio of 6:0, which corresponds to Lewatit CNP LF(H^{+}). Figure 21.7 illustrates the graph of rhenium desorption from Lewatit CNP LF(H^+) cationite according to the ratios.

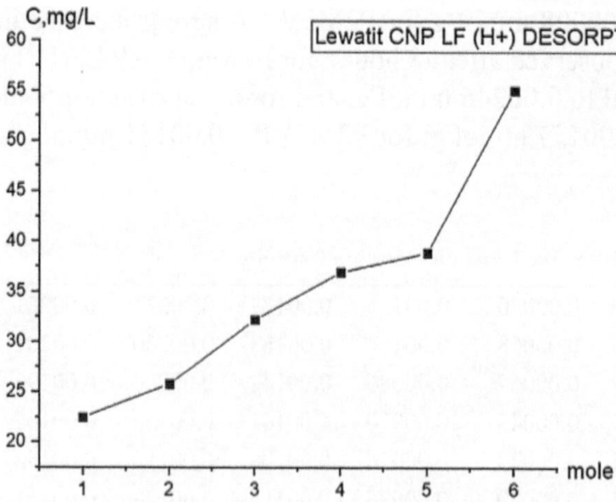

FIGURE 21.7 Graph of rhenium desorption from Lewatit CNP LF(H^+) cationite according to the ratios.

From the graph of cationite desorption, it can be seen that the minimum desorption value corresponds to 1 mole of Lewatit CNP LF(H^+) with indicators of 22.49 mg/liter. Further, with an increase in desorption indicators to 6 moles of Lewatit CNP LF(H^+), the indicator rises to 55.05 mg/liter. The data indicate that those cationite that interacted with P2M5VP anionite show a lower desorption concentration than a separate Lewatit CNP LF(H^+) cationite.

Figure 21.8 illustrates the graph of rhenium desorption from P2M5VP anionite depending on the molar ratio.

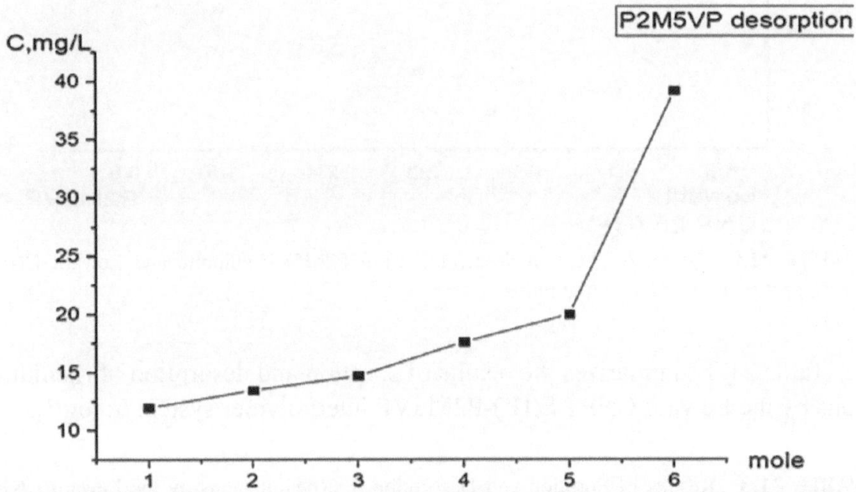

FIGURE 21.8 Graph of rhenium desorption from P2M5VP anionite depending on the molar ratio.

From the graph of anionite desorption, it can be seen that the minimum desorption value corresponds to 1 mole of P2M5VP with indicators of 11.99 mg/L, further with an increase in desorption indicators to 6 mole of P2M5VP with an indicator of 39.2 mg/L.

It follows from the graph that in the Lewatit CNP LF(H^+) interpolymer system, has a high sorption, which indicates desorption data.

Figure 21.9 illustrates the graph of rhenium desorption from P2M5VP anionite and Lewatit CNP LF(H^+) cationite by initial ionite ratios.

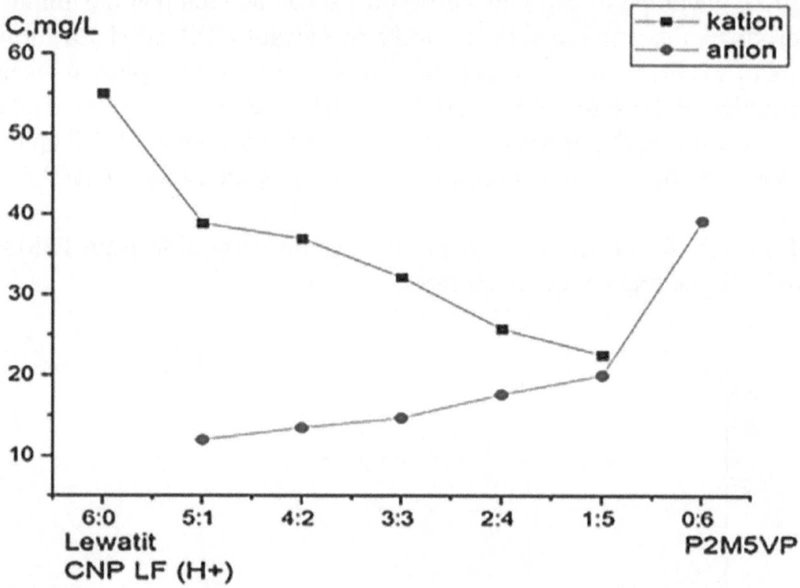

FIGURE 21.9 Graph of rhenium desorption from P2M5VP anionite and Lewatit CNP LF(H^+) cationite by initial ionite ratios.

Table 21.3 summarizes the results of sorption and desorption of rhenium Ions by the Lewatit CNP LF(H^+)-P2M5VP interpolymer system in mg/L.

TABLE 21.3 Results of Sorption and Desorption of Rhenium Ions by the Lewatit CNP LF(H^+)-P2M5VP Interpolymer System in mg/L

Ratios	Residual Concentration in Solution	Concentration of Rhenium Ions After Desorption from Cationite	Concentration of Rhenium Ions After Desorption From Anionite	∑ Desorption From Lewatit CNP LF(H+) and P2M5VP	∑(C Sorpti C Desorptic
6:0	44.93	55.05	–	55.05	99.98
5:1	48.45	38.86	11.99	50.85	99.3
4:2	49.47	36.95	13.48	50.43	99.9
3:3	53.09	32.21	14.7	46.91	100
2:4	55.05	25.78	17.62	43.4	98.45
1:5	55.25	22.49	19.98	42.47	97.72
0:6	60.80		39.2	39.2	100

Figure 21.10 illustrates the graph of rhenium desorption from P2M5VP anionite and Lewatit CNP LF(H^+) cationite by moles.

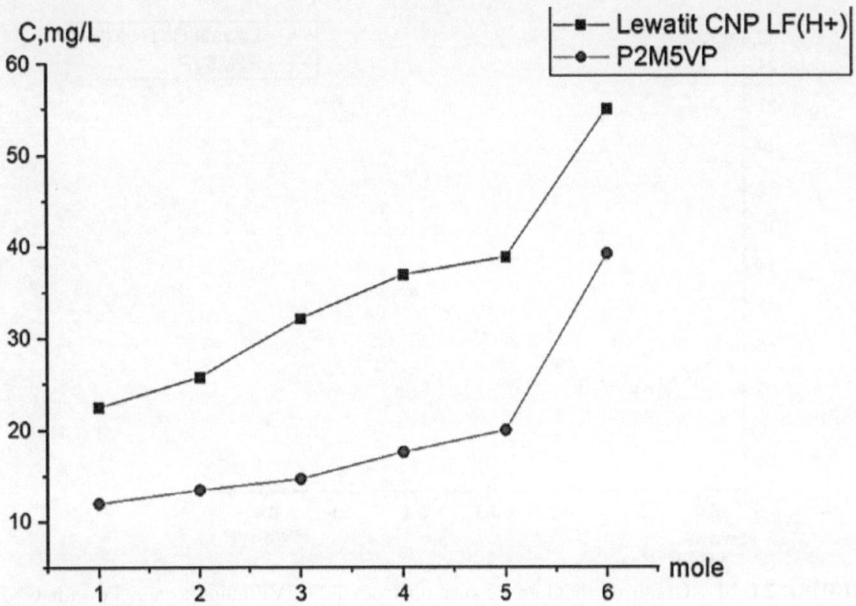

FIGURE 21.10 Graph of rhenium desorption from P2M5VP anionite and Lewatit CNP LF(H^+) cationite by moles.

Table 21.4 summarizes the results of desorption per 1 mole of ionite ions.

TABLE 21.4 Results of Desorption Per 1 Mole of Ionite Ions

Ratios	Desorption from Lewatit CNP LF(H+)	Desorption per 1 mole of Lewatit CNP LF(H+)	Desorption from P2M5VP	Desorption per 1 Mole of P2M5VP
6:0	55.05	9.17		
5:1	38.86	7.77	11.99	11.99
4:2	36.95	9.23	13.48	6.74
3:3	32.21	10.73	14.7	4.9
2:4	25.78	12.89	17.62	4.405
1:5	22.49	22.49	19.98	3.99
0:6			39.2	6.53

Figure 21.11 illustrates the graph of rhenium desorption from P2M5VP anionite and Lewatit CNP LF(H^+) cationite by ratios per 1 mol.

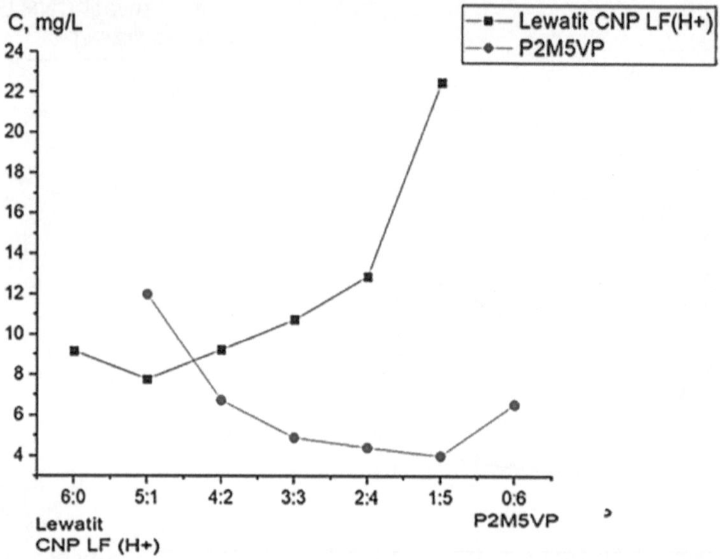

FIGURE 21.11 Graph of rhenium desorption from P2M5VP anionite and Lewatit CNP LF(H^+) cationite by ratios per 1 mol.

21.4 CONCLUSION

The sorption of rhenium ions by an intergel system consisting of a weakly acidic hydrogel, Lewatit CNPLF(H^+), containing functional acid carboxyl groups –COOH, and a weakly basic hydrogel, poly-2-methyl-5-vinylpyridine (gP2M5VP), was studied.

1. In the process of remote interaction of hydrogels in the Lewatit CNPLF(H^+) – P2M5VP intergel system, the specific electrical conductivity and pH of aqueous solutions change [13]. Comparing the dependence of pH and χ on the ratio of Lewatit CNPLF(H^+) – P2M5VP shows that the dependence of the specific electrical conductivity is significantly different; the maximum values of electrical conductivity at ratios of 6:0 and 5:1 coincide with the pH dependent on the ratio of components.
2. The sorption process was studied with the Lewatit CNPLF(H^+) – P2M5VP intergel system in an aqueous medium, during which the specific electrical conductivity and pH of aqueous solutions were measured, and pH peaks were observed at ratios of 6:0 and 5:1 in the Lewatit CNPLF(H^+) – P2M5VP intergel systems. From the data obtained, most of the rhenium is extracted within 72 hours of

desorption; at a ratio of 6:0, Lewatit CNPLF(H^+) – P2M5VP, 55 mg/l is extracted. The results obtained indicate the possibility of creating polymer intergel systems selective to rhenium ions for a highly efficient adsorption technology for the extraction of rhenium ions from industrial solutions [10].

3. The rhenium content in the aqueous medium after sorption showed the maximum value in the ratios of 6:0 and 5:1 at 48 hours of remote interaction.

ACKNOWLEDGMENTS

The work was financially supported by the Committee of Science of the Ministry of Education and Science of the Republic of Kazakhstan. IRN AP14870002 "Development of Modern Methods of Separation and Extraction of Rare Earth Metals from Concentrates and Technological Solutions of Sulfuric Acid Leaching."

KEYWORDS

- **electrical conductivity**
- **hydrogels**
- **interpolymer systems**
- **Lewatit CNP LF(H+)**
- **pH**
- **poly-2-methyl-5-vinylpyridine**
- **rhenium**

REFERENCES

1. Spedding, F. K. H., & Daan, A. K. H. (2005). *Redkozemel'nyyemetally–Metallurgiya.* 610 p.
2. Jumadilov, T. K. (2014). Electrochemical and conformational behavior of intergel systems based on the rare crosslinked polyacid and polyvinylpyridines. In *Proceedings of the International Conference of Lithuanian Chemical Society "Chemistry and Chemical Technology"* (pp. 226–229). Kaunas, Lithuania.
3. Kopishev, E. E., & Reva, Yu. I. (2011). Teoriya ehffekta Dzhumadilova. *Vestnik Almatinskogo Universiteta Energetikii Svyazi, 2*(13), 19–25.

4. Jumadilov, T., Shaltykova, D., & Suleimenov, I. (2013). Anomalous ion exchange phenomenon. In *Austrian-Slovenian Polymer Meeting*. Slovenia. Abstract S5, 51.
5. Dzhumadilov, T. K., Erzhan, B., & Nurbekova, M. A. (2011). Nekotorye osobennosti vzaimodejstviya polimernyh gidrogelej s nizkomolekulyarnymi solyami. *Him. zhurn. Kazahstana, 3*, 92–95.
6. Ahylbekova, M. A., Eskalieva, G. K., Akimov, A. A., Zhunusbekova, N. M., & Dzhumadilov, T. K. (2016). Vzaimovliyanie fiziko-khimicheskikh parametrov azotsoderzhashchikh gidrogelej v vodnoi srede. *Himicheskij zhurnal Kazahstana, 1*, 135–146.
7. Bekturov, E. A., & Jumadilov, T. K. (2009). New approaches hydrogels remote interaction study. *Proceedings of NAS RK, 1*, 86–87.
8. Bekturov, E. A., Jumadilov, T. K., & Korganbayeva, Zh. K. (2010). Remote interaction in polymer systems. *KazNU Herald, 59*, 108–110.
9. Donia, A. M., Atia, A. A., Amer, T. E., El-Hazek, M. N., & Ismael, M. H. (2011). Selective separation of Uranium(VI), Thorium(IV), and Lanthanum(III) from their aqueous solutions using a chelating resin containing amine functionality. *Journal of Dispersion Science and Technology, 32*, 1673–1681.
10. Totkhuskyzy, B., Yskak, L. K., Saparbekova, I. S., Myrzakhmetova, N. O., Jumadilov, T. K., & Gražulevicius, J. V. (2020). Features of the extraction of yttrium and lanthanum with an intergel system based on hydrogels of polyacrylic acid and poly-4-vinylpyridine. *Chemistry Series, 1*(97), 60–67. https://doi.org/10.31489/2020Ch1/60-67.
11. Saparbekova, I. S., Suberlyak, O. V., Yskak, L. K., Malimbayeva, Z. O., Myrzakhmetova, N. O., & Dzhumadilov, T. K. (2019). Some features of the remote interaction of KU 2-8 cation exchanger with AB-17 anion exchanger. In *Proceedings of the International Scientific and Technical Conference "Modern Technologies for the Production and Processing of Polymer Materials"* (p. 87).
12. Yskak, L. K., Dzhumadilov, T. K., Myrzakhmetova, N. O., & Suberlyak, O. V. (2019). Features of distance interaction and mutual activation of hydrogel polymethacrylic acid and anionite AV-17. In *Proceedings of the International Symposium on Specialty Polymers* (p. 106). E. A. Buketov Karaganda State University, Institute of Polymer Materials and Technology.
13. Totkhuskyzy, B., Dzhumadilov, T. K., & Gražulevicius, J. V. (2019). Some peculiarities of the interaction of scandium and yttri ions with activated hydrogels. In *Proceedings of the International Symposium on Specialty Polymers* (p. 104). E. A. Buketov Karaganda State University, Institute of Polymer Materials and Technology.
14. Busev, A. I., Tiptsova, V. G., & Ivanov, V. M. (1978). *Guide to Analytical Chemistry of Rare Elements*. 270 p.

CHAPTER 22

Natural Organic Substances' Transformation in Ukraine's Surface Water Under Solar Radiation

PETRO LINNIK,[1] YAROSLAV IVANECHKO,[1] VLADYSLAV ZHEZHERYA,[1] and ROSTYSLAV LINNIK[2]

[1]*Department of Freshwater Hydrochemistry, Institute of Hydrobiology of the National Academy of Sciences, Kyiv, Ukraine*

[2]*Faculty of Chemistry, Taras Shevchenko Kyiv National University, Kyiv, Ukraine*

ABSTRACT

The article summarizes the results of the authors' long-term research on the most important groups of natural organic substances in surface water bodies of Ukraine. The main attention is focused on the research of humic substances (HS), carbohydrates, and protein-like compounds (PLC). The concentrations of the mentioned groups of organic substances in the water bodies under study are within a fairly wide range: HS – 1.2–120.5 mg/L, carbohydrates – 0.4–4.9 mg/L, and PLC – 0.03–1.41 mg/L. A case study of the Dnipro cascade reservoirs showed that the share of HS in the total balance of dissolved organic matter was 52.8–76.5%, while carbohydrates and protein compounds were 7.2–9.8% and 1.1–3.2%, respectively. The studied groups of natural organic substances are mainly high molecular weight compounds. Thus, among the different fractions of HS, the fraction with a molecular weight of 20–5 kDa dominates. The relative content of high-molecular fractions (>5.0 kDa) reached 43.6–59.6% of HS_{total}. Carbohydrates and protein-like compounds are characterized by a wide range of molecular weight (1–70 kDa). Polysaccharides dominate among

Advanced Topics in Polymer Chemistry and Materials Science: Tamara Tatrishvili and Marc Jean Médard Abadie (Eds)

carbohydrates, the share of which reaches 67.7% to 83.7%. The weight-average molecular weight (M_w) for HS and PLC, which varied over a wide range – from 2.3 to 11.4 kDa and 11.6 to 43.0 kDa, respectively, has been calculated. In this case, average values of M_w accounted for 4.6–7.4 kDa and 22–30 kDa, respectively. It is characteristic of both humic acids and fulvic acids that M_w varies in a wide range of values (6.9–12.2 and 1.8–7.3 kDa, respectively). The studied groups of organic substances are distinguished by different resistance to the effects of water environment factors. HS undergoes transformation mainly under the effect of solar radiation, while carbohydrates and protein compounds are organic substances that are easily oxidized. Their decomposition occurs mainly due to temperature increases and the activation of biological processes, in particular bacterial activity. Data on the ratio of substances with various molecular weights in the studied groups of organic compounds in different seasons are presented.

22.1 INTRODUCTION

Dissolved organic matter (DOM) in surface natural waters represents a wide spectrum of organic compounds differing in their structure and properties [1–3, 4]. It is highly diverse in its component composition. DOM includes substances from various classes and groups – humic substances (HS), carbohydrates, lipids, amino acids, nucleic acids, and the products of their decomposition, lower fatty acids, urea, light organic compounds, phenols, and many other components [1, 3]. These substances differ in their structure, properties, molecular size, origin, and content, which to a large extent depends on a number of factors, in particular on the presence of HS sources within the catchment area, the trophic state of water bodies, the primary production–decomposition ratio, and also on surface runoff.

Natural organic matter in surface water bodies is the main source of organic carbon and some nutrients, including nitrogen, phosphorus, silicon, iron, etc. It significantly affects the chemical composition of surface waters and determines the quality of the aquatic environment as a habitat for aquatic organisms [3, 4]. The most common group of DOM is HS, which mainly belong to terrigenous organic compounds by their origin. A certain portion of HS can be formed in the reservoirs themselves (the so-called "water humus"), but their share in the overall balance is much lower than that which enters with surface runoff. An important role is also played by exometabolites – organic substances that are released by phytoplankton and higher aquatic vegetation during growth and decay. First of all, these are carbohydrates and protein-like compounds (PLC). These organic substances are especially common in surface water bodies with high bio productivity.

Thus, the DOM of natural surface water bodies, in terms of its origin, can be subdivided into allochthonous and autochthonous substances. The first group of substances comes with the surface runoff, whereas the second group is formed as a result of the vital activity of aquatic organisms, including algae, higher aquatic plants, fish, etc. [5]. Moreover, under conditions of warming, the share of autochthonous organic substances in the general balance of DOM can significantly increase [5]. It is important to note that the prolonged period of warmth and abundant solar radiation affects not only the DOM cycle but also its component composition.

Among natural organic substances of surface waters, HS are the most widely distributed [1–3, 6]. Their contribution to the total balance varies from 50% in rivers to 90% in swamps [1, 2, 6]. For the most part, these substances belong to compounds of terrigenous origin because they are removed from the soil complex and enter rivers and water bodies with surface runoff. The concentration of carbohydrates and PLC depends to a large extent on the intensity of biota development and on their influx from plant and animal remnants, and is of considerable interest as well [7, 8].

Data on the various groups of organic substances' contribution to the total balance of river water's DOM are given in monograph [2]. These data suggest that in rivers, the contribution of HS to the total content of carbon in organic matter is not less than 50%. The above groups of organic compounds play an important role in the detoxification of the aquatic environment by binding metal ions into complexes and organic xenobiotics through the formation of adducts.

Over a long period of time, studies have been conducted not only on the component composition of DOM but also on the molecular weight distribution of individual groups of compounds of DOM, as indicated above. On the one hand, this allows for evaluating the ratio of high-molecular and low-molecular compounds, and therefore their ability to complex. On the other hand, based on these results, we can discuss the potential bioavailability of DOM with different molecular weights for assimilation by hydrobionts. In addition, information on seasonal changes in the molecular weight distribution of organic compounds in each of the DOM groups makes it possible to find out how actively they occur at different times of the year.

The transformation of natural organic substances under the effect of solar radiation and the activation of biological processes, in particular, the strengthening of microbiological activity, deserves special attention [5, 9].

In this article, the main focus is on summarizing the results of our many years of research, which are related to the study of the transformation of natural organic substances under the effect of aquatic environment factors. These studies are based on the results of a gel-chromatographic study of the molecular-weight distribution of HS, carbohydrates, and protein compounds in Ukraine's various water bodies (rivers, reservoirs, lakes, including small lakes in urban areas).

22.2 EXPERIMENTAL METHODS AND MATERIALS

The studies of the component composition of DOM, carried out during the periods of 1992–1998 and 2011–2019, covered the reservoirs of the Dnipro Cascade, the rivers of the Prypyat River basin, the Danube River, the Desna River, the Pivdennyi Bug River, the Seret River, the Lyutsymir and Velyke Chorne lakes (the Shatsk lake group), as well as the lakes of Kyiv City (Tel'bin and Verbnoye lakes, and also lakes included in the Opechen system).

Water samples were taken mainly from the surface layer (~0.5 m) using the Ruttner water sampler or the modified sampler-glass [10]. Suspended matter was isolated by the method of membranous filtration. In this case, a water sample of 1.0–1.5 L volume was filtered through the "Synpor" nitrocellulose membranous filters (Czech Republic) or through the "Fioroni" filters (China) with 0.40 and 0.45 μm openings, respectively.

The component composition of DOM was studied using the methods of ion exchange and gel chromatography. The method of ion exchange chromatography was applied to separate and extract individual groups of dissolved organic substances. Accordingly, 0.5–1 L of the natural water filtrate was successively passed through glass columns filled with diethylaminoethyl (DEAE) cellulose and carboxymethyl (CM) cellulose ion exchangers manufactured by the SERVA firm. Elution was carried out using solutions of 0.3 mol/L NaOH, 0.02 mol/L H_2SO_4 (three-step elution from the column with DEAE cellulose), and 0.1 mol/L HCl (elution from the column with CM cellulose). The columns have the following parameters: length – 27.5 cm, diameter – 2.5 cm, sorbent layer – about 4.5 cm, free volume – 12.5 ml. After the column separation, three fractions of DOM were obtained as follows: acidic fraction with HS predominance, basic fraction with PLC predominance, and neutral fraction with carbohydrates predominance. The performed manipulations were described previously [11]. During the process of adsorption, the concentration of HS significantly increased (mainly by a factor of 25–50 depending on the type of water body, and also depending on the concentration of HS in water). In some cases, the obtained concentrate of HS was separated into the fractions of humic acids (HA) and fulvic acids

(FA). This procedure was carried out in acidic media at pH 1.5–2.0 and during the process of heating to ≈ 50°C for several hours. The formation of the precipitate of HS lasted for one day. After that, it was separated by centrifugation. The fraction of FA was in a slightly acidic solution, the pH value of which was closely similar to that registered in the initial natural water. The precipitate of HA was dissolved in an alkaline solution. In its total solution, pH was brought to its initial value.

To separate the above-mentioned groups of organic substances into fractions with various molecular weights, glass columns filled with TOYOPEARL HW-50F and HW-55F gels (Japan) and pre-calibrated by substances with known molecular weights (polyethylene glycols – 0.6, 1.0, 2.0, 15.0, and 20.0 kDa; insulin – 5.8 kDa; albumin – 68.0 kDa; dextran – 70 kDa; and glucose – 0.18 kDa) were used. The free volume of the columns was determined using high-molecular blue dextran (2000 kDa, LOBA CHEMIE, Austria). The column parameters were as follows [12, 13]: HS (gel HW-50F) – length 82.0 cm, diameter 2.8 cm, gel layer height 62.0 cm, and the free volume (V_0) 160 mL; carbohydrates and PLC (gel HW-55F) – length 81.0 cm, diameter 2.8 cm, gel layer height – 60.5 cm, V_0 – 138 mL. Phosphate buffer solution (0.025 mol/L) with pH 7.0 was used as the eluent. After gel-chromatographic separation, the fractions of 15 mL volume were collected into glass tubes using the DOMBIFRAK collector (Ukraine).

The HS concentration was measured by various methods, in particular by the methods of spectrophotometry and fluorescence, and also using the method of photometry [14]. The indirect method of determining HS in the filtrate of natural water and in the fractions after their gel-chromatographic separation was also used. It is based on measuring the water color index using the dichromate-cobalt scale [15]. The concentration of protein-like compounds (PLC) and carbohydrates was assessed by the photometric method using, respectively, Folin's and anthrone reagents [16, 17].

22.3 RESULTS AND DISCUSSION

22.3.1 THE CONCENTRATIONS AND DISTRIBUTION PATTERNS OF VARIOUS GROUPS OF NATURAL ORGANIC SUBSTANCES IN THE UKRAINE'S SURFACE WATER BODIES

The HS, carbohydrates, and PLC concentrations in the water bodies of Ukraine, located in different physical and geographical zones, vary widely: 1.2–120.5, 0.4–4.9, and 0.03–1.41 mg/L, respectively (Figure 22.1). HS predominates in the overall balance of DOM. Thus, their share in the Dnipro cascade's reservoirs is 52.8–76.5%, while the share of carbohydrates does not exceed 7.2–9.8%, and PLC is 1.1–3.2%. Somewhat higher values of protein compounds (9.49%) and organic substances – bacterial metabolites (17.68%) content are observed for the upper section of the Beiyun River (PRC) [3].

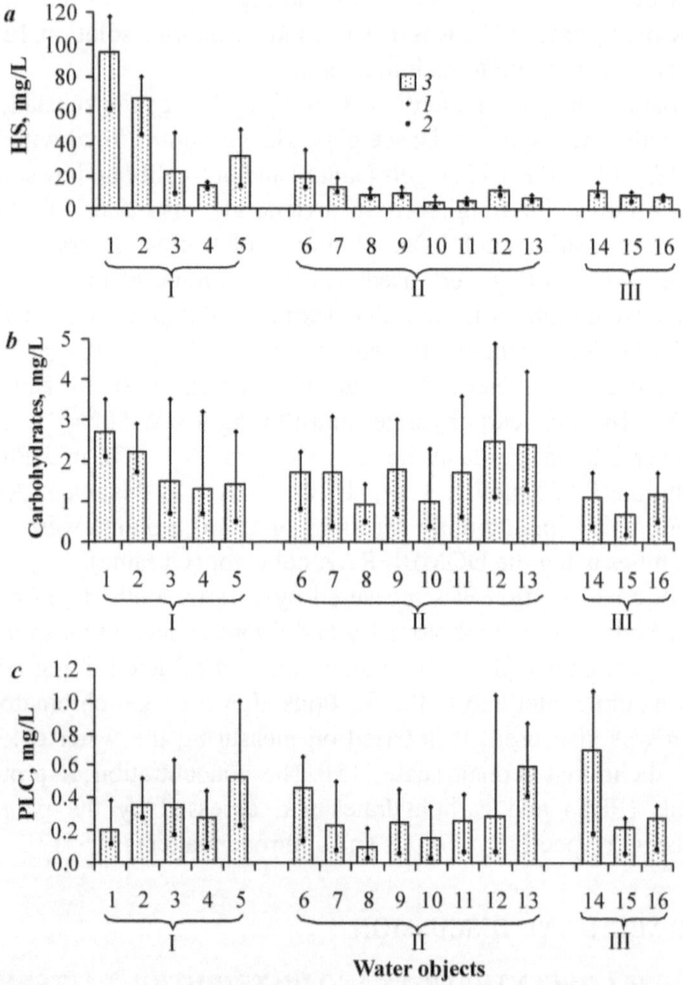

FIGURE 22.1 Limit (*1, 2*) and averaged (*3*) concentrations of HS (a), carbohydrates (b) and PLC (c) in the water bodies from different physical and geographical zones of Ukraine: I – mixed forests zone (1, 2 – the Stvyga and Prypyat' rivers, 3, 4 – the Lyutsymir and Velyke Chorne lakes, 5 – the Kyiv Reservoir); II – forest-steppe zone (6 – the Kaniv Reservoir, 7–10 –the Desna, Pivdennyi Bug, Ros,' and Seret rivers, 11 – the Ternopil Reservoir, the Seret basin, 12 – the upper Kytaivsky Pond, Kyiv City, 13 – Tel'bin Lake, Kyiv City); III – steppe zone (14 – the Kakhovka Reservoir, 15 – the Kiliya Danube delta, 16 – the Sasyk Reservoir, the Danube basin).

The highest concentration of HS is registered in the river waters of the Prypyat River basin because they are supplied mainly by the swamp waters of Polissya, which are enriched by these natural organic acids. Rather high concentrations of HS are registered in the upper reservoirs of the Dnipro River, including the Kyiv and Kaniv reservoirs, as noted previously [18]. As we move down the cascade of the Dnipro reservoirs, concentrations of HS decrease significantly, as can be observed in the Kakhovka Reservoir, which closes the cascade. The average content of this group of organic substances in this reservoir is almost three times lower than in the Kyiv Reservoir [6].

In small and medium rivers located beyond the zone of Polissya, which are unaffected by swamp waters, the content of HS is essentially lower. This is supported by the results of long-term investigations carried out by Nataliia Osadcha [19]. From north to south, the concentration of HS decreases significantly, which is distinctly registered for the reservoirs of the Dnipro River and also for the river waters of some basins. The relative content of HS in the total balance of DOM in the studied water bodies ranges on average from 56.0% to 82.4% [20]. As we might expect, FA contributed significantly to the total content of HS [21]. In Ukraine's surface water, their relative content attained 80.8%–94.8%. Low concentrations of HA were explained by their lower solubility compared to FA. In addition, with their molecular weight being higher, they are easily adsorbed by the suspended matter particles and sediments [22]. As an example, it is appropriate to provide data on the ratio of HA and FA in different types of natural water (Figure 22.2) [1].

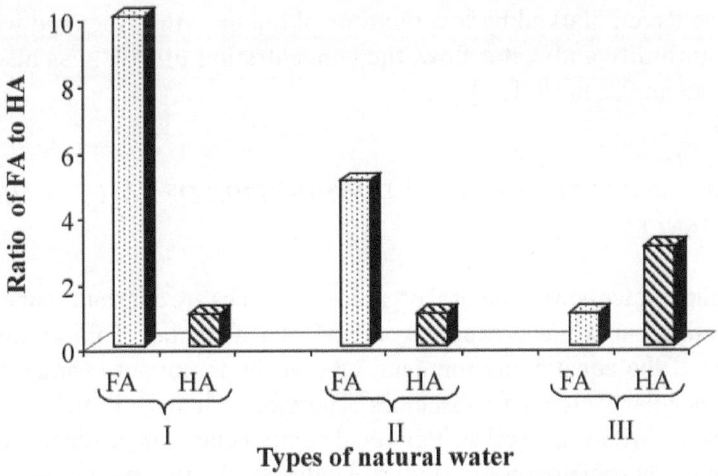

FIGURE 22.2 The ratio of fulvic acids (FA) to humic acids (HA) in various types of natural water: I – low water colority, II – high water colority, III – interstitial solution.

It can be verified that in water bodies with low water colority, the ratio of FA to HA is as much as 10:1, meaning FA dominates. Many researchers estimate that in surface waters, FA makes up the majority of HS, while HA makes up only 10% of HS [1, 20–22]. The above-mentioned ratio is significantly reduced in water bodies with high water colority, and FA:HA changes to 5:1. FA also prevails in these waters, but the share of HA increases significantly. The interstitial solutions of sediments are dominated by HA, and the ratio of FA to HA is 1:3.

The lowest carbohydrate concentrations were observed in water bodies of the Steppe zone (Figure 22.1). In the Prypyat River basin, the carbohydrate content is rather high despite elevated HS concentration. It is believed that in waters with high HS content, phytoplankton growth is inhibited. This, in turn, affects the carbohydrate concentration, which in such water bodies is mainly low. On the other hand, HS and carbohydrates are believed to be the main components of turf formed in swamps [23]. During the flood period, they enter rivers with surface runoff [24].

The concentration of PLC in surface water depends on the trophic status. This fact is supported by the average concentrations of protein compounds in water bodies of Ukraine located in different physical and geographic zones (Figure 22.1). The effect of physical and geographic zones on the concentration of PLC is unlikely to be distinct. For example, in the water of the Prypyat River basin with a high color index, the protein compounds content is low, probably due to the rather high HS concentration inhibiting phytoplankton growth [24]. In the Kiliya delta of the Danube River, marked by low phytoplankton growth associated with high water turbidity and water flow, the concentration of PLC was also rather low – about 0.2 mg/L [25].

22.3.2 MOLECULAR WEIGHT DISTRIBUTION OF NATURAL ORGANIC SUBSTANCES

In surface water bodies, natural organic substances are constantly undergoing transformation, which is caused by the effect of a number of abiotic and biotic factors of the aquatic environment. One of the important characteristics is the molecular weight of substances. Therefore, changes in their molecular weight as a whole, as well as individual components, can be detected, first of all, by gel-chromatographic studies. Undoubtedly, the resistance of certain groups of natural organic substances to such transformation is not the same and depends on their properties. A vivid example of this can be HS, carbohydrates, compounds of protein nature, etc., the resistance of which to the action of environmental factors is significantly different.

22.3.2.1 HUMIC SUBSTANCES

HS are characterized by macromolecular or supramolecular structures due to the association of relatively small heterogeneous components, and also due to the formation of large and very large aggregations in aquatic solutions [21, 26–28]. The molecular weight of HS varies over a wide range [26–31]. Data on the limiting values of the molecular weight of HS, as well as on their dominant fractions, are contradictory [1, 32]. These values depend on many factors, including the origin of HS, their concentration in natural water, and the methods of isolation and identification of individual fractions of HS. For example, the use of the method of ultra-filtration has shown that in the river water relationship between the fractions of HS was as follows: >100 kDa – 9.0%, 100–10 kDa – 15.0, 10–1 kDa – 63.0, and <1 kDa – 13.0% [1].

Studies of the molecular-weight distribution of humic substances seem important, first of all, from the ecological point of view, as the assimilation of these compounds by phytoplankton and aquatic vegetation greatly depends on their molecular weight. There is a general opinion that high-molecular fractions of HS cannot penetrate biological membranes due to their size [27, 33]. A number of publications [34] state that such penetration is possible only for molecules with a molecular weight of up to 5.0 kDa. As an example, data on the molecular weight distribution of HS in surface water bodies of Ukraine are shown in figure 22.3. It can be verified that the fraction with a molecular weight of 20.0–5.0 kDa dominates. The relative content of this fraction amounted to 41.0–58.4% of the HS_{total} concentration. The total content of high-molecular fractions (>5.0 kDa) reached 43.6–59.6% of HS_{total}.

For greater clarity Figure 22.4 provides the data about the weight-average molecular weight (M_w) for HS in each of the studied objects. The values of M_w varied over a wide range – from 2.3 to 11.4 kDa. In this case, their average values accounted for 4.6–7.4 kDa. It is characteristic both for HA and FA that M_w varies in a wide range of values (6.9–12.2 and 1.8–7.3 kDa, respectively). The fact that draws attention is that M_w of FA in the studied water objects turned out to be rather high.

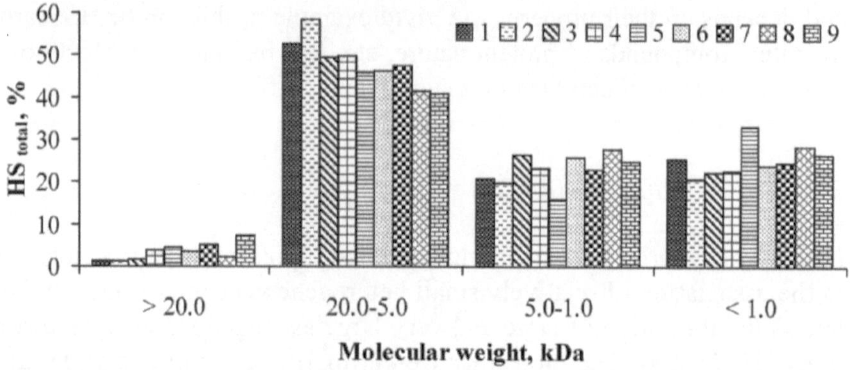

FIGURE 22.3 Ratio of humic substances fractions with different molecular weight (averaged data) in studied water bodies according to results of spectrophotometric measurements of their concentrations. Water bodies: Lutsymir Lake (1), Velyke Chorne Lake (2), Desna River (3), Ros' River (4), Kaniv Reservoir, Desenka branch of the Kaniv reservoir (5), upper Kytaiv pond (6), Pivdennyi Bug River (7), Seret River (8), and Ternopil' Reservoir (9).

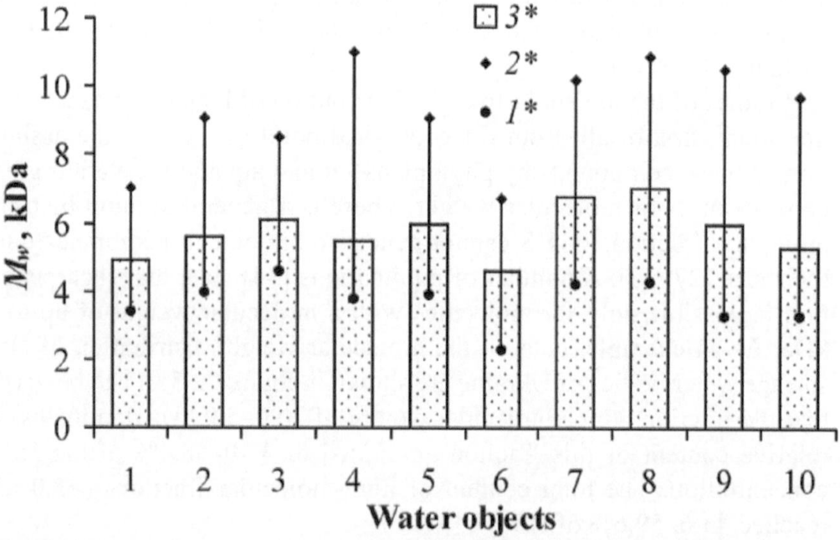

FIGURE 22.4 Limit (*1**, *2**) and averaged (*3**) values of the weight-average molecular weight (M_w, kDa) of HS in the Ukraine's surface water bodies: 1, 2 – the Lyutsymir and Velyke Chorne lakes, 3 – Desna River, 4 – Ros' River, 5 – Pivdennyi Bug River, 6 – Seret River, 7 – Girskyi Tikych River, 8 – the Kaniv reservoir, 9 – the Ternopil reservoir, 10 – the upper Kytaivskyi Pond.

Rather high values of M_w for HS in river and swamp waters are reported in the review article by I.V. Perminova and co-authors (5.2–9.5 kDa and

7.3–16.1 kDa, respectively) [35]. At the same time, the M_W of FA in swamp waters is within 10.8–13.1 kDa. Therefore, it can be assumed that these were FA associates, which are usually characterized by higher values of M_w [36–38]. It is often believed that these compounds have comparatively low molecular weight values. However, this does not apply to FA associates. Apparently, the value of M_w for HS in general and for FA in particular is greatly influenced by their source of origin, concentration in the water of one or another water body and water pH [34, 39, 40].

The molecular weight of HS fractions varies with their concentration in natural water. With an increase in the concentration of HS in water, the proportion of the high-molecular weight fraction (>5.0 kDa) increases noticeably, and this occurs both under natural conditions (Figure 22.5) and in the process of their concentration on a column with DEAE-cellulose when extracted from natural water (Figure 22.6). The rivers of the Pripyat River basin are characterized by the highest HS concentrations. Therefore, they are usually dominated by high-molecular fractions with a molecular weight > 5.0 kDa – from 64.6% to 81.8% of HS_{total}. The concentration of HS in the river waters of Desna and Ros' is incomparably lower than in the rivers of the Pripyat River basin. However, as the concentration of HS in them increases, the share of high-molecular weight fractions (>5.0 kDa) goes up noticeably.

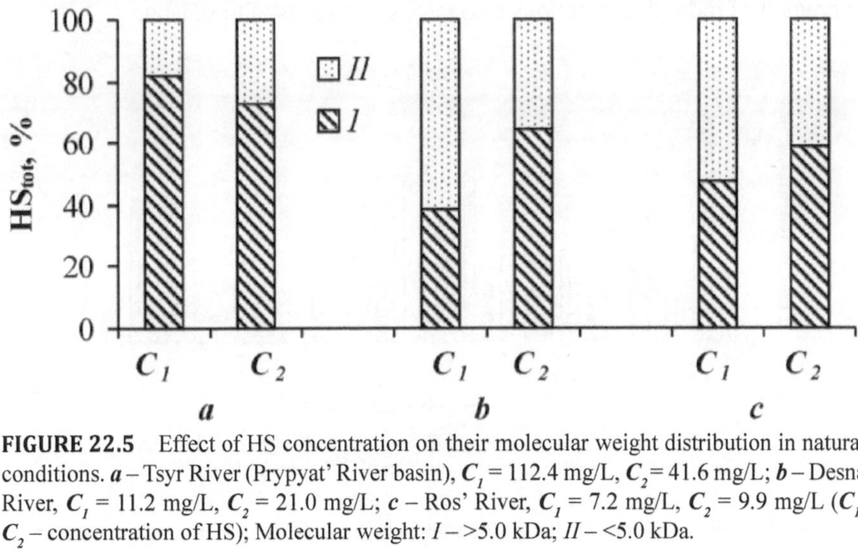

FIGURE 22.5 Effect of HS concentration on their molecular weight distribution in natural conditions. ***a*** – Tsyr River (Prypyat' River basin), C_1 = 112.4 mg/L, C_2= 41.6 mg/L; ***b*** – Desna River, C_1 = 11.2 mg/L, C_2 = 21.0 mg/L; ***c*** – Ros' River, C_1 = 7.2 mg/L, C_2 = 9.9 mg/L (C_1, C_2 – concentration of HS); Molecular weight: *I* – >5.0 kDa; *II* – <5.0 kDa.

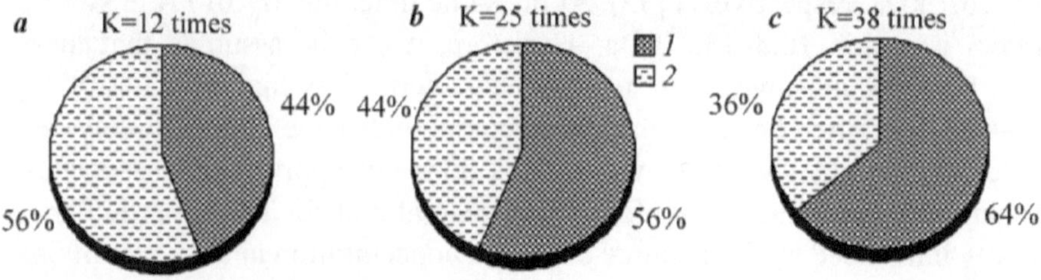

FIGURE 22.6 The effect of HS concentration on their molecular-weight distribution in natural water from the Desenka branch (Kaniv reservoir) during extraction. K – concentration coefficient; *1* – >5.0 kDa; 2 – <5.0 kDa.

22.3.2.2 CARBOHYDRATES

Fractionation of carbohydrates in terms of their molecular weight makes it possible to assess the contribution of their individual fractions to the total balance and to establish the relationship between high molecular and low molecular weight compounds. This allows for a better understanding of their transformation in the natural aquatic environment.

It has been found that in Ukraine's surface water bodies, the molecular weight of carbohydrates varies over a wide range – from <1.0 to >70.0 kDa (Figure 22.7). It is known that mainly polysaccharides subjected to further transformation are released by phytoplankton and higher aquatic plants. As a consequence, the content of individual fractions also varies from season to season.

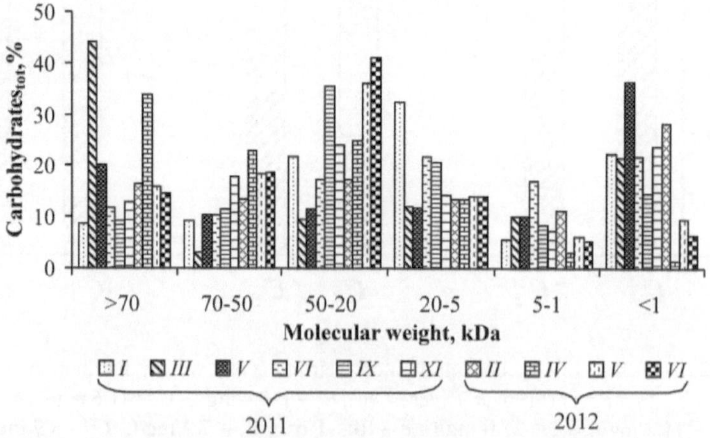

FIGURE 22.7 Ratio of carbohydrates' fractions with various molecular weights in water of upper Kytaivskyi pond (Kyiv City) for 2011–2012;% to Carbohydrates$_{total}$ (roman numerals are months).

High molecular compounds (polysaccharides) prevail. The average annual content of the fractions with a molecular weight >70.0 kDa accounted for 10.8–20.9%, 70.0–50.0 kDa – 10.7–23.4%, 50.0–20.0 kDa – 13.7–25.9%, 20.0–5.0 kDa – 7.7–19.5%, and 5.0–1.0 kDa – 5.8–13.7% of the total concentration of dissolved carbohydrates. The contribution of the fraction of low molecular carbohydrates (<1.0 kDa) accounted for 16.3–32.3%. It is known that this fraction contains simple sugars – monosaccharides and disaccharides. Thus, carbohydrates were represented mainly by polysaccharides. Their contribution ranged in total from 67.7% to 83.7%.

A wide range of molecular weights of carbohydrates is typical for these organic compounds, as they are characterized by a polymer structure. For example, it has been found that in the products of bacterial degradation of green algae, the molecular weight of carbohydrates varied within 0.7–200.0 kDa [41]. In swamps, about 15% of dissolved organic carbon was represented by carbohydrates with a molecular weight >100 kDa [42].

22.3.2.3 PROTEIN-LIKE COMPOUNDS

The findings of gel-chromatographic studies have shown that PLC registered in Ukraine's surface water bodies as carbohydrates are characterized by a wide range of values for their molecular weight – from <1.0 to >70.0 kDa (Figure 22.8) [43]. Such a pattern of the distribution of PLC among various fractions differing in their molecular weight was reported previously in studies of this group of organic compounds in the Dnipro reservoirs [44].

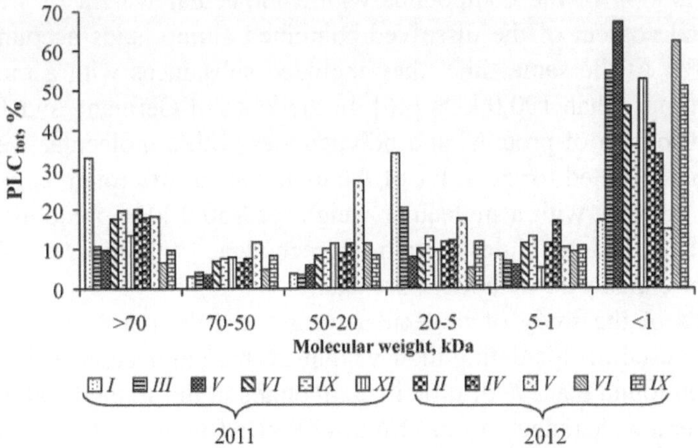

FIGURE 22.8 Ratio of protein-like compound (PLC)' fractions with various molecular weights in water of upper Kytaivskyi pond (Kyiv City) for 2011–2012;% to PLC_{total} (roman numerals are months).

Among various fractions of PLC, the fraction with a molecular weight of <1 kDa dominates. In the total balance of PLC, the maximal total content of the fractions with molecular weights of 5.0–1.0 and <1.0 kDa attains 27–76% (Figure 22.8).

A similar situation was observed in the other water bodies under study. The contribution of the above-mentioned fractions accounted for 39.3–57.9% of PLCtot. Previously, it has been shown that in summer, in the Dnipro reservoirs, the contribution of PLC with a molecular weight of ≤5.0 kDa attains 74.5–90.2% of their total content in the water [44]. Data on the seasonal dynamics of the relative content of the PLC fractions with a molecular weight of ≤5.0 kDa in the studied water bodies are characterized. In some of them, the contribution of these fractions essentially increases in spring and summer, which is probably caused by the intensification of the processes of decomposition of high-molecular protein compounds under the effect of extracellular hydrolytic enzymes of animal, algal, and bacterial origin. However, in some cases, the content of these fractions significantly decreases during the same period, which is explained by the assimilation of the compounds with a rather low molecular weight by hydrobionts during the process of their growth [1, 45]. Since excretion, decomposition, and assimilation processes in water bodies occur simultaneously, seasonal changes in the relative content of these fractions of PLC with a molecular weight of ≤5.0 kDa are difficult to reveal.

A wide range of the molecular weights of protein compounds is discussed in the literature. For example, in the water of the Delaware estuary (USA), the contribution of the compounds with a molecular weight of <1.0 kDa to the total content of the dissolved combined amino acids accounted for about 60%. At the same time, they included substances with a molecular weight of more than 100.0 kDa [46]. In the lakes of Germany and Japan, a high contribution of proteins and polypeptides with a molecular weight of <5.0 kDa accounted for 55–74% of the total content of protein compounds [47–49]. Proteins with a molecular weight of >50.0 kDa predominated in the products of bacterial destruction of green algae [41]. Thus, the obtained results correlated well with literature data.

Results of the study of molecular-weight distribution of the dissolved PLC were used in calculating their weight-average molecular weight (M_w). It has been found that M_w of protein compounds in the studied water bodies varied over a wide range – from 11.6 to 43.0 kDa (Figure 22.9). The average values of M_w are in limits 22–30 kDa.

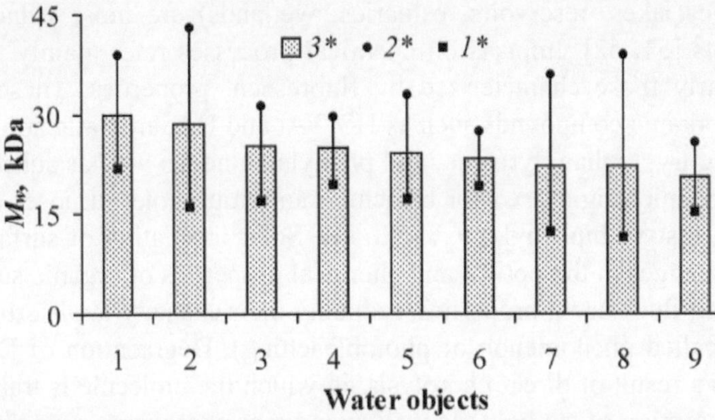

FIGURE 22.9 Limit (*1**, *2**) and averaged (*3**) values of the weight-average molecular weight (M_w, kDa) of PLC in the Ukraine's surface water bodies: 1 – the Pivdennyi Bug River, 2 – the Ternopil reservoir, 3 – the Desenka branch (the Kaniv reservoir), 4 – the Lyutsymir Lake, 5 – the Seret River, 6 – the Velyke Chorne Lake, 7 – the upper Kytaivskyi Pond, 8 – the Ros' River, 9 – the Desna River.

At some time, based on result of fluorescent investigations in lakes of China the dissolved PLC were represented by the compounds with a lowe M_w (about 3.0 kDa) [50]. The fractional composition of these compounds was as follows: >3.0 kDa – 12.4–18.6%, 3–2 kDa – 44.3–54.7%, 2–1 kDa – 16.8–21.5%, and <1.0 kDa – 15.6–16.1%. As it is evident, the contribution of the fraction with the molecular weight of 3–2 kDa was maximal.

The M_w values of PLC also varied from season to season, which was probably conditioned by the influence of the above mentioned processes. It is not inconceivable that high-molecular compounds prevail during the process of PLC excretion. However, their decomposition is accompanied by the formation of large amounts of substances with a relatively low molecular weight, which influences the value of M_w. It essentially decreases. On the other hand, assimilation of protein compounds with a lower molecular weight by hydrobionts results in the predominance of high molecular compounds in aquatic environment, which, in its turn, results in the increase in the value of M_w.

22.3.3 EFFECT OF THE SOLAR RADIATION AND BIOLOGICAL ACTIVITY ON THE RATIO VARIOUS FRACTIONS OF NATURAL ORGANIC SUBSTANCES IN SURFACE WATER OBJECTS

Solar radiation can significantly affect the physical, chemical, and biological processes occurring in surface water, and water bodies with slow water exchange (lakes, reservoirs, estuaries, wetlands) are most vulnerable to its effects [51, 52]. Impacts on chemical processes refer mainly to DOM, particularly those characterized by fluorescent properties. These include natural organic compounds such as HS, FA, and HA, aromatic amino acids, including tryptophan, tyrosine, and phenylalanine, as well as substances of anthropogenic origin used for bleaching and household purposes (diamino stilbene, distyrylbiphenyl, etc.) [53, 54]. Solar irradiation of surface water causes changes in the optical and chemical properties of organic substances exhibiting fluorescent properties, reducing their capacity to absorb sunlight (the so-called phenomenon of photobleaching). Degradation of DOM can occur as a result of direct photolysis, in which the molecule is transformed after absorption of sunlight, as well as indirect photolysis, whereby natural organic substances (photosensitizers) absorb a photon, producing a number of reactive intermediate species that can transform molecules [55, 56].

Solar radiation experiments show a shift in the composition of natural organic matter (NOM). Due to irradiation, the concentration of high molecular weight (HMW) fractions decreases, and that of the low molecular weight (LMW) fractions increases [52, 54, 57, 58].

Carbohydrates and PLC belong to easily oxidized DOM. They undergo bacterial destruction at increased water temperatures and activation of microbiological activity. However, there is evidence that carbohydrates are also photodegraded in the presence of HS [59].

22.3.2.1 HUMIC SUBSTANCES

HS belong to stable organic compounds of natural waters, as they experience little bacterial degradation in contrast to other organic substances, such as carbohydrates, PLC, etc. However, it is known that HS strongly absorb ultraviolet (UV) radiation from sunlight and undergo photochemical changes [60–62]. The HS ability to absorb ultraviolet light decreases when their concentration decreases. As a result of photochemical oxidation, the transformation of high-molecular fractions of HS into compounds with a lower molecular weight occurs [60].

The findings of our studies have shown that HS undergo the greatest transformation in the spring and summer periods when the intensity of photochemical processes reaches its maximum. There is an increase in the share of the HS fraction with a molecular weight not exceeding 5 kDa in general and < 1 kDa in particular during this period (Figure 22.10).

The share of HS in Kaniv Reservoir with a molecular weight of ≤5 kDa noticeably rises in summer by 2.8 and 2.4 times, respectively, compared to the winter and autumn periods. This is attributed to the manifestation of various factors, including the effect of solar radiation. The share of the HS fraction with a molecular weight of ≤5 kDa in Ternopil Reservoir is highest in the spring and summer periods. It is assumed that the effect of photochemical processes on the transformation of HS is more noticeable in this small water body than in Kaniv Reservoir.

22.3.2.2 CARBOHYDRATES

Carbohydrates occupy an important place in the total balance of natural organic substances of surface waters. According to the available data, carbohydrates in river waters account for about 10–23%, and in lake waters—for 20.0–24.5% of C_{org} [1, 13, 63]. Polysaccharides, i.e., compounds of polymeric nature, most often prevail [51, 64–66]. As has been already mentioned above, Ukraine' surface water bodies are also dominated by polysaccharides [13, 66].

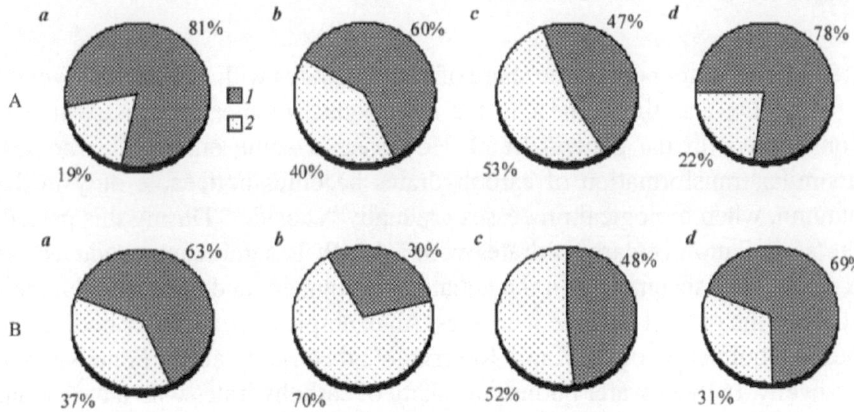

FIGURE 22.10 Ratio of HS' fractions with various molecular weights (1, >5 kDa; 2, ≤5 kDa);% of total content in waters of (A) upper part of the Kaniv Reservoir (Dnipro River basin) and (B) Ternopil Reservoir (Seret River basin) in different seasons; (a–d) winter, spring, summer and autumn, respectively, 2011–2013 and 2016–2017 [51].

Molecular weight distribution of carbohydrates in water bodies under study is characterized by some special features [67]. In some of them, the transformation of HMW carbohydrates is noticeable in spring and summer. As an example of such water bodies, (Figure 22.11) are the data on the molecular weight distribution of carbohydrates in the water of the Kytaivskyi pond and the Pivdennyi Bug River in different seasons.

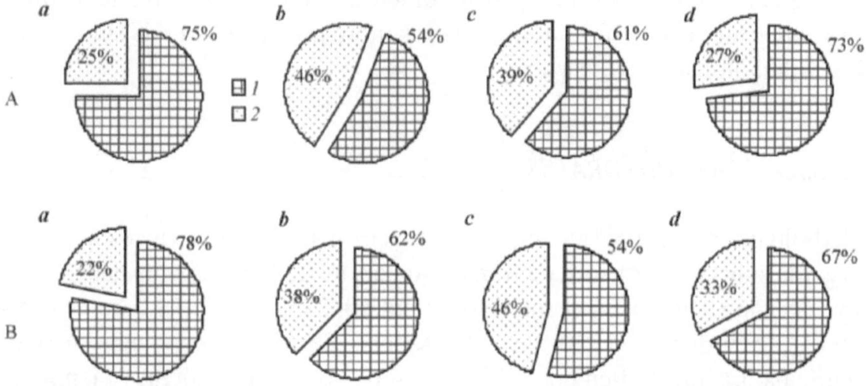

FIGURE 22.11 Ratio of carbohydrates' fractions with various molecular weights (1, > 5 kDa; 2, ≤ 5 kDa);% of total content in waters of (A) Kytaivskyi pond (Kyiv City) and (B) Pivdennyi Bug River in different seasons; (a–d) winter, spring, summer and autumn, respectively, 2011–2013 and 2016–2017.

In these water bodies, the share of carbohydrates with a molecular weight of ≤ 5 kDa was the highest in the spring and summer seasons, which is consistent with the general trend. However, in some other water bodies, a similar transformation of carbohydrates becomes noticeable only in the autumn, when biological processes gradually "subside." During this period, the assimilation of carbohydrates with a LMW is significantly reduced; as a result, they accumulate in the aquatic environment and, accordingly, their share in the overall balance increases. An example of such water bodies can be the Obolon Bay of the Kaniv Reservoir and Horikhuvatskyi Pond No. 5 in Kyiv city. In these water bodies, the share of carbohydrates with a molecular weight of ≤ 5 kDa increases to 38–48% compared to 8–25% and 23–29% in other seasons.

At the same time, in some water bodies, in particular, in rivers, no noticeable changes in the molecular weight distribution of carbohydrates were found (Figure 22.12). The largest share of compounds with a molecular weight of ≤ 5 kDa was detected in the summer-autumn or winter-spring periods of the year.

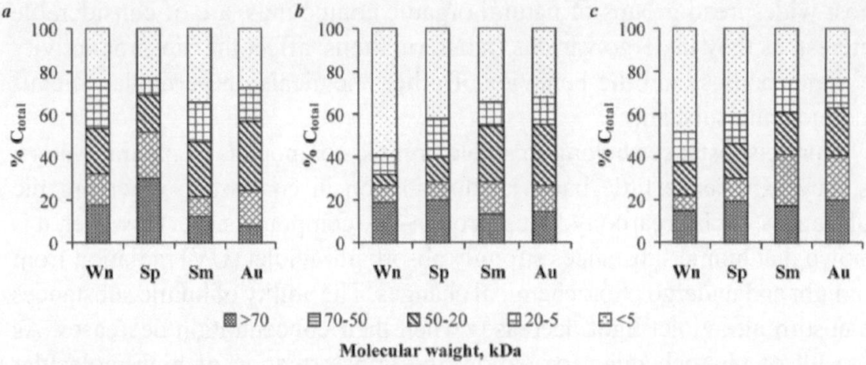

FIGURE 22.12 Mass share of carbohydrate fractions with different molecular weights,% of their total content in waters of (a) the Desna River (mouth), (b) the Seret River (downstream of the Ternopil' Reservoir), and (c) the Ros' River (Bilotserkovs'ke Reservoir) in different seasons of the year, 2011–2013. C_{total} – total content of carbohydrates in the water; Wn, Sp, Sm, Au – winter, spring, summer, and autumn accordingly.

Since carbohydrates and protein compounds belong to easily oxidized organic substances, the main role in their transformation is played by biological processes, first of all, with the participation of bacteria, the activity of which increases significantly with increasing temperature. Therefore, under the conditions of global warming, we should expect the activation of destructive processes in the transformation of natural organic compounds. At the same time, the appearance of organic compounds with a LMW in natural conditions cannot always be detected, since they become biologically active and are assimilated by hydrobionts. This is the reason why seasonal changes in the molecular mass distribution of carbohydrates and protein compounds are not always clearly manifested. Nevertheless, the results of our many years of observations indicate that these transformations are taking place, and in the conditions of modern climate changes, they will take place even more actively.

22.4 CONCLUSION

Dissolved organic substances of surface water bodies are an important component of the chemical composition of natural water, undergoing noticeable seasonal changes in their component composition and significantly affecting the state of the aquatic environment as a habitat for aquatic organisms. In this situation, humic substances and carbohydrates, as the most widespread groups of natural organic compounds, are of considerable interest, as they undergo various transformations, affect the bio-productivity of water bodies, and the behavior of other chemicals, in particular, metals and biogenic substances.

Humic substances belong to stable organic compounds of natural waters, as they experience little bacterial degradation in contrast to other organic substances, such as carbohydrates, protein-like compounds, etc. However, it is known that humic substances strongly absorb ultraviolet (UV) radiation from sunlight and undergo photochemical changes. The ability of humic substances to absorb ultraviolet light decreases when their concentration decreases. As a result of photochemical oxidation, the transformation of high-molecular fractions of humic substances into compounds with a lower molecular weight occurs. Most often, this transformation occurs in the spring-summer period when photochemical oxidation processes are activated. The mass fraction of substances with a molecular weight $\leq$ 5.0 kDa increases maximally during this period to 50–60% of the total content of humic substances.

Carbohydrates and protein-like compounds belong to easily oxidized DOM. They undergo bacterial destruction at increased water temperatures and activation of microbiological activity. However, there is evidence that carbohydrates are also photodegradable in the presence of HS. The largest share of carbohydrates with a molecular weight $\leq$ 5 kDa is observed in autumn (38–48%), when biological processes and bioaccumulation of low molecular weight fractions gradually decrease.

Regardless of which factors of the water environment exert the greatest influence on the transformation of natural organic substances, the formation of compounds with a LMW leads to their greater bioavailability and, as a result, significantly affects the development and functioning of aquatic organisms that assimilate these compounds. At the same time, the complexing ability of natural organic substances with respect to metal ions changes, as well as the toxicity of the water environment. We can argue about a number of other potential changes in the chemical composition of natural water in surface water ecosystems. Under conditions of modern warming, these changes will intensify, as the intensity of photochemical and biological processes increases.

ACKNOWLEDGMENTS

The authors wish to thank Dr. Sc. Nataliia Semeniuk for editing the English translation.

KEYWORDS

- **carbohydrates**
- **dissolved organic matter**
- **humic substances**
- **molecular-weight distribution**
- **protein-like compounds**
- **solar radiation**
- **surface water objects**

REFERENCES

1. Findlay, S. E. G., & Sinsabaugh, R. L. (2003). *Aquatic Ecosystems: Interactivity of Dissolved Organic Matter*. Academic Press.
2. Thurman, E. M. (1985). *Organic Geochemistry of Natural Waters*. Kluwer Academic Publishers Group.
3. Zhang, L., Sun, Q., Peng, Y., Zhao, H., Liu, H., You, Y., & Zhang, Y. (2021). Components and structural characteristics of dissolved organic matter in the overlying water of the Beiyun River. *Energy, 221*, Article 119921. https://doi.org/10.1016/j.energy.2021.119921.
4. Williams, C. J., Frost, P. C., Morales-Williams, A. M., Larson, J. H., Richardson, W. B., Chiandet, A. S., & Xenopoulos, M. A. (2016). Human activities cause distinct dissolved organic matter composition across freshwater ecosystems. *Global Change Biology, 22*(2), 613–626.
5. Mash, H., Westerhoff, P. K., Baker, L. A., Nieman, R. A., & Nguyen, M. L. (2004). Dissolved organic matter in Arizona reservoirs: Assessment of carbonaceous sources. *Organic Geochemistry, 35*, 831–843.
6. Linnik, P. N., Ivanechko, Ya. S., Linnik, R. P., & Zhezherya, V. A. (2013). Humic substances in surface waters of Ukraine. *Russian Journal of General Chemistry, 83*(13), 2715–2730.
7. Sirenko, L. A., & Kozitskaya, V. N. (1988). *Biologicheski aktivnye veshchestva vodorosley i kachestvo vody* [Biologically active substances of algae and water quality]. Naukova Dumka Press. [In Russian].
8. Osadchyy, V., Nabyvanets, B., Linnik, P., Osadcha, N., & Nabyvanets, Yu. (2016). *Processes Determining Surface Water Chemistry*. Springer International Publishing.
9. Chen, M., Price, R. M., Yamashita, Y., & Jaffé, R. (2010). Comparative study of dissolved organic matter from groundwater and surface water in the Florida coastal Everglades using multi-dimensional spectrofluorometry combined with multivariate statistics. *Applied Geochemistry, 25*(6), 872–880.

10. Zhezherya, V. A. (2012). Modyfikovany batometr-sklyanka [Modified water sampler-glass]. Patent 75995, MPK51 (2012.01) G 01 N 1/00, Ukraine. Declared 27.04.2012; Published 25.12.2012, Bulletin No. 24. [In Ukrainian].
11. Linnik, P. N., Zhezherya, V. A., Ivanechko, Ya. S., & Linnik, R. P. (2014). Humic substances and their role in the migration of metals in high-colored surface waters: The case study of the Pripyat' River basin. *Russian Journal of General Chemistry, 84*(13), 2572–2587.
12. Linnik, P. N., Zhezherya, V. A., Linnik, R. P., Ignatenko, I. I., & Zubenko, I. B. (2015). Metals in surface water of Ukraine: The migration forms, features of distribution between the abiotic components of aquatic ecosystems, and potential bioavailability. *Russian Journal of General Chemistry, 85*(13), 2965–2984.
13. Linnik, P. N., & Ivanechko, Ya. S. (2014). Dissolved carbohydrates in the surface water bodies of Ukraine. *Hydrobiological Journal, 50*(6), 87–107.
14. Popovich, G. M. (1990). Sorption concentration and spectrophotometric determination of fulvic and humic acids in waters. Author's abstract of PhD thesis. Kyiv.
15. Nabyvanets, B. I., Osadchyy, V. I., Osadcha, N. M., & Nabyvanets, Yu. B. (2007). *Analytical Chemistry of Surface Water*. Naukova Dumka Press.
16. Semenov, A. D. (1977). *Manual on the Chemical Analysis of the Surface Water on Land.* Gidrometeoizdat Press. [In Russian].
17. Debeiko, E. V., Ryabov, A. K., & Nabivanets, B. I. (1973). Direct photometric determination of dissolved proteins in natural waters. *Gidrobiologicheskiy Zhurnal, 9*(6), 109–113. [In Russian].
18. Linnik, P. N., & Vasilchuk, T. A. (2005). Role of humic substances in the complexation and detoxification of heavy metals: Case study of the Dnieper reservoirs. In I. V. Perminova, K. Hatfield, & N. Nertkorn (Eds.), *Use of Humic Substances to Remediate Polluted Environments: From Theory to Practice* (pp. 135–154). NATO Science Series IV: Earth and Environmental Science. Springer Press.
19. Osadcha, N. M. (2011). Regularities of humic substance migration in surface water of Ukraine. Doctoral thesis, Kyiv.
20. Linnik, P. N., Ivanechko, Ya. S., Linnik, R. P., & Zhezherya, V. A. (2013). Humus substances of surface waters and the peculiarities of their distribution among various fractions. *Hydrobiological Journal, 49*(5), 90–111.
21. Rigobello, E. S., Campos, S. X., de Azevedo, E. R., Dantas, A. D. B., & Vieira, E. M. (2017). Comparative characterization of humic substances extracted from freshwater and peat of different apparent molecular sizes. *Revista Ambiente & Água–An Interdisciplinary Journal of Applied Science, 12*(5), Article 774, 12 pp.
22. Nguyen, H. V. M., Hur, J., & Shin, H. S. (2022). Humic acids and fulvic acids: Characteristics, sorption of hydrophobic organic contaminants, and formation of disinfection by-products during chlorination. In A. Makan (Ed.), *Humus and Humic Substances–Recent Advances* (Chapter 1, pp. 1–19). IntechOpen.
23. Sachse, A., Babenzien, D., Ginzel, G., Gelbrecht, J., & Steinberg, C. E. W. (2001). Characterization of dissolved organic carbon (DOC) in a dystrophic lake and an adjacent fen. *Biogeochemistry, 54*, 279–296.
24. Osypenko, V. P., Vasylchuk, T. O., & Yevtukh, T. V. (2011). Comparison of the content of carbohydrates and protein-like substances in the water of the rivers of the Prypyat River basin and in the Kyiv reservoir depending on the concentration of humic substances. *Gidrologiya, Gidrokhimiya i Gidroekologiya, 1*(22), 179–184. [In Ukrainian].

25. Osadcha, N. M. (1993). Role of organic compounds in the process of copper (II) transformation in water bodies used for fishery. Author's abstract of PhD thesis. Kyiv.
26. Perminova, I. V. (2000). Analysis, classification, and prediction of the properties of humus acids. Author's abstract of Doctoral thesis. Moscow.
27. Ghabbour, E. A., & Davies, G. (2003). *Humic Substances: Nature's Most Versatile Materials*. Taylor and Francis, Inc.
28. Fischer, T. (2017). Humic supramolecular structures have polar surfaces and unipolar cores in native soil. *Chemosphere, 183*, 437–443. https://doi.org/10.1016/j.chemosphere.2017.05.125.
29. Linnik, P. N., & Vasilchuk, T. A. (2004). Humic substances of natural waters and their importance for aquatic ecosystems: A review. *Hydrobiological Journal, 40*(3), 79–101.
30. Osadcha, N. M. (2010). Polydispersity of humic substances of surface water of the Dnipro River Basin. *Naukovi pratsi UkrNDGMI, 259*, 145–170. [In Ukrainian].
31. Kulikova, N. A., Perminova, I. V., & Kudryavtsev, A. V. (2010). A comparative study of molecular weight distribution of water-soluble humic substances, humic acids, and fulvic acids extracted from sod-podzolic soils. *Moscow University Soil Science Bulletin, 65*(4), 155–158.
32. Conte, P., & Piccolo, A. (1999). Conformational arrangement of dissolved humic substances: Influence of solution composition on association of humic molecules. *Environmental Science & Technology, 33*(10), 1682–1690.
33. Tercero Espinoza, L. A., ter Haseborg, E., Weber, M., & Frimmel, F. H. (2009). Investigation of the photocatalytic degradation of brown water natural organic matter by size exclusion chromatography. *Applied Catalysis B: Environmental, 87*, 56–62.
34. Linnik, P. N., & Nabivanets, B. I. (1986). *Forms of Metals Migration in Surface Fresh Waters*. Gidrometeoizdat Press.
35. Perminova, I. V., Frimmel, F., Kudryavtsev, A. V., Kulikova, N. A., Abbt-Braun, G., Hesse, S., & Petrosyan, V. S. (2003). Molecular weight characteristics of humic substances from different environments as determined by size exclusion chromatography and their statistical evaluation. *Environmental Science & Technology, 37*, 2477–2485.
36. Makharadze, T., & Makharadze, G. (2021). Measurement of complex formation process of nickel (II) with freshwater fulvic acids using the solubility method. *Fine Chemical Engineering, 2*(2), 54–61.
37. Makharadze, G., Supatashvili, G., & Makharadze, T. (2018). New version of calculation of stability constant of metal–fulvate complexes on the example of zinc fulvate. *International Journal of Environmental Science and Technology, 15*, 2165–2168.
38. Makharadze, T., & Makharadze, G. (2020). Investigation of complex formation process of copper with macromolecular organic substances isolated from natural waters. *Organic Chemistry Plus, 1*(1), 1–6.
39. Varshal, G. M., Koshcheeva, I. Y., Sirotkina, I. S., Velukhanova, T. K., Intskirveli, L. N., & Zamokina, N. C. (1979). The study of organic substances in surface waters and their interaction with metal ions. *Geochemistry, 4*, 598–607. [In Russian].
40. Makharadze, G., & Makharadze, T. (2014). Method of calculation of stability constants of fulvic complexes on the example of copper. *Journal of Chemical and Chemical Engineering, 8*, 108–111.
41. Akiyama, T. (1972). Chemical composition and molecular weight distribution of dissolved organic matter produced by bacterial degradation of green algae. *Geochemical Journal, 6*, 93–104.

42. Satoh, Y., Shoji, S., Satoh, H., & Takahashi, M. (1987). Dissolved organic matter in colored water from mountain bog pools in Japan. I. Seasonal changes in the concentration and molecular weight distribution. *Archiv für Hydrobiologie, 110*, 589–603.
43. Linnik, P. N., & Ivanechko, Ya. S. (2015). Dissolved protein-like substances in surface water bodies of various types. *Hydrobiological Journal, 51*(2), 85–104.
44. Linnik, P. N., & Vasil'chuk, T. A. (1997). Nitrogenous organic matter in the water of the Dnieper reservoirs. *Hydrobiological Journal, 33*(5), 22–29.
45. Berman, T., & Bronk, D. A. (2003). Dissolved organic nitrogen: A dynamic participant in aquatic ecosystems. *Aquatic Microbial Ecology, 31*, 279–305.
46. Coffin, R. B. (1989). Bacterial uptake of dissolved free and combined amino acids in estuarine waters. *Limnology and Oceanography, 34*(3), 531–542.
47. Hama, T., & Handa, N. (1980). Molecular weight distribution and characterization of dissolved organic matter from lake waters. *Archiv für Hydrobiologie, 90*(1), 106–120.
48. Hama, T., & Handa, N. (1983). The seasonal variation of organic constituents in a eutrophic lake, Lake Suwa, Japan. Part II. Dissolved organic matter. *Archiv für Hydrobiologie, 98*(4), 443–462.
49. Steinberg, C. (1977). Schwer abbaubare, stickstoffhaltige gelöste organische Substanzen im Schöhsee und in Algenkulturen. *Archiv für Hydrobiologie* (Suppl.), *53*, 48–158.
50. Yue, L., Wu, F., Liu, C., et al. (2006). Relationship between fluorescence characteristics and molecular weight distribution of natural dissolved organic matter in Lake Hongfeng and Lake Baihua, China. *Chinese Science Bulletin, 51*(1), 89–96.
51. Linnik, P., Osadchyi, V., & Osadcha, N. (2023). Photochemical processes in surface water bodies and their potential impacts on the chemical composition of water: A review. *Lakes & Reservoirs: Research & Management, 28*, e12436. https://doi.org/10.1111/lre.12436.
52. Slavik, I., Kostrowski, D., & Uhl, W. (2023). Effect of solar radiation on natural organic matter composition in surface waters and resulting impacts on drinking water treatment. *Environmental Technology, 44*(11), 1549–1565.
53. Mostofa, K. M. G., Wu, F., Liu, C. Q., Vione, D., Yoshioka, T., Sakugawa, H., & Tanoue, E. (2011). Photochemical, microbial, and metal complexation behavior of fluorescent dissolved organic matter in aquatic environments. *Geochemical Journal, 45*, 235–254.
54. Paul, A., Dziallas, K., Zwirnmann, E., Gjessing, E. T., & Grossart, H. P. (2012). UV irradiation of natural organic matter (NOM): Impact on organic carbon and bacteria. *Aquatic Sciences, 74*(3), 443–454.
55. Vähätalo, A., Carena, L., & Vione, D. (2022). Photochemical reactions in inland waters. In T. Mehner & K. Tockner (Eds.), *Encyclopedia of inland waters* (2nd ed., Vol. 1, pp. 105–117). Elsevier.
56. Vione, D., & Scozzaro, A. (2019). Photochemistry of surface fresh waters in the framework of climate change. *Environmental Science & Technology, 53*(14), 7945–7963.
57. Gowland, D. C. A., Robertson, N., & Chatzisymeon, E. (2021). Photocatalytic oxidation of natural organic matter in water. *Water, 13*, Article 288, 21 pp. https://doi.org/10.3390/w13030288.
58. Milstead, R. P., Horvath, E. R., & Remucal, C. K. (2023). Dissolved organic matter composition determines its susceptibility to complete and partial photooxidation within lakes. *Environmental Science & Technology, 57*(32), 11876–11885.

59. Grzybowski, W. (2009). Terrestrial humic substances induce photodegradation of polysaccharides in the aquatic environment. *Photochemical and Photobiological Sciences, 8*, 1361–1363.
60. Catalán, N., Obrador, B., Felip, M., & Pretus, J. L. I. (2013). Higher reactivity of allochthonous vs. autochthonous DOC sources in a shallow lake. *Aquatic Sciences, 75*, 581–593.
61. Coelho, C., Guyot, G., ter Halle, A., & Richard, C. (2011). Photoreactivity of humic substances: Relationship between fluorescence and singlet oxygen production. *Environmental Chemistry Letters, 9*, 447–451.
62. Chróst, R. J., Münster, U., Rai, H., Albrecht, D., Witzel, P. K., & Overbeck, J. (1989). Photosynthetic production and exoenzymatic degradation of organic matter in the euphotic zone of a eutrophic lake. *Journal of Plankton Research, 11*(2), 223–242.
63. Sachse, A., Henrion, R., Gelbrecht, J., & Steinberg, C. E. W. (2005). Classification of dissolved organic carbon (DOC) in river systems: Influence of catchment characteristics and autochthonous processes. *Organic Geochemistry, 36*, 923–935.
64. Sverre, M. M. (1995). Release of extracellular products by phytoplankton with special emphasis on polysaccharides. *Science of the Total Environment, 165*, 155–164. https://doi.org/10.1016/0048-9697(95)04549-g.
65. Cole, J. J., McDowell, W. H., & Likens, G. E. (1984). Sources and molecular weight of "dissolved" organic carbon in an oligotrophic lake. *Oikos, 42*(1), 1–9.
66. Vasil'chuk, T. A., & Linnik, P. N. (1998). Carbohydrates in water of the Dnieper reservoirs. *Hydrobiological Journal, 34*(2–3), 212–218.
67. Linnik, P. N., Zhezherya, V. A., & Linnik, R. P. (2021). Potential transformations of dissolved organic substances and their complexes with metals in surface waters under solar radiation. *Russian Journal of General Chemistry, 91*(13), 2931–2942.

CHAPTER 23

The Role of Various Groups of Natural Organic Substances in Complexation in Ukraine's Surface Waters

PETRO LINNIK,[1] VLADYSLAV ZHEZHERYA,[1] and ROSTYSLAV LINNIK[2]

[1]*Department of Freshwater Hydrochemistry, Institute of Hydrobiology of the National Academy of Sciences, Kyiv, Ukraine*

[2]*Faculty of Chemistry, Taras Shevchenko Kyiv National University, Kyiv, Ukraine*

ABSTRACT

The paper deals with the findings of the authors' long-term research on the role of various groups of dissolved organic matter in metal complexation in Ukraine's surface water bodies. It has been found that humic substances, carbohydrates, and protein-like substances play an important role in metal complexation in surface waters. Humic substances, as a rule, play the main role in complexation since they prevail in content among other groups of dissolved organic matter. The share of metal complex compounds with carbohydrates increases with a decrease in humic substances concentration and an increase in carbohydrates concentration. This is typical for high bio productivity water bodies with low humic substances concentration and is mainly observed in summer and autumn. The greatest transformation of high-molecular weight complex metal compounds to low-molecular weight complexes, which become bioavailable to aquatic organisms, also occurs during these seasons. The share of the labile fraction as potentially bioavailable varied widely from a few percent to almost 100%, and on average was within 15–84% of the dissolved metals concentration. It should be noted that

Advanced Topics in Polymer Chemistry and Materials Science: Tamara Tatrishvili and Marc Jean Médard Abadie (Eds)

the share of the labile fraction of metals increases in water bodies with low HS concentration that are subject to significant human impact. The share of anionic complexes with humic substances was, on average, 41.6–72.8% (Al), 42.0–71.8% (Cu), 40.5–65.7% (Zn), 59.5–72.8% (Pb), and 52–82% (Fe) of dissolved metal concentration. The share of the investigated metals in neutral complexes with carbohydrates was 9–57% of dissolved metal concentration. The smallest share is made up of metal complexes with protein compounds (cationic fraction). The fraction of cationic complexes may also include free (hydrated) ions and hydroxocomplexes. It was found that complex metal compounds with humic substances were mainly represented by complexes with a molecular weight of less than 5 kDa, with their share varying within 42–74% of the total metal concentration in complexes with HS. The molecular weight distribution of metals' neutral complexes with carbohydrates has a slightly different nature. Most of the specified complex compounds are high molecular weight, and the share of complexes with a molecular weight ≤ 5.0 kDa, as a rule, does not exceed 50% of their total content in the neutral complexes' composition. The molecular weight distribution of the metal complexes with PLC wasn't investigated due to their low concentrations in the water of the studied water bodies.

23.1 INTRODUCTION

In surface water bodies, metals migrate in dissolved, colloidal, and suspended forms [1–5]. In this case, the suspended form can be represented by metals contained in the crystalline lattice of mineral particles of the suspended substance or adsorbed on their surface, as well as metals in the composition of biota and detritus. Metals in the dissolved state can be in the form of aqua- and hydroxocomplexes, and also in the form of complex compounds with inorganic (hydrocarbonate, sulfate, chloride ions, etc.) and organic ligands, including primarily humic substances (HS), carbohydrates, protein-like compounds (PLC), and many other chemical substances, which are the products of aquatic organisms' metabolism [1, 6]. The dissolved forms of metals are characterized by high migration ability and bioavailability. At the same time, the suspended forms of metals are inaccessible to hydrobionts [7].

From an environmental point of view, there is insufficient data on the total concentration of metals in the surface waters. These data don't allow for assessing metals' potential toxicity and biological role in aquatic ecosystems, which depend on the physical and chemical state of metals in the natural environment [8–10]. The total concentration of metals in the water makes it impossible to assess their mobility and possible transformations of their coexisting forms under the effect of the environmental factors and processes occurring in water bodies [3, 11]. The state of metals in surface waters depends on hydrolysis, complexation, adsorption on the surface of suspended solids, and sedimentation. Complexation is accompanied by an increase in the migration activity of metals as a result of their transformation into the dissolved state. Adsorption on suspended particles is accompanied by a decrease in the migration activity of metals that settle to the bottom with suspended solids when the water velocity decreases. The degree of adsorption of complex metal compounds differs significantly from the adsorption of their free ions. Thus, information on the total concentration of metals in the water can be used only in assessing the degree of water body contamination and its spatial and temporal changes.

The metals' bioavailability and toxicity in surface waters depend to a large extent on the relationship between their coexisting forms. The dissolved metal forms are considered more bioavailable for assimilation by hydrobionts. However, the dissolved form of the metals also includes some portion of colloidal particles. Mainly, the term "metal bioavailability" implies the portion of metal available to aquatic organisms [12]. To assess this form of metals, it is essential to know the most widely distributed dissolved forms of metals in surface water (Figure 23.1). The metals in the labile fraction are more bioavailable than those in the form of complexes with dissolved organic substances or those in colloidal and suspended forms [9, 13].

Thus, metals' labile fractions include primarily their free ions, hydroxocomplexes, and complexes with inorganic ligands (in the scheme, they are indicated by the dotted line). The metal complexes' lability with natural organic compounds largely depends on the strength of bonds, the resistance of organic substances to environmental factors, and also on the molecular weight of their individual fractions. High-molecular metal complexes can't penetrate through the biological membrane.

An important coexisting form of metals in natural surface waters is represented by complex compounds with various groups of dissolved organic matter (DOM) and different molecular weight (Figure 23.1). Complexation leads primarily to an increase in the metals solubility, to a change in the ratio of their oxidized and reduced forms, to a decrease in the metals toxicity, as well as to a change in their bioavailability and ability to adsorb on suspended substances [14–16].

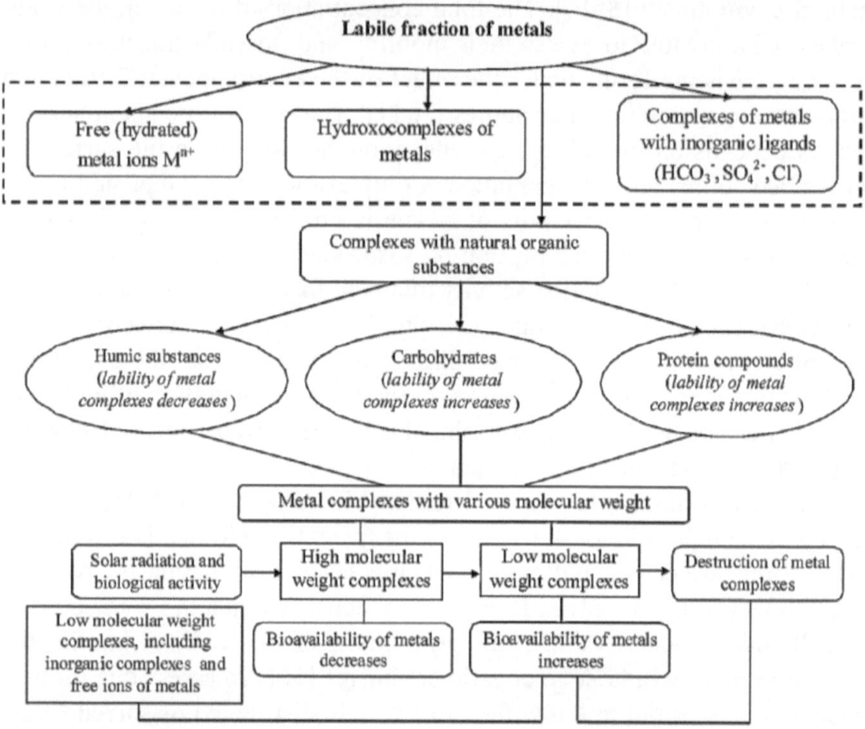

FIGURE 23.1 Coexisting forms of dissolved metals in natural surface water and their potential lability and bioavailability.

Complexation with HS as the dominant group of DOM in surface waters has been studied to the greatest extent [3, 14, 17–23]. Often, it is with HS that the predominant part of dissolved metals is associated, which was also established for surface water bodies of Ukraine [24]. The research findings described in recent publications indicate that not only HS, but also other groups of DOM are involved in the complexation, in particular carbohydrates, protein-like compounds (PLC), and other chemical substances that are metabolic products of phytoplankton, higher aquatic plants, and microorganisms [25, 26, 27]. The proportion of autochthonous organic compounds, including polysaccharides, can reach 25% of the total DOM content in fresh waters during intensive algae growth (algal bloom) [28]. In particular, this is true of eutrophic water bodies, where the HS concentration in water is rather low. At the same time, it is known that carbohydrates and protein compounds belong to the group of readily oxidized substances. An increase in water temperature and microbial activity leads to the transformation of high-molecular compounds into compounds with a lower molecular weight, which are bioavailable to aquatic organisms. Metal complexes with such low molecular weight organic compounds become bioavailable.

The article summarizes the results of the authors' long-term research on the role of various groups of DOM in metal complexation in Ukraine's surface water bodies.

23.2 EXPERIMENTAL METHODS AND MATERIALS

23.2.1 REAGENTS, EQUIPMENTS, METHODS

Standard solutions of Al (MCO 0535:2003 – 10.0 mg/mL), Fe (MCO 0518:2003 – 1.00 mg/mL), Mn (MCO 0521:2003 – 1.00 mg/mL), Cu (MCO 0524:2003 – 10.0 mg/mL), Zn (MCO 0032:1998 – 1.00 mg/mL), and Pb (MCO 0525:2003 – 1.00 mg/mL) were used in this study. The reference and model solutions were prepared using double-distilled water. Solutions of KOH (0.3 mol/L) and H_2SO_4 (1.0 mol/L) were prepared by diluting 45% and 98% solutions of high-purity grade, respectively. Chrome Azurol S, 1,10-phenanthroline, luminol (Fluka Chemie, Germany), and H_2O_2 (30%) solutions were also used.

Unico UV 2800 spectrophotometer, KFK-2 photo colorimeter, chemiluminescence photometer, pH meter pH 150MI (Russia), a device for membrane filtration, a device for UV treatment with a DRT-1000 mercury-quartz lamp, glass columns for ion-exchange and gel chromatography research, and a collector of fractions "Dombifrak" were used.

Spectrophotometry, chemiluminescence, ion-exchange and gel chromatography, membrane filtration, and UV treatment have been applied to investigate the coexisting forms of metals in surface waters.

23.2.2 SAMPLING AND SAMPLE PREPARATION

Water samples were taken in surface water bodies of various types: reservoirs of the Dnipro cascade (Kyivs'ke, Kanivs'ke (upper section, including the Desenka arm), Dniprovs'ke, Kakhovs'ke); small reservoirs (Yurpils'ke, Girskyy Tikych River, Chorna Kamyanka village; Middle Bilotserkivs'ke, Ros' River, Bila Tserkva City; Ternopils'ke, Seret River, Ternopil City); rivers Tsyr, mouth; Prypyat,' Svalovychi village (Prypyat' River basin); Desna, mouth; Pivdennyy Bug, Khmelnitsky city; Seret, Ternopil City;

Kiliya Danube Delta, Kiliya arm; lakes (Lyutsymir, Chorne Velyke of the Shatsk group, Volyn' region); Yordans'ke and Verbne, within the Kyiv city; the upper Kytayivskyy and second Gorikhuvatskyy ponds, within the Kyiv city) using Ruthner bathometer or modified bathometer-bottle [29].

To separate suspended solids, a freshly collected water sample was passed through a nitrocellulose membrane filter with a pore diameter of 0.4 μm (Czech Republic) under the pressure of about 2×10^5 Pa. The extraction of HS from the natural water filtrate and their concentration by at least 25–50 times was achieved by ion-exchange chromatography using a column filled with diethylaminoethylcellulose (DEAE-cellulose) manufactured by SERVA. HS was eluted from the column using a 0.3 mol/L KOH solution. The elution rate was about 1.0 mL/min. To separate HS into humic acids (HA) and fulvic acids (FA) fractions, the resulting HS concentrates were acidified to pH 1.5–2.0 and heated to approximately 50°C for several hours. Then, the resulting HA precipitate was allowed to settle for a day and centrifuged at 5000 rpm for 15 minutes. The solution above the precipitate containing FA was decanted. The HA precipitate was dissolved in 3–5 mL of 0.3 mol/L KOH solution. The pH value of the thus obtained solutions of FA and HA was adjusted to the original natural water value.

The molecular weight distribution of HS, FA, HA, and their complex compounds with metals was studied by gel chromatography. For this purpose, a glass column filled with TOYOPEARL HW-50F gel (Japan) was used. The column length was 82 cm, diameter – 2.8 cm; gel layer height – 62 cm; column free volume (V_0) – 160 mL.

The ion-exchange chromatography was also used in separating carbohydrates from the other organic matter, including HS. Water filtrates were passed successively through two columns filled with ion exchangers. The first column contained DEAE-cellulose, which was used for the extraction of HS (acidic group), whereas the second column contained carboxymethyl cellulose (CM) – for PLC (basic group). Carbohydrates were mainly in the neutral fraction, that is, in the water filtrate that passed through the columns with the above-mentioned ion exchangers.

The molecular weight distribution of carbohydrates and protein compounds, and their complexes with metals was also studied by gel chromatography using a glass column filled with TOYOPEARL HW-55F gel (Japan). Column parameters: length – 81.0 cm, diameter – 2.8 cm, gel layer height – 60.5 cm, V_0 – 138 mL.

The columns were calibrated using solutions of polyethylene glycols with molecular weight of 1.0, 2.0, 15.0, and 20.0 kDa, insulin – 5.8 kDa, albumin – 68.0 kDa, dextran – 70 kDa, and glucose (0.18 kDa). The concentrations of polyethylene glycols, proteins, dextran, and glucose were 2.0, 1.0, and 0.5 mg/mL, respectively. The eluent was 0.025 mol/L phosphate buffer solution with pH 7.0. The V_0 value was measured using a solution of bludextran (2000 kDa).

The sequence of work on columns with ion-exchange celluloses and gels is outlined in earlier publications [30].

23.2.3 PHOTOCHEMICAL OXIDATION OF ORGANIC SUBSTANCES

The metals concentration in the obtained groups of DOM was measured after their destruction by UV irradiation in an acidic medium, pH ≈ 1.0–1.5, with the addition of 0.1 mL of 30% H_2O_2. The UV treatment was accomplished in 50-mL quartz glasses for 2.0–2.5 hours using a DRT-1000 mercury-quartz lamp.

23.2.4 METHODS FOR DETECTING METALS CONCENTRATION

The Al(III) and Fe(III) concentrations were measured photometrically using the Chrome Azurol S [31] and o-phenanthroline [32] reagents. Mn(II), Cu(II), and Cr(III) concentrations were detected using chemiluminescent analysis methods [33–35]; Zn(II) and Pb(II) content was determined by anodic stripping voltammetry [36].

23.3 RESULTS AND DISCUSSION

The dissolved metals concentration in Ukraine's surface water bodies is within a wide range: $Al_{dissolved}$ – 4.2–580 μg/L, $Fe_{dissolved}$ – 4.9–1306, $Mn_{dissolved}$ – 5.2–1990, $Cu_{dissolved}$ – 2.3–57.9, $Zn_{dissolved}$ – 5.4–130.0, $Pb_{dissolved}$ – 0.7–28.6, and $Cr_{dissolved}$ – 2.7–48.0 μg/L. [24]. The maximum values are usually observed in water bodies with a high content of DOM, in particular HS. However, concentrations of some metals (Mn, Fe) increase significantly due to their migration from the bottom sediments of surface water bodies under dissolved oxygen deficiency and the formation of anaerobic conditions in the bottom layer.

23.3.1 LABILE FRACTION OF METALS

The labile fraction of metals includes free (hydrated) ions, hydroxocomplexes, and inorganic complex compounds, as well as organic complexes that are capable of dissociation depending on the environmental conditions, as already mentioned (*see* Figure 23.1). This form of metals is the most bioavailable and toxic to living organisms because it is able to penetrate the cell membrane due to its size. High-molecular organic compounds and metal complexes with them can't be transported through the cell membrane and therefore belong to indifferent compounds. However, such compounds can be assimilated by aquatic organisms from natural water after their destruction into low molecular weight compounds. Microorganisms, as well as solar radiation, can contribute to such destruction.

The share of the labile fraction of some metals in the studied water bodies is given in Figure 23.2. The largest share of the labile fraction is typical of Mn(II), as it is the least capable of binding with natural organic ligands. The degree of Mn(II) binding into complexes increases significantly in water with high water color values. This is confirmed by the results of studies of the rivers of the Pripyat River basin, as well as the Kyiv reservoirs, which differ from other reservoirs of the Dnieper cascade in terms of high concentrations of HS. The labile fraction of Cu(II) in the studied water bodies was the lowest, due to its active binding in complexes with natural organic compounds. Other metals, in particular Al(III), Fe(III), Zn(II), Pb(II), and Cr(III), like Cu(II), migrate in water mainly in the form of complex compounds. It should be noted that the share of the labile fraction of metals increases in water bodies with low HS concentration, as well as in rivers and reservoirs that are subject to significant anthropogenic impact (the so-called water bodies of the urban area).

It is important to note that the share of the labile fraction also significantly depends on the methods of metal detection. The most reliable results can be obtained using anodic stripping voltammetry and chemiluminescence methods. Overestimated results of labile Al(III) and Fe(III) concentration are obtained when using photometric methods.

23.3.2 PATTERNS OF THE METALS DISTRIBUTION AMONG COMPLEX COMPOUNDS WITH NATURAL ORGANIC SUBSTANCES

The HS, carbohydrates, and protein-like compounds are potential natural organic ligands that bind the metal ions into complexes. This is confirmed by

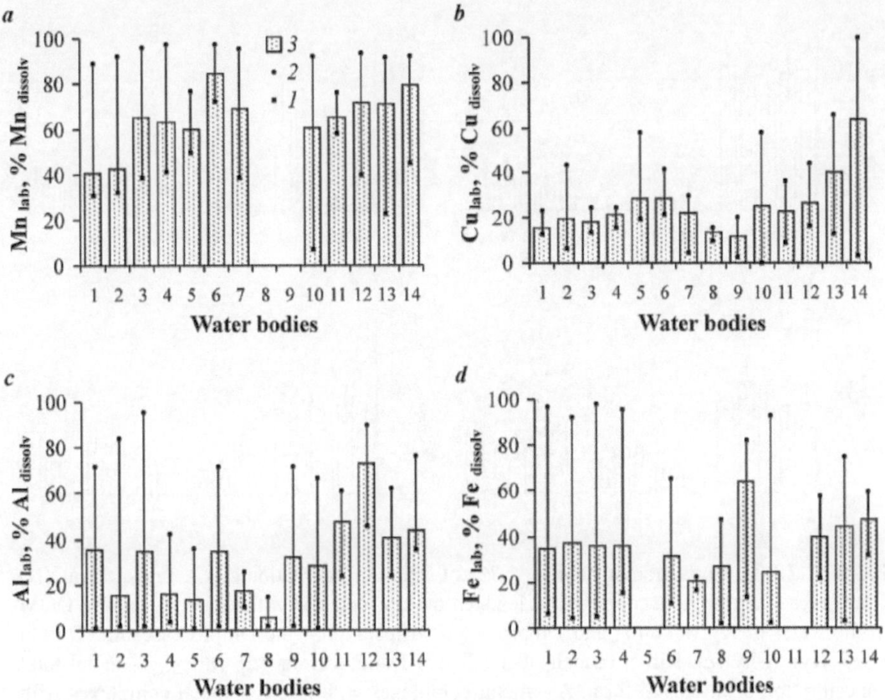

FIGURE 23.2 Share of the labile fraction of manganese (***a***), copper (***b***), aluminum (***c***) and iron (***d***) in Ukraine's surface water bodies. 1 – Prypyat River, 2, 3, 4, 5 – Kyivs'ke, Kanivs'ke, Dniprovs'ke, Kakhovs'ke reservoirs of the Dnipro cascade, 6, 7, 8, 9 – Danube, Desna, Teteriv, Girskyi Tikych rivers, 10 – upper Kytaivskyi Pond (Kyiv City), 11, 12, 13 – lakes Tel'bin, Verbne and lakes belonging to the Opechen' system (Kyiv City), 14 – Lybid' river (Kyiv City). Here and in Figure 23.3 and 23.4 – *1*, *2* – limit values, *3* – average values.

the convincing findings of our long-term studies of the component composition of DOM and the distribution of metals among complex compounds with various groups of natural organic substances. As an example, the data on the distribution of metals among anionic, cationic, and neutral complexes in various types of water bodies are given in Figure 23.3.

For this purpose, water bodies characterized by different HS concentrations in the water were selected. So, in particular, the HS concentration in the water of the Kyiv reservoir is 15.5–45.3 mg/L, of the Dniprovske reservoir – 9.4–18.2 mg/L, of Telbin Lake – 5.2–8.2 mg/L, and of the Kiliya Danube Delta – 5.0–7.3 mg/L.

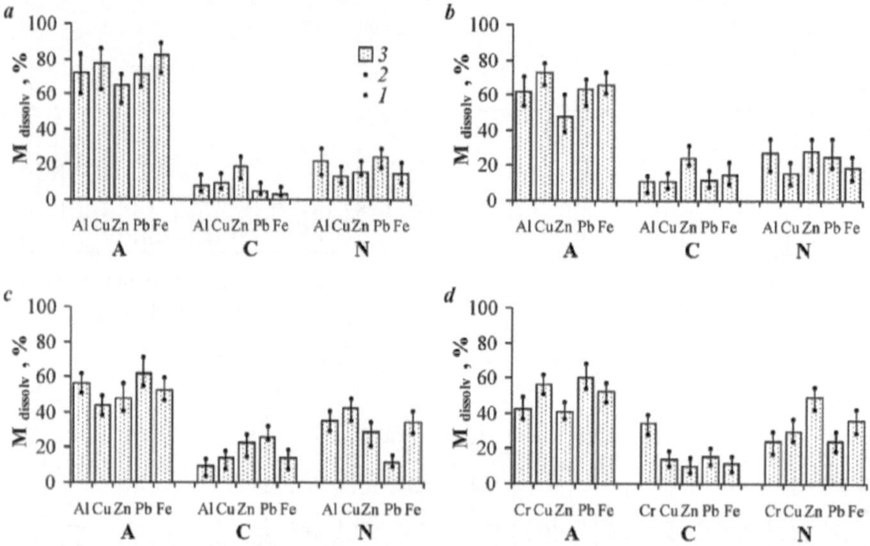

FIGURE 23.3 Mass fraction ($M_{dissolved}$,%) of anionic (A), cationic (C), and neutral (N) complexes of aluminium, copper, zinc, lead, chromium, and iron with various groups of DOM in the water of Kyivs'ke (***a***) and Dniprovs'ke (***b***) reservoirs (the Dnipro cascade), Tel'bin Lake, Kyiv City (***c***), Kiliya branch of the Kiliya Danube Delta (***d***). $M_{dissolved}$ – metal total concentration of dissolved form. A – metal complexes with HS, C – metal complexes with protein-like compounds, N – metal complexes with carbohydrates.

A number of previous studies of the metal complexes in surface waters [37, 38] have revealed the key role of the HS, as the most widespread group of the DOM, in metal binding. The predominant part of the dissolved metals in the water of the studied water bodies was found in the acidic (anionic) group of the DOM. In particular, the share of anionic complexes was, on average, 41.6–72.8% (Al), 42.0–71.8% (Cu), 40.5–65.7% (Zn), 59.5–72.8% (Pb), and 52–82% (Fe). It can be verified that with a decrease in the HS concentration, the share of anionic complexes of metals decreases. At the same time, the share of metals' neutral complexes with carbohydrates is increasing. This is especially true for small water bodies in an urbanized area, such as Tel'bin Lake (Figure 23.3). Carbohydrates concentration in the water of this lake in the summer reaches the maximum values – almost 4.0 mg/L. At the same time, it is known that carbohydrates in surface waters are able to bind metal ions into complexes [39, 40, 41]. In addition, there is evidence that polysaccharides can be used as adsorbents for the removal of metals from polluted waters [42]. If in the water of the Kyivs'ke reservoir the share of neutral complexes of metals with carbohydrates is on average 13.5–24% of $M_{dissolv}$, then in the water of Lake Telbin it reaches 12.5–35.4% of $M_{dissolv}$. The smallest share is made up of metal complexes with protein compounds. This is caused by the low content of PLS in the water of the studied water bodies. In addition, the fraction of cationic complexes may include free (hydrated) ions and hydroxocomplexes of metals with a positive charge.

The summarized data on the seasonal dynamics of the neutral complexes make it possible to conclude that the share of the complexed metal increases markedly in summer and autumn, when the carbohydrate concentration rises [43]. To some extent, this is confirmed by the findings of the Kanivs'ke Reservoir research (Figure 23.4). On the other hand, the share of zinc and lead neutral complexes, on the contrary, is noticeably higher in winter. It should be noted that such seasonal dynamics are sometimes poorly expressed, and it can be difficult to explain.

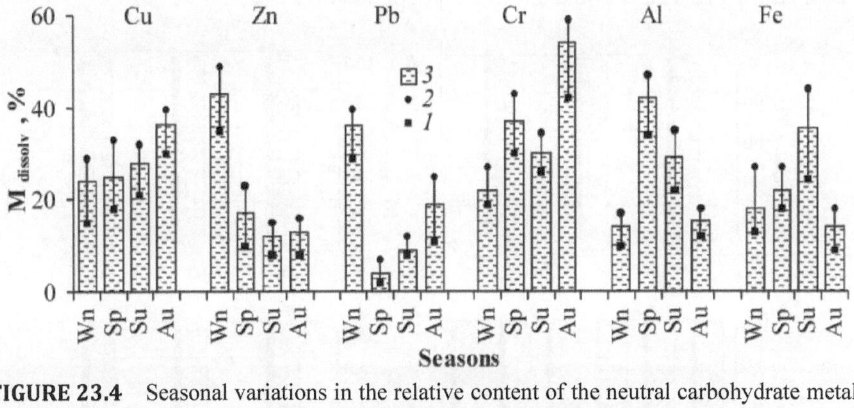

FIGURE 23.4 Seasonal variations in the relative content of the neutral carbohydrate metal complexes (% $M_{dissolv}$) in the upper part of the Kanivs'ke reservoir.

23.3.3 MOLECULAR WEIGHT DISTRIBUTION OF METAL COMPLEXES WITH VARIOUS GROUPS OF DOM

23.3.3.1 METAL COMPLEXES WITH HUMIC SUBSTANCES

The molecular weight distribution of anionic complexes requires special attention since HS plays a key role in binding metal ions in surface water bodies. The obtained research results are important for understanding the role of different fractions of HS in complexation, as well as for assessing the potential bioavailability for hydrobionts of complex metal compounds with different molecular weights.

The generalized data on the molecular weight distribution of the anionic metal complexes are given in figure 23.5. The distribution of metals among complex compounds with HS has some features depending on the concentration of this group of DOM in water. The majority of metals in the water of the Tsyr River were found in the composition of high molecular fractions of HS (> 20 and 20–5 kDa) – 58.4% of Al_{anion}, 58.7% of Fe_{anion}, and 83.8% of Cu_{anion}. This is due to the high concentration of HS, as a result of which their composition is dominated by high-molecular fractions. As the HS concentration decreases, the share of high-molecular complexes of metals with the above mentioned ligands also decreases. This is already becoming noticeable for the Kanivs'ke Reservoir, where the share of high molecular weight metal complexes (> 20 and 20–5 kDa) does not exceed 37–45% of M_{anion} (37.6% of Al_{anion}, 45.0% of Fe_{anion}, 44.0 of Cu_{anion}, 37.0% of Zn_{anion}, 38.3% of Pb_{anion}, and 41.2% of Cr_{anion}. HS concentration in the water of Tsyr River and Kanivs'ke Reservoir were 112,5 and 26,5 mg/L respectively.

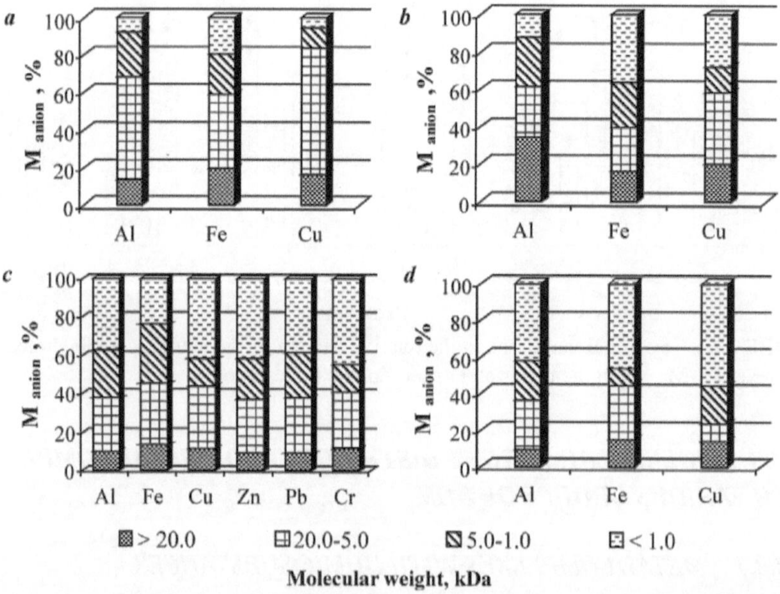

FIGURE 23.5 The share of anionic complexes of metals with different molecular weights (% M_{anion}) in the water of the Tsyr River (***a***), the Pripyat River (***b***), the upper part of the Kanivs'ke reservoir (***c***), and the Kiliya Danube Delta, Skhidnyi arm (***d***). $C_{HS:}$ ***a*** – 112.5 mg/L, ***b*** – 46.8 mg/L, ***c*** – 26.5 mg/L, ***d*** – 6.4 mg/L.

A similar distribution of metals among different fractions of HS is typical for other Ukraine's water bodies where the major fraction of the studied metals was detected in the HS fractions with the relatively low molecular weight (<1 and 1–5 kDa): 51.6–70.3% of $Al_{anionic}$, 55.6–74.0% of $Fe_{anionic}$, and 41.8–65.8% of $Cu_{anionic}$ ($Al_{anionic}$, $Fe_{anionic}$, and $Cu_{anionic}$ – total metal concentration in complexes with HS). For example, the data on the distribution of these metals among anionic complexes with HS in the water of the Pivdennyy Bug River and Desna River are given (Figure 23.6).

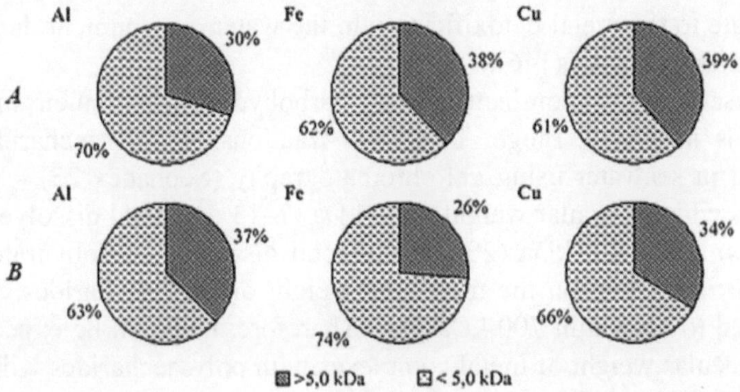

FIGURE 23.6 Relative content (on average values) of various metals' anionic complexes with HS in water of the Pivdennyy Bug River (***A***) and the Desna River (***B***), in% to the total concentration of each of the metals in the mentioned complexes' composition.

A significant part of dissolved metals in surface waters migrates as part of their complexes with HS. The transformation of DOM, including HS, under the effect of solar radiation is inevitably accompanied by changes in the molecular weight of complex metal compounds with them. This can be confirmed by the example of the Kanivs'ke Reservoir, where the transformation of humic substances takes place in summer due to the impact of solar radiation. The share of Fe(III) complexes with HS of relatively low molecular weight (≤ 5.0 kDa) increased to 63% in summer compared to the winter-spring period, when their share was 43% and 35%, respectively (Figure 23.7).

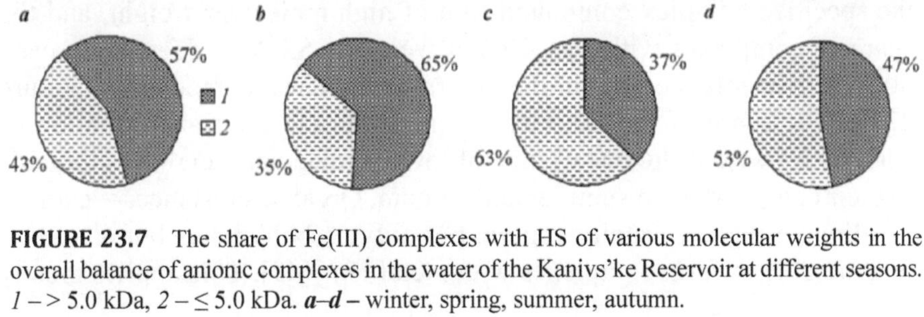

FIGURE 23.7 The share of Fe(III) complexes with HS of various molecular weights in the overall balance of anionic complexes in the water of the Kanivs'ke Reservoir at different seasons. *1* – > 5.0 kDa, *2* – ≤ 5.0 kDa. ***a–d*** – winter, spring, summer, autumn.

23.3.3.2 METAL COMPLEXES WITH CARBOHYDRATES

Polysaccharides, as protein and lipids are important components of extracellular polymeric substances secreted by microalgae cells. Polysaccharides make up from 45 to 95% of extracellular polymeric substances. An important characteristic of carbohydrates is their polyfunctionality, which usually results in several possible coordination sites for metal atoms [44, 45]. They contribute to the metal detoxification in the water environment due to their complexing properties [46].

Polysaccharides dominate among carbohydrates, and their molecular weight is in a wide range. Thus, two fractions of polysaccharides were detected in seawater using gel chromatography (Sephadex-25) – polysaccharides with molecular weight < 4.0 kDa (7–13% of total dissolved carbohydrates) and > 4.0 kDa (20–33% of total dissolved carbohydrates) [47]. There are also data on the molecular weight of polysaccharides, which is estimated to be within 700 kDa [48]. Therefore, it should be expected that the molecular weight of metal complexes with polysaccharides will also be characterized by a wide range of values.

Data on the molecular weight distribution of carbohydrates and Cu(II) complexes with carbohydrates in the water of the upper part of the Kanivs'ke Reservoir are given as an example in Figure 23.8. As expected, carbohydrates and their complexes with Cu(II) are characterized by a wide range of molecular weights. Polysaccharides were dominant, and the share of low molecular weight compounds increased only in autumn. Accordingly, complex compounds of Cu(II) with carbohydrates are also characterized by a wide range. The share of complex compounds with lower molecular weight (5–1 kDa and < 1 kDa) in the overall balance of neutral complexes increased noticeably in spring and autumn (43% and 53%, respectively).

The molecular weight distribution of metal complexes with carbohydrates (neutral fraction) follows a slightly different pattern. Most of the specified complex compounds are of high molecular weight, and the share of complexes with a molecular weight ≤ 5.0 kDa does not exceed 50% of their total content in the neutral complexes' composition (Figure 23.9). The share of the latter can increase due to the transformation of carbohydrates and the occurrence of lower molecular weight compounds in their composition in summer and autumn. Organic substances—exometabolites of algae—usually contain not only carbohydrates but also PLC that are able to form complexes with metal ions [49]. This applies mostly

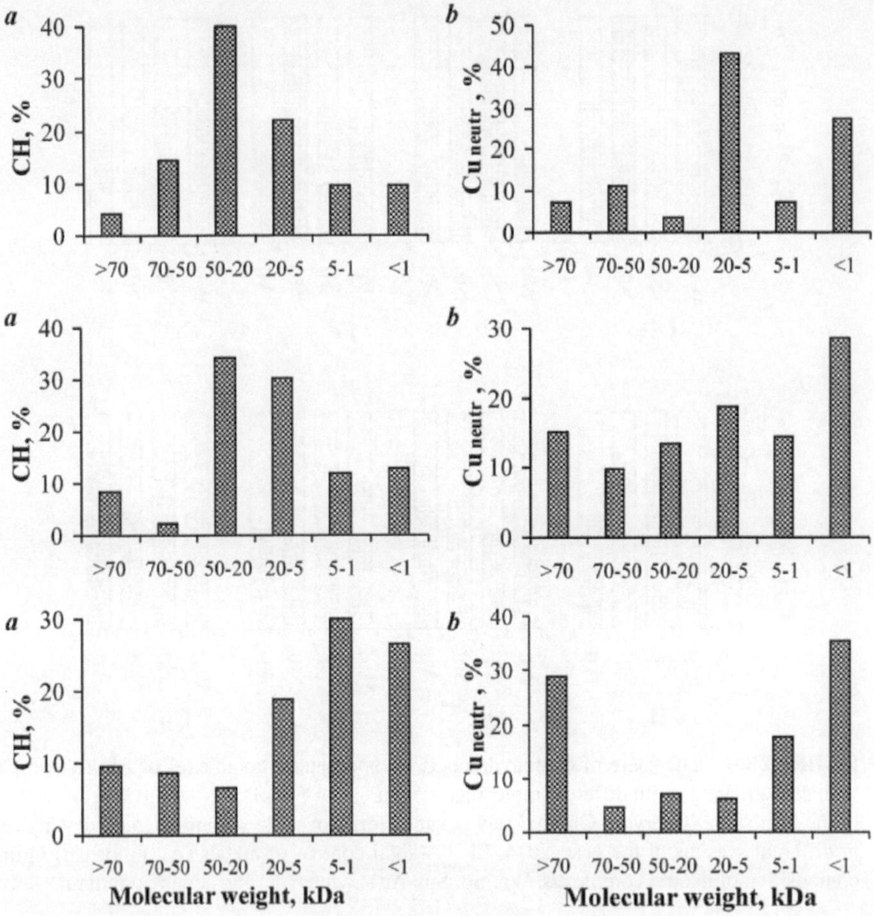

FIGURE 23.8 Molecular weight distribution of carbohydrates (***a***) and Cu(II) complexes with carbohydrates (***b***) in the water of the upper part of the Kanivs'ke Reservoir in different seasons of 2016. ***A*** – winter, ***B*** – spring, ***C*** – autumn. C_{CH}: ***A*** – 0.90 mg/L; ***B*** – 1.05 mg/L; ***C*** – 1.70 mg/L. C_{CH} – concentration of carbohydrates. $C_{Cu(neutral)}$: ***A*** – 10.3 µg/L; ***B*** – 12.4 µg/L; ***C*** – 7.5 µg/L.

to cyanobacteria [50]. The molecular weight range of PLC is known to be very wide [50, 51]. However, the molecular weight distribution of the corresponding metal complexes with these ligands hasn't been investigated due to their low concentrations in the water of the studied water bodies.

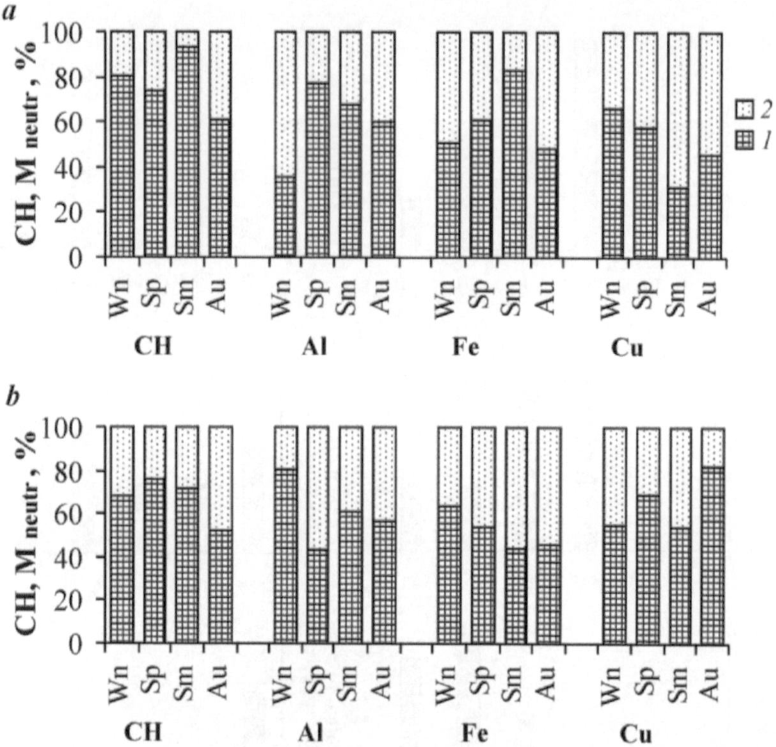

FIGURE 23.9 The share of carbohydrates (CH) and neutral complexes of Al, Fe, and Cu with carbohydrates with different molecular weight (1 – > 5.0 kDa, 2 – ≤ 5.0 kDa) in waters of the Kanivs'ke Reservoir, Obolon' Bay (***a***) and Horikhovats'kyy Pond no. 5, Kyiv City (***b***) in different seasons of the year, 2016. M_{neutr} – total content of metals (Al, Fe or Cu) in the composition of neutral complexes. Wn, Sp, Sm, Au – winter, spring, summer, autumn.

23.4 CONCLUSION

An important form of existence for metals in natural surface waters is complex compounds with various groups of dissolved organic matter (DOM) and different molecular weights. It is the complexation that contributes to the increase in the metals' migration mobility and their preferential presence in a dissolved state. This significantly affects metal toxicity and potential bioavailability. Among the dissolved organic substances in surface waters, humic substances (HS), carbohydrates, and protein-like substances can play an important role in the complexation of metals. HS play a primary role in complexation due to the fact that they dominate among other groups of DOM in most surface water bodies of Ukraine. However, in water bodies with a relatively low content of HS and high bioproductivity, the binding of metal ions into complexes also occurs with other DOM groups, particularly carbohydrates and, to a much lesser extent, protein-like compounds (PLC). It is known that carbohydrates and PLC belong to the group of readily oxidized substances. An increase in water temperature and microbial activity leads to the transformation of high-molecular compounds into compounds with a lower molecular weight, which are bioavailable to aquatic organisms. Metal complexes with such low molecular weight organic compounds become bioavailable.

The labile fraction of metals is the most bioavailable and toxic to living organisms because it is able to penetrate the cell membrane due to its size. This fraction includes free (hydrated) ions, hydroxocomplexes, and inorganic complex compounds, as well as organic complexes that are capable of dissociation under the influence of environmental conditions. The share of the labile fraction of metals increases in water bodies with low HS concentration, as well as in rivers and reservoirs that are subject to significant anthropogenic influence. Among the studied metals, the largest share of the labile fraction is observed for manganese, since it is characterized by the least complex binding. The share of the labile fraction varied widely from a few percent to almost 100%, and on average was within 15–84% of the dissolved metals concentration.

Metals in surface waters are mostly bound into complexes by humic substances. The share of anionic complexes was, on average, 41.6–72.8% (Al), 42.0–71.8% (Cu), 40.5–65.7% (Zn), 59.5–72.8% (Pb), and 52–82% (Fe). It was found that the share of anionic complexes of metals decreases when HS concentration decreases too. At the same time, the share of metals' neutral complexes with carbohydrates is increasing. The share of the investigated metals in neutral complexes with carbohydrates was 9–57%. The highest share of the neutral complexes is observed in water bodies which are characterized by a high bioproductivity and relatively low HS concentrations. An increase in the share of metal complexes with carbohydrates is usually observed in the spring-summer period. The smallest share consists of metal complexes with protein compounds, attributed to the low PLS content in the studied water bodies. Additionally, the cationic complexes may include free (hydrated) ions and positively charged metal hydroxocomplexes.

Complex compounds of metals with humic substances were mainly represented by complexes with a molecular weight of less than 5 kDa, the share of which varies between 42 and 74% of the total metal concentration in complexes with HS. It has been found that the proportion of metal complexes with humic substances with a molecular weight of less than 5 kDa can increase in summer, probably due to photochemical processes in surface waters. For example, the share of Fe(III) complexes with HS of relatively low molecular weight ($\leq$ 5.0 kDa) increased to 63% in summer compared to the winter-spring period, when their share was 43% and 35%, respectively. The molecular weight distribution of metals' neutral complexes with carbohydrates has a slightly different character. Most of the specified complex compounds are high molecular weight, and the share of complexes with a molecular weight $\leq$ 5.0 kDa, as a rule, does not exceed 50% of their total content in the neutral complexes' composition. The share of the latter can increase due to the carbohydrates' transformation and the occurrence of compounds with a lower molecular weight in summer and autumn. The molecular weight distribution of the metal complexes with PLC wasn't investigated due to their low concentrations in the water of the studied water bodies.

ACKNOWLEDGMENTS

The authors wish to thank Dr. Sc. Nataliia Semeniuk for editing the English translation.

KEYWORDS

- **carbohydrates**
- **complex compounds**
- **complexation**
- **dissolved organic matter**
- **humic substances**
- **metals**
- **molecular-weight distribution**
- **protein-like compounds**
- **surface water bodies**

REFERENCES

1. Town, R. M., & Filella, M. (2002). Size fractionation of trace metal species in freshwaters: Implications for understanding their behaviour and fate. *Reviews in Environmental Science and Bio/Technology, 1*, 277–297.

2. Namieśnik, J., & Rabajczyk, A. (2010). The speciation and physico-chemical forms of metals in surface waters and sediments. *Chemical Speciation and Bioavailability, 22*(1), 1–24.
3. Linnik, P. N., & Nabivanets, B. I. (1986). Forms of metals migration in fresh surface waters. Gidrometeoizdat Press.
4. Lu, J. Y., Chakrabarti, C. L., Back, M. H., Sekaly, A. L. R., Gregoire, D. C., & Schroeder, W. H. (1996). Speciation of some metals in river surface water, rain and snow, and the interactions of these metals with selected soil matrices. *Journal of Analytical Atomic Spectrometry, 11*, 1189–1201.
5. Pokrovsky, O. S., & Shirokova, L. S. (2013). Diurnal variations of dissolved and colloidal organic carbon and trace metals in a boreal lake during summer bloom. *Water Research, 47*(2), 922–932.
6. Tipping, E. (2004). *Cation Binding by Humic Substances*. Cambridge University Press.
7. Bungala, S., Machiwa, J., & Shilla, D. (2021). Distribution and speciation of heavy metals in water and sediments of the coastal marine areas of Tanzania. *Journal of Environmental Protection, 12*, 734–754.
8. Benson, N. U., Anake, W. U., & Olanrewaju, I. O. (2013). Analytical relevance of trace metal speciation in environmental and biophysicochemical systems. *American Journal of Analytical Chemistry, 4*, 633–641.
9. De Paiva Magalhães, D., da Costa Marques, M. R., Baptista, D. F., & Buss, D. F. (2015). Metal bioavailability and toxicity in freshwater. *Environmental Chemistry Letters, 13*(1), 69–87.
10. Väänänen, K., Leppänen, M. T., Chen, X. P., & Akkanen, J. (2018). Metal bioavailability in ecological risk assessment of freshwater ecosystems: From science to environmental management. *Ecotoxicology and Environmental Safety, 147*, 430–446.
11. Warren, L. A., & Haack, E. A. (2001). Biogeochemical controls on metal behavior in freshwater environments. *Earth-Science Reviews, 54*, 261–320.
12. Peakall, D., & Burger, J. (2003). Methodologies for assessing exposure to metals: Speciation, bioavailability of metals, and ecological host factors. *Ecotoxicology and Environmental Safety, 56*, 110–121.
13. Guéguen, C., Clarisse, O., Perroud, A., & McDonald, A. (2011). Chemical speciation and partitioning of trace metals (Cd, Co, Cu, Ni, Pb) in the lower Athabasca River and its tributaries (Alberta, Canada). *Journal of Environmental Monitoring, 13*, 2865–2872.
14. Osadchyy, V., Nabyvanets, B., Linnik, P., Osadcha, N., & Nabyvanets, Yu. (2016). *Processes determining surface water chemistry*. Springer International Publishing.
15. Adams, W., Blust, R., Dwyer, R., Mount, D., Nordheim, E., Rodriguez, P. H., & Spry, D. (2020). Bioavailability assessment of metals in freshwater environments: A historical review. *Environmental Toxicology and Chemistry, 39*(1), 48–59.
16. Markich, S., Brown, P. L., Batley, G. E., Apte, S. C., & Stauber, J. L. (2001). Incorporating metal speciation and bioavailability into water quality guidelines for protecting aquatic ecosystems. *Australasian Journal of Ecotoxicology, 7*(2), 109–122.
17. Glaus, M. A., Hummel, W., & Vanloon, L. R. (2000). Trace metal-humate interactions. I. Experimental determination of conditional stability constants. *Applied Geochemistry, 15*(7), 953–973.
18. Pandey, A. K., Pandey, S. D., & Misra, V. (2000). Stability constants of metal humic acid complexes and its role in environmental detoxification. *Ecotoxicology and Environmental Safety, 74*, 195–200.

19. Town, R. M., & Filella, M. (2000). Dispelling the myths: Is the existence of L1 and L2 ligands necessary to explain metal ion speciation in natural waters? *Limnology and Oceanography, 45*(6), 1341–1357.
20. Eshwar, M., Srilatha, M., Rekha, K. B., & Sharma, S. H. K. (2017). Complexation behavior of humic and fulvic acids with metal ions and their assessment by stability constants. *International Journal of Pure and Applied Bioscience, 5*(6), 899–907.
21. Makharadze, T., & Makharadze, G. (2021). Measurement of complex formation process of nickel (II) with freshwater fulvic acids using the solubility method. *Fine Chemical Engineering, 2*(2), 54–61.
22. Makharadze, T., & Makharadze, G. (2020). Investigation of complex formation process of copper with macromolecular organic substances, isolated from natural waters. *Organic Chemistry Plus, 1*(1), 1–6.
23. Dinu, M. I. (2015). Interaction between metal ions in waters with humic acids in gley–podzolic soils. *Geochemistry International, 53*(3), 265–276.
24. Linnik, P. N., Zhezherya, V. A., Linnik, R. P., Ignatenko, I. I., & Zubenko, I. B. (2015). Metals in surface water of Ukraine: The migration forms, features of distribution between the abiotic components of aquatic ecosystems, and potential bioavailability. *Russian Journal of General Chemistry, 85*(13), 2965–2984.
25. Strmečki, S., Plavšić, M., Steigenberger, S., & Passow, U. (2010). Characterization of phytoplankton exudates and carbohydrates in relation to their complexation of copper, cadmium and iron. *Marine Ecology Progress Series, 408*, 33–46.
26. Gouvêa, S. P., Vieira, A. A. H., & Lombardi, A. T. (2005). Copper and cadmium complexation by high molecular weight materials of dominant microalgae and of water from a eutrophic reservoir. *Chemosphere, 60*, 1332–1339.
27. Mostofa, K. M. G., Liu, C. Q., Feng, X., Yoshioka, T., Vione, D., Pan, X., & Wu, F. (2013). Complexation of dissolved organic matter with trace metal ions in natural waters. In K. M. G. Mostofa, T. Yoshioka, A. Mottaleb, & D. Vione (Eds.), *Photobiogeochemistry of Organic Matter* (pp. 769–849). Springer-Verlag.
28. Lamelas, C., Wilkinson, K. J., & Slaveykova, V. I. (2005). Influence of the composition of natural organic matter on Pb bioavailability to microalgae. *Environmental Science & Technology, 39*, 6109–6116.
29. Zhezherya, V. A. (2012). Patent 75995 Ukrayina, MPK51 (2012.01) G 01 N 1/00. Modyfikovany batometr-sklyanka [Modified water sampler-glass]. *Bulletin N 24*.
30. Linnik, P. N., Zhezherya, V. A., Ivanechko, Y. S., & Linnik, R. P. (2014). Humic substances and their role in migration of metals in the high colored surface waters: The case study of rivers of the Pripyat' River basin. *Russian Journal of General Chemistry, 84*(13), 2572–2587.
31. Savranskii, L. I., & Nadzhafova, O. Yu. (1992). Spectrophotometric investigation of a complex of copper, iron and aluminum with Chrome Azurol S in the presence of cationic and nonionogenic surfactants. *Zhurnal Analiticheskoy Khimiyi, 47*(9), 1613–1617.
32. Nabyvanets, B. Y., Osadchyy, V. I., Osadcha, N. M., & Nabyvanets, Yu. B. (2007). *Analytical Chemistry of Surface Waters*. Naukova Dumka Press.
33. Nabivanets, B. I., & Linnik, P. N. (1981). *Kinetic Methods for the Analysis of Natural Waters*. Naukova Dumka Press.
34. Linnik, P. N. (2003). Complexation as the most important factor in the fate and transport of heavy metals in the Dnieper water bodies. *Analytical and Bioanalytical Chemistry, 376*(3), 405–412.

35. Linnik, P. N., Leshchinskaya, A. A., & Nabivanets, B. I. (1989). Methodology for investigating coexisting forms of chromium in natural waters. *Hydrobiological Journal, 25*(2), 91–96.
36. Linnik, P. N., & Nabivanets, Yu. B. (1988). Application of stripping voltammetry method for determination of free and bound zinc and lead ions in natural waters. *Gidrobiologichesky Zhurnal, 24*(1), 68–71.
37. Hiraide, M. (1992). Heavy metals complexed with humic substances in fresh water. *Analytical Sciences, 8*, 453–459.
38. MacCarthy, P., & Perdue, E. M. (1991). Complexation of metal ions by humic substances: Fundamental considerations. In *Interactions at the Soil Colloid–Soil Solution Interface* (pp. 469–489). Springer.
39. Alekseev, Yu. E., Garnovskii, A. D., & Zhdanov, Yu. A. (1998). Complexes of natural carbohydrates with metal cations. *Russian Chemical Reviews, 67*, 649–669.
40. Linnik, P. N., Zhezherya, V. A., & Linnik, R. P. (2017). Role of neutral fraction of dissolved organic matter in the migration of metals in surface waters: II. Neutral metal complexes in water bodies of different types. *Russian Journal of General Chemistry, 87*(13), 3233–3243.
41. Kaplan, D., Christiaen, D., & Arad, S. (1987). Chelating properties of extracellular polysaccharides from *Chlorella* spp. *Applied and Environmental Microbiology, 53*(12), 2953–2956.
42. Joly, N., Ghemati, D., Aliouche, D., & Martin, P. (2020). Interaction of metal ions with mono- and polysaccharides for wastewater treatment: A review. *Natural Products Chemistry & Research, 8*(3), Article 373, 17 pp.
43. Linnik, P. N., Zhezherya, V. A., & Zhezherya, T. P. (2021). Features of the hydrochemical regime of water bodies of urbanized areas in summer: I. Biogenic and organic substances. *Russian Journal of General Chemistry, 91*(13), 2805–2816.
44. Allscher, T., Klüfers, P., & Mayer, P. (2008). Carbohydrate-metal complexes: Structural chemistry of stable solution species. In B. Fraser-Reid, K. Tatsuta, & J. Thiem (Eds.), *Glycoscience* (Chapter 4.7, pp. 1077–1139). Springer-Verlag.
45. Hartinger, C. G., Nazarov, A. A., Ashraf, S. M., Dyson, P. J., & Keppler, B. K. (2008). Carbohydrate-metal complexes and their potential as anticancer agents. *Current Medicinal Chemistry, 15*(25), 2574–2591.
46. Babiak, W., & Krzemińska, I. (2021). Extracellular polymeric substances (EPS) as microalgal bioproducts: A review of factors affecting EPS synthesis and application in flocculation processes. *Energies, 14*, Article 4007, 23 pp.
47. Sakugawa, H., & Handa, N. (1983). Chemical studies of dissolved carbohydrates in seawater. Part 1. The concentration and separation of dissolved carbohydrates. *Journal of the Oceanographical Society of Japan, 39*, 279–288.
48. Magaletti, E., Urbani, R., Sist, P., Ferrari, C. R., & Cicero, A. M. (2004). Abundance and chemical characterization of extracellular carbohydrates released by the marine diatom *Cylindrotheca fusiformis* under N- and P-limitation. *European Journal of Phycology, 39*, 133–142.
49. Tu, X., Xu, P., Zhu, Y., Mi, W., & Bi, Y. (2023). Molecular complexation properties of Cd^{2+} by algal organic matter from *Scenedesmus obliquus*. *Ecotoxicology and Environmental Safety, 263*, Article 115378, 10 pp.

50. Guo, T., Yang, Y., Liu, R., & Li, X. (2017). Enhanced removal of intracellular organic matters (IOM) from *Microcystis aeruginosa* by aluminum coagulation. *Separation and Purification Technology, 189*, 279–287.
51. Linnik, P. N., & Ivanechko, Y. S. (2015). Dissolved protein-like substances in surface water bodies of various types. *Hydrobiological Journal, 51*(2), 85–104.

CHAPTER 24

Hexagonal 2D-Materials Intercalation by Magnetic Clusters

SHIO MAKATSARIA,[1,2] LEVAN CHKHARTISHVILI,[1,3] OTAR TSAGAREISHVILI,[3] ROIN CHEDIA,[3,4] VAKHTANG GABUNIA,[4] SHALVA KEKUTIA,[5] JANO MARKHULIA,[5] VLADIMER MIKELASHVILI,[5] TAMAR DUNDUA,[6] and IRMA JINIKASHVILI[7]

[1]*Department of Engineering Physics, Georgian Technical University, Tbilisi, Georgia*

[2]*Deltamed Georgia, Official Representatives of Siemens Healthcare Diagnostics in Georgia, Tbilisi, Georgia*

[3]*Semiconducting and Powder Composite Materials Laboratory, Ferdinand Tavadze Metallurgy and Materials Science Institute, Tbilisi, Georgia*

[4]*Petre Melikishvili Institute of Physical and Organic Chemistry, Ivane Javakhishvili Tbilisi State University, Tbilisi, Georgia*

[5]*Vladimer Chavchanidze Institute of Cybernetics, Georgian Technical University, Tbilisi, Georgia*

[6]*Faculty of Agricultural Sciences and Biosystems Engineering, Georgian Technical University, Tbilisi, Georgia*

[7]*Department of Medical Chemistry, Tbilisi State Medical University, Tbilisi, Georgia*

Advanced Topics in Polymer Chemistry and Materials Science: Tamara Tatrishvili and Marc Jean Médard Abadie (Eds)

ABSTRACT

Iron(0) pentacarbonyl $Fe(CO)_5$ is widely used to obtain magnetic materials containing ultra-dispersed iron for different functional applications. By its impregnation and subsequent transformation, it becomes possible to obtain nano-zero valent iron and iron oxides supported in various (organic and inorganic) matrices. Previously, we carried out the chemical transformation of $Fe(CO)_5$ impregnated in porous sorbents and biopolymers. The aim of this work was to intercalate magnetic clusters, such as ferromagnetic free iron Fe and its ferrimagnetic oxide – magnetite Fe_3O_4, in hexagonal 2D materials, such as graphene G, graphene oxide GO, reduced graphene oxide rGO, and hexagonal boron nitride h-BN. The structural similarity of these compounds determines the similarity of their chemical properties. h-BN:Fe, h-BN:Fe_3O_4, rGO:Fe, and rGO:Fe_3O_4 composites were obtained by the thermal treatment of h-BN–$Fe(CO)_5$, h-BN–$Fe(CO)_5$–H_2O, GO–$Fe(CO)_5$, and GO–$Fe(CO)_5$–H_2O systems in an autoclave at 200°C. h-BN doped with Fe or Fe_3O_4 can serve as delivery agents for boron-^{10}B isotope atoms to cancer cells. Nano powders of pure h-BN and h-BN:Fe and h-BN:Fe_3O_4 composites were obtained, and their chemical and phase compositions, crystal structure, and morphology were investigated. In the case of thermal treatment of the GO–$Fe(CO)_5$–H_2O system in an autoclave, rGO was obtained from GO, on which magnetite was deposited, with a particle size of 20–60 nm. In this case, exfoliation of GO and formation of bulky rGO:Fe_3O_4 composite powder does not occur, as it happens with vacuum-thermal treatment of GO–$Fe(NO_3)_3$ complex at 130°C. Fourier- Transform-infrared spectra of the obtained nanocomposites were recorded. Magnetization curves at room temperature were also measured for composites containing a magnetic phase, and the numerical values of their main parameters (saturation and remnant magnetizations and coercive force) were established. It is concluded that the nanocomposite powder of hexagonal boron nitride doped with magnetite nanoclusters h-BN:Fe_3O_4 will be an effective nano-agent for delivering boron-10 (^{10}B) isotope atoms to tumor cells in Boron Neutron Capture Therapy (BNCT).

24.1 INTRODUCTION

Magnetic nanoparticles delivering therapeutically active agents consist of two components: a chemically functional constituent (biomolecules such as peptides, aptamers, antibodies, as well as chemical therapy drugs, nucleic acids, radionuclides, etc.) and magnetic materials (iron, nickel, cobalt, their oxides, etc.). Among them, nanopowder boron compounds, and in particular, hexagonal boron nitride (h-BN) doped with ferromagnetic clusters, may find interesting applications in cancer BNCT (Boron Neutron Capture Therapy) [1]. The physical mechanism of BNCT involves accumulating a critically high concentration of boron-10 (^{10}B) isotope atoms in the tumor cells and then irradiating the patient with an appropriate dose of thermal neutrons. Such accumulation can be enhanced by applying an external magnetic field targeting the tumor, directing the ^{10}B isotope atoms contained in boron nitride magnetic particles.

Exfoliation of h-BN-like 2D materials such as graphite, several-layer graphene and its oxides, molybdenum disulfide, tungsten disulfide, metal carbides, carbonitrides, and so-called MXenes are well studied. It is possible to conduct this by vacuum, ultrasound, microwave, thermal, intercalation, and other methods. In the case of h-BN, the simplest method of separation and activation is to heat the material to a high temperature (900–1000°C) in air. As a result of partial oxidation, the B–OH functional groups are formed. This helps to increase the distance between neighboring hexagonal BN layers and, in this way, facilitates the intercalation of other phases.

The structural similarity of h-BN or "white graphite" with several-layer graphene makes it possible to obtain their composites by similar methods. This circumstance was also taken into account when developing the methods of depositing component phases on h-BN, graphene, and graphene oxide surfaces. We suggested [2, 3] several methods of deposition and intercalation of magnetic clusters on hexagonal boron nitride, which were based on different chemical processes: (1) Obtaining h-BN–Fe composite by reduction of $FeSO_4$ through sodium borohydride; (2) Preparation of h-BN–Fe composite by reduction of h-BN–Fe_2O_3 composite with hydrogen according to the scheme: Fe^{+3} ? $Fe(OH)_3$? FeOOH ? Fe_2O_3 ? Fe; (3) Obtaining the h-BN–Fe_3O_4 composite by co-precipitation of iron(II) and iron(III) compounds in the presence of boron nitride on the basis of chemical reaction: $FeSO_4$ + $2FeCl_3$ + $8NH_4OH$? Fe_3O_4 + $(NH_4)_2SO_4$ + $6NH_4Cl$ + $4H_2O$; (4) Iron deposition on h-BN using iron(O) pentacarbonyl; and (5) Obtaining h-BN–Fe_3O_4 composite using iron(O) pentacarbonyl.

Due to the structural-morphological features of hexagonal boron nitride, graphene, and its oxides, it is possible to intercalate them with magnetic clusters or coat their surface using the same chemical processes. These methods are used in works [4, 5]. Recently, a lot of attention has been paid to obtaining hybrid composites containing graphene oxide and studying their biocidal properties. Biocidal properties can be enhanced by coating nanosized metal particles or their compounds (especially oxides) [6–11].

Previously, the ability of graphene oxide to exfoliate in a vacuum was used to intercalate nanosized metals and their oxides into it. Based on graphite, graphite foil, and its wastes, graphene and its oxides were obtained. A new method of obtaining composites containing nanosized metallic silver and copper, as well as copper, iron, and titanium oxides, based on vacuum-thermal exfoliation of graphene oxide complexes at 100–130°C, was proposed in [12–14].

The present work aims to develop methods for obtaining composites of the above-mentioned 2D materials with a hexagonal structure doped with magnetic clusters and their characterization.

24.2 EXPERIMENTAL

In this section, we describe the starting materials, the technological equipment used, and the measuring apparatus, as well as the methods of synthesizing 2D-material–magnetic nanocomposites consisting of graphene (G), graphene oxide (GO), reduced graphene oxide (rGO), or hexagonal boron nitride (h-BN) on one hand, and elemental iron (Fe) and magnetite (Fe_3O_4) on the other hand.

Organic compounds and polymers (carbohydrates, glucose, sucrose, cellulose, polyvinyl alcohol, polyvinylpyrrolidone, etc.) were used as a carbon source (both for carbidization and as a reducing agent), which pyrolysis produced so-called carbon black (activated amorphous carbon). Natural flake graphite and its powder were used to obtain GO, rGO, and finally, G. Graphite flake, natural, 325 mesh – 99.98% (on a metallic basis), was purchased from Alfa Aesar. By grinding graphite foil, a powder with a particle size of 140 μm was obtained, which was also used to obtain GO.

Amorphous boron and boric acid were used to obtain boron nitride. Ammonia, ammonium chloride and ammonium hydroxide were used as nitrogen sources. As iron sources, its compounds such as $FeSO_4{\cdot}7H_2O$, $FeCl_3{\cdot}6H_2O$, $Fe(NO_3)_3{\cdot}9H_2O$ and $Fe(CO)_5$ were used. Their reduction was performed with sodium borohydride $NaBH_4$ purchased from Sigma Aldrich. $KMnO_4$, $NaNO_3$, H_2SO_4 (98%) and HCl (37%) were purchased from Sigma Aldrich. An inert environment was created in the reaction area using nitrogen and argon.

The degree of purity of chemical reagents and solvents used for synthesis reached 99.00–99.99%. Reagents purchased from Sigma Aldrich were used without prior purification. Powders' morphology and microstructure were studied with a SEM (Scanning Electron Microscope) JEOL–JSM 6510 LV equipped with an energy dispersive analyzer (Dispersive Micro-X-ray Spectral Analyzer X-MaxN, Oxford Instruments), through which the elemental composition of magnetic composites was determined. XRD (X-Ray Diffraction) patterns were obtained with DRON–3M (Cu Kα, Ni filter, 2°/min) and XZG–4 (Cu Kα, $\lambda = 1.5418$ Å) diffractometers. Then, the powder particle sizes were determined by the Scherer method. Thermal treatment (<1500°C) of GO and rGO was implemented in a high-temperature vacuum furnace Kejia. High-temperature vacuum furnaces STG–100–17 (1700°C) and Kejia were used for boron nitride synthesis and iron(III) chloride intercalation. In the same furnaces, metal oxides deposited on boron nitride were reduced in a hydrogen flow.

For the ultrasound treatment and homogenization of suspensions, an ultrasonic cleaner (45 kHz) and JY92–IIDN Touch Screen Ultrasonic Homogenizer (20–25 kHz, 900 W) were used. GO and rGO particle sizes were determined by a photon correlation nanoparticle size analyzer, Winner 802 DLS, and Malvern Instruments Mastersizer. The specific surface area of the composite powder was measured using a Micromeritics Gemini VII Instrument.

FTIR (Fourier Transform InfraRed) spectra were recorded with an Agilent Cary 630 spectrometer (350–5000 cm^{-1}).

VSM (Vibrating Sample Magnetometer) Lake Shore 7300 was used for the general characterization of the magnetic state of the obtained material by measuring its magnetic properties, namely, magnetization. Information on saturation and remnant magnetizations, coercive force, etc., was obtained. Superparamagnetic properties of nanostructured materials doped with magnetic particles were also determined. Several versions of the chemical synthesis method are elaborated for G, GO, rGO, h-BN, and their composites with Fe and Fe_3O_4.

The synthesis of GO was carried out by modifying the method known in the literature. The method is based on adding concentrated sulfuric acid to a mixture of expanded graphite and potassium permanganate powders. We used graphite plate powder instead of expanded graphite powder. The synthesis was carried out as follows: 1.5 g of finely ground $KMnO_4$ was added to 0.5 g of graphite plate powder, and the mixture was placed in a 250 ml flask. To the mixture cooled to 0°C, 10 ml of 98% H_2SO_4 was added in one portion. The mixture was stirred for 1 h. After that, the temperature of the reaction mixture was increased to 40°C and stirred again for 1.5 hour. 50

g of ice was added to the flask and heated to 70°C. Two drops of 30% H_2O_2 were added to the reaction mixture.

The resulting yellow suspension was stirred for 1 h. Separation of GO from the suspension was performed by centrifugation.

For vacuum exfoliation of GO and obtaining rGO, 0.5 g of GO plates were cut into small pieces (2–4 mm) and placed in a 1 l glass flask. After vacuuming the flask, it was heated at a rate of 5°C/min to 220–250°C. In this temperature range, there was a volumetric exfoliation of the GO plates and the formation of a fluffy black powder. Vacuuming was continued for 1 h, and then the flask was cooled under vacuum. The bulk density of the obtained powder reached 20–30 mg/ml.

All the operations of obtaining rGO–Fe composite described below were carried out in a fume cupboard with full compliance with safety rules because iron carbonyl is a strong poisonous substance. 1.5 g of graphene oxide was placed in a Teflon test tube, and 1.5 ml of iron (O) pentacarbonyl was added. The test tube was placed in a 0.5 l iron reactor – Teflon lined autoclave. The autoclave was heated at a rate of 50°C/min and kept at 230°C for 2 h. After that, the autoclave was connected to the vacuum system and vacuumed for 2 h at 120–140°C (2–4 mmHg). A black magnetic powder was obtained, which was stored in a desiccator. The process of obtaining the rGO–Fe_3O_4 composite was carried out by a similar route, with the difference that water was additionally introduced into the autoclave. The composition of the reaction mixture was rGO – 0.3 g, $Fe(CO)_5$ – 0.3 g, and H_2O – 0.1 ml. When obtaining G from rGO, the initial rGO contained 15–17 wt.% oxygen because it consists of various types of oxygen-containing functional groups. Their removal was performed by heating the rGO to high temperatures (>1000°C) in a vacuum under an inert gas or hydrogen. Graphene was obtained through heating at 1000–1200°C for 5 h in a vacuum. Graphene obtained by this method contained 2–5% oxygen. Graphene with a defective structure was obtained by a chemical method.

For obtaining G–Fe_3O_4 composite using iron(O) pentacarbonyl, 0.3 g of fluffy graphene powder is placed in teflon test tube and 0.3 ml of iron(O) pentacarbonyl was added. The mixture was stirred with a magnetic stirrer under argon for 1 hour. 0.1 mL of water was added to the reaction mixture, and then the test tube was placed in a 0.5 L iron reactor in the form of a Teflon-lined autoclave. The reaction was carried out at 230°C for 2 hours. After that, the autoclave was connected to the vacuum system and vacuumed for 2 hours at 120–140°C (2–4 mmHg). A black magnetic powder was obtained, which was stored in a desiccator.

The exfoliation and activation of boron nitride are often used to obtain various inorganic and polymeric composites. During exfoliation, there is a partial breakdown of the crystal lattice and its functionalization, as a result of which it is easier to intercalate boron nitride with different compounds. Currently, many methods for the exfoliation of 2D compounds have been developed. Among them, the thermal exfoliation method is the simplest in terms of technical performance. Thermal exfoliation and activation of h-BN were carried out at 900–1000°C in air, as a result of which its surface was functionalized. The process was carried out as follows: 15 g of h-BN was placed in a corundum jar, which was heated in air for 2 hours. The mass increase due to the oxidation of the sample was 16–18%. After the sample cooled, the activated h-BN was milled in a nanomill, and the powder was boiled in ethyl alcohol to remove the boron oxide formed during the process. Before starting the intercalation with iron compounds, the powder was preground in an ultrasonic homogenizer.

To obtain h-BN–Fe composite, 100 ml of 50% alcohol solution, 5 g of activated h-BN powder, and 5 g of $FeSO_4·7H_2O$ were placed in 250 ml three-necked flask equipped with thermometer, gas tube, and dropping funnel. The reaction mixture was cooled with ice water. Argon was pumped into flask, and after 30 min of stirring, a 0.5 M $NaBH_4$ solution in ethanol was added drop-wise. Molar ratio Fe^{+2}:$NaBH_4$ = 1:10. Reduction time was 120 min. The obtained black precipitate was attached to the bottom of the flask with a magnet and washed three times with ethanol. Then, the suspension was filtered under an argon stream and dried in a vacuum (2–4 mmHg) at 70°C for 4 hours.

The method of obtaining the h-BN–Fe_3O_4 composite by co-precipitation of iron(II) and iron(III) compounds in the presence of activated h-BN is based on the chemical reaction: $FeSO_4 + 2FeCl_3 + 8NH_4OH \rightarrow Fe_3O_4 + (NH_4)2SO_4 + 6NH_4Cl + 4H_2O$. For its realization, 2.8 g of activated h-BN, 0.7 g of $FeSO_4·7H_2O$, and 1.3 g of $FeCl_3·6H_2O$ were placed in a 250 ml three-necked flask equipped with a gas tube, thermometer, and dropping funnel. Argon was pumped into the flask, and 100 ml of freshly distilled water was added. The suspension was heated to 70°C, and after 30 minutes, 15 ml of 25% ammonium hydroxide (NH_4OH) solution was added dropwise. The resulting black suspension was stirred for 1 hour at 75°C. The reaction mixture was cooled to room temperature. The obtained black precipitate was attached to the bottom of the flask with a magnet and washed three times with water and ethanol. Then, it was filtered in an argon stream, and the obtained precipitate was washed again with anhydrous ethanol. The wet mass was dried in a vacuum at 120°C (6 hours, 2–4 mmHg). The brownish-black powder of h-BN–Fe_3O_4 was obtained and kept in a desiccator.

24.3 G-, GO- AND RGO-MATRIX COMPOSITES

Let's start with considering the processes of obtaining G and its reducing from GO. GO consists of oxygen-containing functional groups. FTIR data showed that GO samples contain OH, C=O, CO, COH, C=C and some other functional groups: OH – 3200–3700, CH_2 – 2924 and 2854, aromatic ring C=C – 1641, C=O – 1744, epoxy C–O – 1213 cm^{-1}, etc. Figure 24.1 shows XRD patterns of GO obtained by the method of oxidation from the system H_2SO_4–$KMnO_4$–graphite.

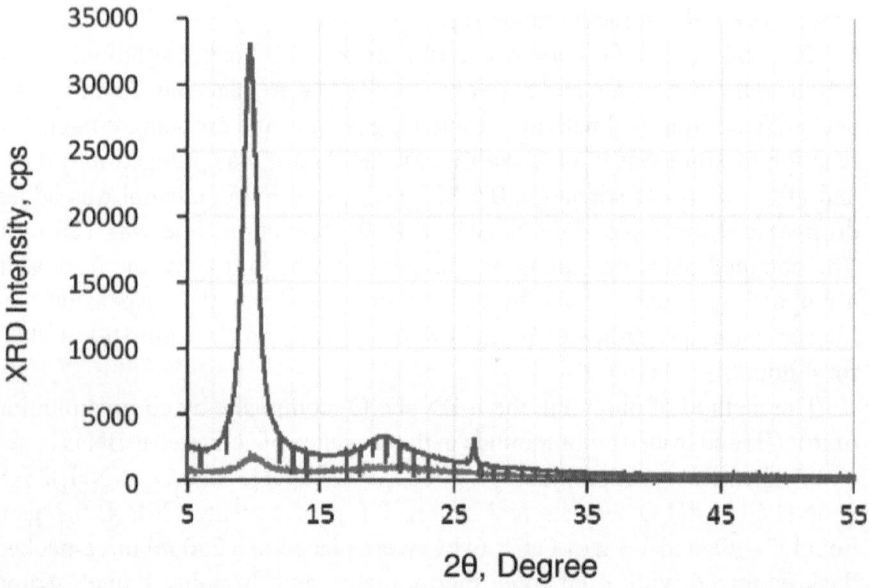

FIGURE 24.1 XRD patterns of GO obtained by oxidation with adding H_2SO_4 to $KMnO_4$–graphite mixture (Upper) and $KMnO_4$ powder to the H_2SO_4–graphite mixture (lower).

When GO is heated, water molecules adsorbed on it are removed. At 200–230°C, it undergoes a fundamental transformation and gives the rGO phase. With increasing temperature (200–1000°C), the GO diffraction maximum at 2θ = 10–12.5° disappears, and new peaks at 2θ = 20.03–26.16° appear, which correspond to rGO and G with a defective structure.

EDX (Energy Dispersive X-ray) analysis determined that the GO sample contained 27–33 w/w% oxygen. When heated to 250°C, due to the partial removal of oxygen-containing functional groups, rGO was formed with an oxygen content of 15–17%. However, as the temperature was raised to 1000°C, its content in the resulting graphene decreased to 3–5%. The geometrical shape of GO plates did not change when heated in air or any gas stream. When samples were heated in a vacuum (0.01 MPa, 220–250°C), instant expansion and the formation of a bulky powder occurred. The thermal treatment of the resulting powder at 1000°C yielded graphene with a defective structure. Graphene obtained by the chemical method had a defective structure with a peak at 2θ = 26.16°. According to the EDX spectrum, graphene obtained under these conditions consisted only of carbon and oxygen in a wt.% ratio ranging approximately from 20:1 to 29:1.

At the next stage of the research, we used the ability of GO to expand in vacuum for the intercalation of iron and iron oxides. The paper [14] describes in detail the methods of receiving biocidal composites containing nanosized particles of metallic silver, copper, and iron and titanium oxides intercalated in rGO. The technology of obtaining composites containing graphene layers is described in the work [15], where the method of exfoliating graphene oxide in vacuum is used as well.

As GO consists of oxygen-containing functional groups and forms complex bonds with metal ions, it undergoes vacuum separation and forms composites containing metal oxides. At the initial stage, we synthesized the GO–$Fe(NO_3)_3$ complex. Its vacuum decomposition led to the formation of the non-magnetic (namely, antiferromagnetic) composite rGO–Fe_2O_3. The composite was obtained in vacuum at a temperature of 120–135°C. According to the EDX analysis, the rGO–Fe_2O_3 composite contained 9.7 wt.% Fe, although the diffractogram did not show the Fe_2O_3 phase because it was amorphous [14]. Nitrogen oxides, oxygen, and water were released during the thermal treatment of the GO–$Fe(NO_3)_3$ complex. A high pressure developed between the layers of GO included in the nitrate complexes, which led to its decomposition into nanosized plates, and, accordingly, a powder with a larger surface area was obtained. It is established that the characteristic peak of GO registered at the 2θ = 10.07–12.30° range corresponds to the (001) plane.

According to the literature, during the reduction of GO to rGO by different methods, the maxima of diffraction peaks appear at different values of 2θ between 20.0 and 26.64° [16, 17]. By XRD analysis, it is established that during the decomposition of the GO–$Fe(NO_3)_3$ complex, rGO was formed with the XRD peak at 2θ = 24–25°.

Partial reduction of the rGO–Fe_2O_3 composite obtained by vacuum exfoliation (400–450°C) with hydrogen resulted in the magnetic composite rGO–Fe_3O_4 and complete reduction of rGO–Fe.

Magnetic composites on graphene and its oxides were obtained using iron(0) pentacarbonyl. The method is based on the decomposition of $Fe(CO)_5$ in a closed system in the presence of GO, rGO, and G. By decomposing iron carbonyl, zero-valent nanoiron was obtained, which was intercalated between the GO (or G) layers, and the rGO–Fe magnetic composite was obtained. If the same process is carried out in the presence of water, the rGO–Fe_3O_4 composite is formed.

Previously, similar technology was used to deposit nano-zerovalent iron and magnetite phases on cellulose, wood, biochar and biopolymers. As described in the experimental section, the rGO–Fe–Fe_3O_4 composite was obtained by the interaction of GO and $Fe(CO)_5$ at 230°C in an autoclave. It is established that GO under pressure does not undergo expansion and the plates retain their original geometric shapes containing the metallic iron phase. It is seen from the XRD pattern, in this case both iron and magnetite phases were formed. GO's diffraction maximum at $2\theta = 11.37°$ disappeared, but new peaks $2\theta = 20.03–24.98°$ appeared instead. They correspond to rGO.

XRD pattern of the rGO–Fe–Fe_3O_4 composite product was compared with that of GO, and not rGO, since the starting material was GO with the addition of iron carbonyl $Fe(CO)_5$, and it is impossible to obtain an XRD pattern of the rGO–$Fe(CO)_5$ mixture because, at room temperature, $Fe(CO)_5$ is a liquid substance (and at the same time very toxic). The absence of a GO maximum in the lower XRD pattern allows us to conclude that it is transformed into rGO and forms the target composite product rGO–Fe–Fe_3O_4.

The rGO–Fe_3O_4 composite was obtained by a similar procedure, with the difference that water is added to the reaction zone. Since the GO does not expand in the autoclave, it is better to use pre-exfoliated rGO to obtain rGO–Fe_3O_4 composite powder.

rGO–Fe_3O_4 consists of separate plates deposited with the magnetite phase, the amount of which in the composite can be varied within wide limits, and it depends on the rGO: $Fe(CO)_5$ molar ratio. The elemental composition of one of the rGO–Fe_3O_4 composite samples determined by EDX analysis is: C – 74.19, O – 22.17, and Fe – 3.64 wt.%. The SEM micrograph of the obtained composite rGO–Fe_3O_4 and EDX mapping of elements distribution in this composite show that iron is evenly distributed in the sample bulk, which is an advantage of the used method compared to the co-precipitation one.

By a similar method, the magnetite phase was impregnated in activated graphene. The relevant procedure was described in the experimental section.

rGO was obtained by vacuum exfoliation of GO, which consists of transparent plates. By heating it in a vacuum at 1000°C, defective graphene of the same morphology was formed. Graphene consists of thin plates, the size of which reaches 1360 nm.

After depositing the magnetite phase on it, the size of the plates was reduced by two times, and the average size reached 710 nm.

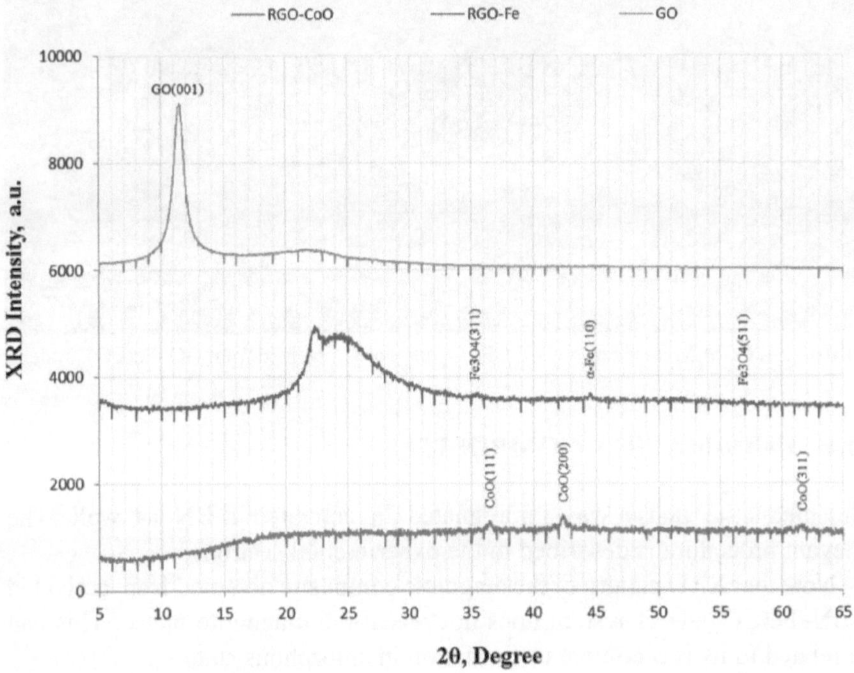

FIGURE 24.2 XRD patterns of GO (Upper), and rGO–Fe–Fe_3O_4 (Middle) and also rGO–CoO composites (Lower) obtained, respectively, by vacuum exfoliation of GO–$Fe(NO_3)_3$ and GO–$Co(NO_3)_2$ complexes at 120–130°C.

Figure 24.2. represents XRD patterns of GO and rGO–Fe–Fe_3O_4 composite (and also rGO–CoO composite for comparison) obtained by the vacuum exfoliation method, while Figure 24.3 shows morphology of the rGO–Fe–Fe_3O_4 composite obtained from GO–$Fe(CO)_5$ system at temperature of 250°C in autoclave.

FIGURE 24.3 SEM image of rGO–Fe–Fe_3O_4 composite obtained from GO–$Fe(CO)_5$ system.

24.4 H-BN-MATRIX COMPOSITES

The magnetite phase was impregnated in activated h-BN as well. The relevant procedure is described in the experimental section.

Note that XRD pattern of the magnetic composite obtained from activated h-BN–$Fe(CO)_5$–H_2O system does not reveal the magnetite phase. This can be related to its low content or formation in amorphous state.

h-BN–Fe composite obtained by $Fe(OH)_3$ deposition on the h-BN surface and reduction in hydrogen flow does not reveal any iron oxide phase in Co-rays, while traces of Fe_2O_3 and Fe_3O_4 are visible in Cu-rays (Figure 24.4). From the SEM image (Figure 24.5), it appears that the magnetite phase was present as filaments wrapped around the boron nitride grains. The magnetite phase was unevenly distributed on the grains of the activated h-BN–Fe_3O_4 composite.

Tables 24.1 and 24.2 and Figure 24.6, respectively, show examples of elemental composition, its statistics, and elements distribution mapping in the obtained h-BN-based composites determined by EDX analysis.

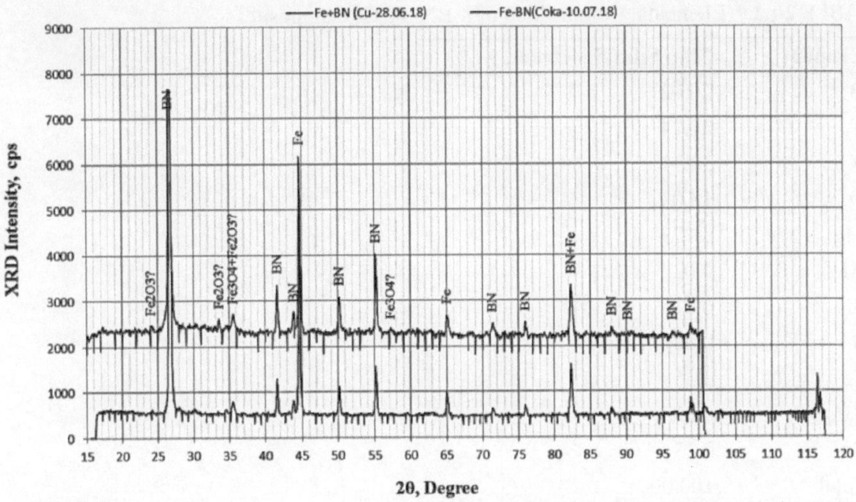

FIGURE 24.4 XRD patterns in Co- (Lower) and Cu-rays (Upper) of h-BN–Fe composite obtained by $Fe(OH)_3$ deposition on h-BN and further reduction by hydrogen at 450°C.

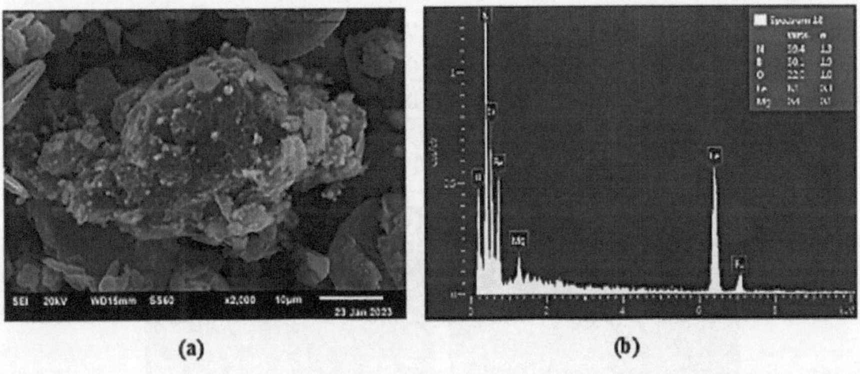

FIGURE 24.5 h-BN–Fe_3O_4 composite obtained in autoclave by decomposition of $Fe(CO)_5$ in presence of h-BN: (a) SEM image and (b) EDX spectrum.

FTIR spectroscopy was used to determine the chemical and structural nature of the obtained h-BN–Fe and h-BN–Fe_3O_4 composites. In the h-BN–Fe composite, strong and broad peaks at 1351 and 762 cm^{-1} were observed, apparently due to B–N stretching and out-of-plane B–N–B bending vibrations, respectively. These two peaks are typical for sp^2 bonds in h-BN, and they provide essential proof of the h-BN presence. Their placement is determined by the material synthesis conditions, crystal structure, composite mixture, physical-chemical properties, etc. The broad absorption bands at

TABLE 24.1 Elemental Composition of h-BN–Fe_3O_4 Composite

Element	Map Sum Spectrum, wt.%
B	20.02
C	26.24
N	34.51
O	13.03
Mg	0.17
Al	0.02
Si	0.01
S	0.07
Fe	5.35
Cu	0.33
Zn	0.23
Total	100.00

TABLE 24.2 Statistics of Elements Distribution in h-BN–Fe_3O_4 Composite

Element, wt.%	B	C	N	O	Mg	Al	Si	S	Fe	Cu	Zn
Maximum	20.02	26.24	34.51	13.03	0.17	0.02	0.01	0.07	5.35	0.33	0.23
Minimum	20.02	26.24	34.51	13.03	0.17	0.02	0.01	0.07	5.35	0.33	0.23
Average	20.02	26.24	34.51	13.03	0.17	0.02	0.01	0.07	5.35	0.33	0.23

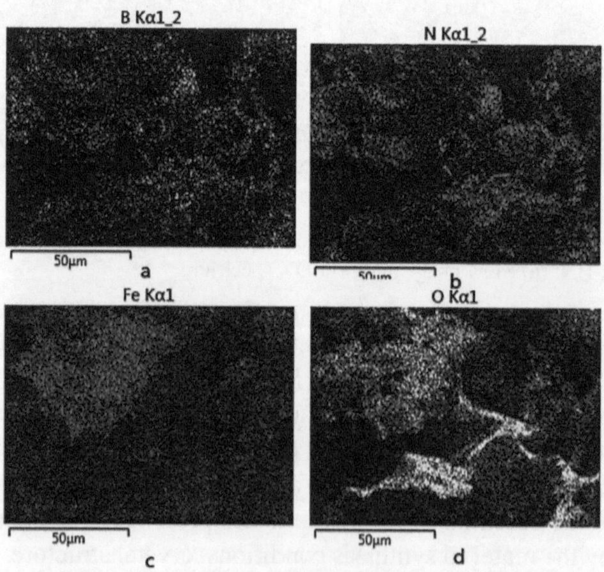

FIGURE 24.6 EDX mapping of elements distribution in h-BN–Fe_3O_4 composite obtained in autoclave by decomposition of $Fe(CO)_5$ in presence of h-BN: (a) B, (b) N, (c) Fe and (d) O.

around 3420–3250 cm^{-1} could be related to B–O–H stretching of absorbed water molecules.

Spectra of another group of h-BN–Fe samples also showed two main peaks characteristic of hexagonal boron nitride at 1356 and 762 and 1349 and 758 cm^{-1}, which could be related to B–N tension and B–N–B bending vibrations, respectively. The fingerprint peak characteristic of maghemite (Fe_2O_3) was not clearly observed, indicating the reduction of iron oxide with hydrogen. It is known that such reduction depends on various factors, including the size of nanoparticles, reaction routes, processing conditions, etc. The broad absorption bands at around 3400–3200 cm^{-1} in these samples could be ascribed to the hydroxyl group –OH vibrations.

In the FTIR spectra of h-BN–Fe_3O_4 composite samples again exhibited characteristic peaks for hexagonal boron nitride: 1366 and 794, and 1349 and 766 cm^{-1}, respectively. In comparison to h-BN–Fe, spectra of these samples reveal extra peaks at 560 and 562 cm^{-1}, which are characteristic of the Fe–O bond, indicating that these materials are composed of h-BN and Fe_3O_4. Note that the exact positions and intensities of the peaks are a combination of both h-BN and magnetite and can vary depending on the concentration of magnetite dopants and their specific crystallographic sites within the h-BN lattice. Another absorption peak at around 3200 cm^{-1} could again be ascribed to the hydroxyl group –OH vibrations.

The magnetic properties, namely magnetization curves, i.e., magnetization as a function of the applied magnetic field, of such composites were studied in [18]. Such curves provide important information about material magnetic parameters, including saturation and remnant magnetizations, as well as coercive force. They should be analyzed by taking into account that the magnetic properties of ferro- and ferrimagnetic powder materials depend on particle size. As is known, saturation and remnant magnetizations increase with crystallinity, and the coercivity variation can be caused by several combined factors, such as atom distribution in the magnetic phase structure and crystallite size and distribution in the composite.

Some of the obtained samples exhibited magnetic behavior characteristic of soft magnetic materials. One among them was close to the behavior of superparamagnetic materials.

Magnetization curves of another group of samples showed the so-called S-shaped hysteresis loops, which are extremely thin due to the absence of remnant magnetization and negligible coercivity. In particular, in one sample, there is actually no coercivity, and therefore, this nanocomposite exhibits superparamagnetic behavior.

24.4 CONCLUSION

In summary, several versions of exfoliation and chemical synthesis methods have been developed to form graphene oxide, reduced graphene oxide, graphene, and hexagonal boron nitride matrix magnetic nanocomposites by coating or intercalating their nanoparticles with ferromagnetic elemental iron or ferrimagnetic iron oxide – magnetite. The structure, morphology, and some physical properties of the hexagonal 2D-material–magnetic nanocomposites obtained in this way are studied. These materials would be prospective for a variety of high-tech applications, in particular, in medicine as controlled drug delivery agents.

ACKNOWLEDGMENT

Sh. Makatsaria acknowledges that this research has been supported by Shota Rustaveli National Science Foundation of Georgia (SRNSFG) Doctoral Educational Programs Grant # PHDF-22–1299.

KEYWORDS

- **BNCT**
- **chemical deposition**
- **exfoliation**
- **graphene**
- **graphene oxide**
- **hexagonal boron**
- **nitride**
- **intercalation**
- **iron**
- **magnetite**
- **nanocomposite**
- **reduced graphene oxide**
- **therapeutic agent delivery**

REFERENCES

1. Makatsaria, S., Chkhartishvili, L., Dekanosidze, S., & Chedia, R. (2023). Nanopowder boron compounds doped with ferromagnetic clusters for BNCT. *International Journal of Advanced Nano Computing and Analysis, 2*(1), 1–12.
2. Chkhartishvili, L., Chedia, R., Tsagareishvili, O., Mirzayev, M., Makatsaria, S., Gogolidze, N., Barbakadze, N., Buzariashvili, M., Lekashvili, O., & Jinikashvili, I. (2022). Preparation of neutron-capturing boron-containing nanosystems. In *Proceedings of the 9th International Conference on Advanced Nano Materials* (pp. 1–15). Victoria, Canada.
3. Chkhartishvili, L., Makatsaria, S., & Gogolidze, N. (2023). Boron-containing fine-dispersive composites for neutron therapy and neutron shielding. In *Proceedings of the International Scientific Practical Conference "Innovations and Modern Challenges–2022"* (pp. 221–226). Tbilisi, Georgia: Publishing House "Tech. Univ."
4. Pis, I., Nappini, S., Bondino, F., Mentes, T. O., Sala, A., Locatelli, A., & Magnano, E. (2018). Fe intercalation under graphene and hexagonal boron nitride in-plane heterostructure on Pt(111). *Carbon, 134*, 274–282.
5. Zhang, W., Wan, W., Zhou, H., Chen, J., Wang, X., & Zhang, X. (2013). *In-situ* synthesis of magnetite/expanded graphite composite material as high rate negative electrode for rechargeable lithium batteries. *Journal of Power Sources, 223*, 119–124.
6. Zhang, W., Yang, Y., Ziemann, E., Be'er, A., Bashouti, M. Y., Elimelech, M., & Bernstein, R. (2019). One-step sonochemical synthesis of a reduced graphene oxide–ZnO nanocomposite with antibacterial and antibiofouling properties. *Environmental Science: Nano, 6*(10), 3080–3090.
7. Ahmed, J. (2019). Use of graphene/graphene oxide in food packaging materials: Thermomechanical, structural and barrier properties. *Reference Module in Food Science*, 1–22. https://doi.org/10.1016/B978-0-08-100596-5.22499-2.
8. Zhu, J., & Lua, A. C. (2021). Antibacterial ultrafiltration membrane with silver nanoparticle impregnation by interfacial polymerization for ballast water. *Journal of Polymer Science, 59*, 2295–2308.
9. Mohanty, F., & Swain, S. K. (2018). Silver nanoparticles decorated polyethylmethacrylate/graphene oxide composite: As packaging material. *Polymer Composites*. https://doi.org/10.1002/pc.24944.
10. Majumder, P., Dutta, K., & Dutta, P. (2019). Synthesis, properties of graphene oxide–metal oxide mixed nanocomposites and their applications–Review. *International Journal of Advanced Science and Engineering, 5*(3), 1032–1039.
11. Yaragalla, S., Bhavitha, K. B., & Athanassiou, A. A. (2021). Review on graphene-based materials and their antimicrobial properties. *Coatings, 11*(1197), 1–18.
12. Dundua, T., Sachaneli, T., Kvartskhava, G., Gamkrelidze, N., Meladze, S., Sarajishvili, K., Japharidze, M., & Jinikhashvili, I. (2021). Preparation of graphene oxide composites containing nanosized silver, copper, and titanium oxide and study of their biocidal properties. In L. Chkhartishvili & M. Chikhradze (Eds.), *Abstracts of the 6th International Conference on Nanotechnology* (p. 28). Tbilisi, Georgia: Publishing House "Tech. Univ."
13. Nadaraia, L., Dundua, T., Gamkrelidze, N., Tsitsishvili, V., Barbakadze, N., & Chedia, R. (2021). Graphite foil waste to graphene: New carbon precursors for synthesis of graphene and its oxides. *Key Engineering Materials, 89*, 68–74.

14. Dundua, T. (2021). Preparation of graphene oxide composites containing nanometals and oxides from graphite foil wastes and study of their biocidal activity. *Nano Studies, 21/22*, 91–110.
15. Nadaraia, L., Jalabadze, N., Khundadze, L., Rurua, L., Japaridze, M., & Chedia, R. (2021). Effects of graphene on morphology, fracture toughness, and electrical conductivity of titanium dioxide. *Diamond and Related Materials, 114*, Article 108319, 1–10.
16. Ruidiaz–Martinez, M., Alvarez, M. A., Lopez–Ramon, M. V., Cruz–Quesada, G., Rivera–Utrilla, J., & Sanchez–Polo, M. (2020). Hydrothermal synthesis of rGO–TiO_2 composites as high-performance UV photocatalysts for ethylparaben degradation. *Catalysts, 10*(520), 1–25.
17. Tripathi, S. N., Saini, P., Gupta, D., & Choudhary, V. (2013). Electrical and mechanical properties of PMMA/reduced graphene oxide nanocomposites prepared via ` polymerization. *Journal of Materials Science, 48*(18), 6223–6232.
18. Makatsaria, S., Kekutia, S., Markhulia, J., Mikelashvili, V., Chkhartishvili, L., & Chedia, R. (2021–2022). Magnetic properties of nanopowder h-BN doped with Fe and Fe3O4 nanoclusters. *Nano Studies, 21/22*, 287–292.

PART VI
Calculations

CHAPTER 25

Analysis of Engineering Properties and Applications of Polymer Composites Using Deep Learning Techniques

LELA MIRTSKHULAVA

Department of Computer Science, Ivane Javakhishvili Tbilisi State University, Tbilisi, Georgia

ABSTRACT

Polymer composites, characterized by their distinct properties, are pivotal for a myriad of applications, yet optimizing their attributes remains complex. Recently, deep learning has positioned itself as an instrumental tool in materials science, adept at deciphering extensive data to forecast properties and potential uses. This paper offers an exhaustive review of the integration of deep learning in evaluating the engineering characteristics and applicability of polymer composites. Discussions encompass facets from data curation to model creation and validation. Through detailed examples, the paper delves into deep learning applications in determining the mechanical, thermal, and electrical properties of polymer composites. Discussions encompass facets from data curation to MT-NN model creation and validation.

25.1 INTRODUCTION

Polymer composites have emerged as an important class of materials with unique combinations of properties that make them suitable for various applications. However, the optimization of their properties and applications is still a challenging task. Deep learning has emerged as a powerful tool

Advanced Topics in Polymer Chemistry and Materials Science: Tamara Tatrishvili and Marc Jean Médard Abadie (Eds)

in materials science, offering a way to extract insights from vast amounts of data and to make predictions of properties and applications [1, 2]. This paper provides a comprehensive review of the use of deep learning in the analysis of engineering properties and applications of polymer composites. The review covers various aspects of the use of deep learning, including data preparation, model development, and validation. It also provides examples of applications of deep learning in the analysis of various properties of polymer composites, including mechanical, thermal, and electrical properties [3–6]. The paper highlights the advantages of deep learning in materials science, including its ability to handle high-dimensional data, learn complex relationships between different variables, and make predictions with high accuracy. The review also identifies the challenges in the use of deep learning in materials science, such as the need for large amounts of high-quality data and the potential limitations of the models [7–12].

Overall, this review provides a valuable resource for researchers and engineers working in the field of polymer composites. It summarizes the current state of the art in the use of deep learning in materials science and identifies areas for future research. The paper also provides references to recent studies that have applied deep learning to the analysis of engineering properties and applications of polymer composites, which can help guide further research in this area. The landscape of materials engineering has been revolutionized by the convergence of advanced computational methods and the vast array of materials, particularly polymers, that offer varied and unique properties. Among these computational methods, deep learning, a subset of machine learning, has been at the forefront, showing incredible promise in deciphering complex patterns and relationships within data. Polymer composites, known for their customizable properties by combining different materials, provide an excellent subject for these computational explorations [13–17].

Polymer composites, being mixtures of polymers and other materials (often fibers or fillers), bring together the best of both worlds, i.e., the flexibility and processability of polymers and the strength and rigidity of reinforcing agents. Their wide range of applications, from aerospace to consumer goods, owes to their tailorability, allowing engineers to design materials with specific desired properties [18, 19].

Deep learning, particularly with neural networks, can model non-linear relationships in large datasets without the need for explicit feature engineering. Given the multifaceted nature of polymer composites, which can have a vast array of constituents leading to a myriad of properties, deep learning becomes an apt tool to map these intricate relationships. The aim here is to delve deep into the potential of deep learning techniques in analyzing the engineering properties of polymer composites. By harnessing the computational prowess of deep learning, one can:

1. Predict the mechanical, thermal, and electrical properties of new polymer composite combinations.
2. Accelerate the material discovery process by identifying promising polymer composite candidates for specific applications.
3. Understand the underlying factors that influence the behavior of polymer composites, thus refining the design process.

25.2 A POLYMER REFERENCE MODEL

25.2.1 DEFINITION AND DEVELOPMENT

A polymer reference model is a theoretical representation of a polymer's structure and behavior based on fundamental principles and experimental data. It serves as a starting point for understanding and predicting various properties of polymers. The model incorporates key features of polymer chains, such as monomer units, bond angles, bond lengths, and the arrangement of side groups.

Some commonly used approaches for creating polymer reference models include:

1. **Molecular Dynamics (MD) Simulations:** This use classical mechanics to model the motion of atoms and molecules over time. In the case of polymers, MD simulations track the movements of individual atoms in the polymer chain, considering interatomic forces and thermal effects. MD simulations can provide insights into polymer conformation, mechanical properties, and thermal behavior.

2. **Monte Carlo Simulations:** These are used to explore the conformational space of polymer chains by randomly sampling different configurations. These simulations are particularly useful for understanding the thermodynamics of polymer systems and predicting phase transitions.

3. **Quantum Mechanics (QM) Calculations:** It provides a more accurate representation of the electronic structure and bonding in

polymers. QM calculations can be applied to study electronic properties, molecular orbitals, and charge distributions in polymers.

4. **Statistical Mechanics:** These are based on statistical ensembles and can provide information about the thermodynamic properties of polymer systems, such as heat capacity, entropy, and free energy.
5. **Density Functional Theory (DFT):** It is a quantum mechanical approach that allows the calculation of electronic properties and energies of polymers with reasonable computational cost. It is often used to study the electronic structure and properties of polymer chains and their interactions with other molecules or surfaces.
6. **Empirical Models:** These use experimental data to establish relationships between various properties of polymers. These models are often simple and computationally efficient, making them practical for large-scale simulations and predictions.

It is important to note that the accuracy and complexity of a polymer reference model depend on the specific application and the level of detail required. Researchers often choose the most appropriate modeling technique based on the research objectives and available computational resources.

25.2.2 POLYMER COMPOSITES – CHALLENGING TASKS

25.2.2.1 BACKGROUND

Polymer composites are materials composed of two or more constituents, typically a polymer matrix reinforced with fibers or particles, resulting in a combination of properties that are superior to those of the individual components. These composites have gained immense popularity in various industries, including aerospace, automotive, construction, and consumer goods, due to their lightweight, high strength-to-weight ratio, corrosion resistance, and other unique properties. As a result, polymer composites have opened up new possibilities for engineering applications, leading to innovations in diverse fields [20–22]. The engineering properties of polymer composites, such as mechanical strength, thermal conductivity, and electrical conductivity, play a crucial role in determining their performance and suitability for specific applications. Understanding and optimizing these properties are essential for the development of advanced materials that meet the growing demands of modern industries.

25.2.2.2 SIGNIFICANCE

Despite the numerous advantages of polymer composites, the process of optimizing their properties and predicting their applications presents significant challenges. Traditionally, materials scientists have relied on experimental testing and simulation techniques, which can be time-consuming, expensive, and limited by the complexity of composite materials.

This is where deep learning comes into the picture. Deep learning, a subset of artificial intelligence, has shown immense promise in various domains, including image and speech recognition, natural language processing, and healthcare. In recent years, researchers have started exploring its potential in materials science, specifically in the analysis of polymer composites. The significance of using deep learning in this context lies in its ability to handle vast amounts of high-dimensional data efficiently and learn complex relationships between different variables. By utilizing deep learning models, researchers can gain valuable insights from extensive datasets and make accurate predictions about the properties and applications of polymer composites. This technology has the potential to accelerate materials discovery and development processes, leading to the creation of new, innovative, and high-performance materials for various industries.

Furthermore, the integration of deep learning with materials science is in line with the ongoing trend of digitalization and Industry 4.0. The ability to predict material properties and performance accurately can reduce the time and cost of materials development, enhance materials selection for specific applications, and even enable the design of tailor-made materials with desired properties.

In conclusion, the use of deep learning in the analysis of engineering properties and applications of polymer composites represents a groundbreaking approach that holds great promise for revolutionizing the field of materials science and advancing technological advancements in numerous industries. By offering new insights, efficient data analysis, and accurate predictions, deep learning can unlock the full potential of polymer composites and pave the way for the development of high-performance materials that meet the demands of the modern world.

25.3 POLYMER PROPERTIES PREDICTED USING DEEP LEARNING TECHNIQUES

25.3.1 PREDICTING POLYMER PROPERTIES

Predicting polymer properties using deep learning techniques has emerged as a promising approach in recent years. Deep learning is a subset of machine learning that utilizes artificial neural networks to automatically learn patterns and representations from large datasets. When applied to polymer properties, deep learning models can analyze complex relationships between the molecular structure of polymers and their resulting properties. Here's how polymer properties are predicted using deep learning techniques:

1. **Data Collection** – A large dataset of polymer samples with known properties. This dataset should encompass various types of polymers, their chemical structures, and corresponding property measurements.
2. **Data Preprocessing:** The collected data may need to undergo preprocessing steps, such as cleaning, normalization, and feature extraction, to ensure that the input is suitable for deep learning models.
3. **Model Architecture:** Deep learning models, like convolutional neural networks (CNNs) or recurrent neural networks (RNNs), are chosen or designed for the specific prediction task. The architecture of the neural network determines how data is processed and information is extracted.
4. **Training:** The deep learning model is trained on the preprocessed data using a process called backpropagation. During training, the model adjusts its internal parameters to minimize the difference between predicted polymer properties and the actual measured properties in the training dataset.
5. **Validation:** After training, the model is validated using a separate dataset to assess its performance on unseen data. This step helps ensure that the model generalizes well and does not overfit the training data.
6. **Testing and Prediction:** Once the model is trained and validated, it can be used to predict polymer properties for new, unseen samples. By inputting the molecular structure or relevant features of a polymer into the trained model, it can provide predictions of various properties, such as mechanical strength, glass transition temperature, or thermal stability.
7. **Model Optimization:** The model's performance can be further optimized through hyperparameter tuning, regularization techniques, or by using more advanced deep learning architectures, such as attention mechanisms or transformer-based models.

8. **Continual Improvement:** The predictive model can be continually improved by incorporating new data, refining the architecture, or using transfer learning techniques to leverage knowledge from related domains.

By utilizing deep learning techniques, researchers and engineers can efficiently analyze vast amounts of data, discover intricate patterns in polymer properties, and make accurate predictions, ultimately advancing the development and understanding of polymer materials for diverse applications.

25.4 ENGINEERING PROPERTIES AND APPLICATIONS OF POLYMER COMPOSITES

25.4.1 ENGINEERING PROPERTIES OF POLYMER COMPOSITES

Polymer composites are materials that consist of a polymer matrix reinforced with fibers, particles, or fillers. The engineering properties of polymer composites are influenced by the type and arrangement of the reinforcement, the nature of the polymer matrix, and the manufacturing process. Here are some of the key engineering properties of polymer composites:

1. **Strength:** Polymer composites can exhibit high strength, especially when reinforced with materials like carbon fibers, glass fibers, or aramid fibers. The fibers carry the majority of the load, while the polymer matrix holds them in place.
2. **Stiffness:** The stiffness of polymer composites is significantly influenced by the stiffness of the reinforcing material and its volume fraction in the composite. Composites with high stiffness are suitable for applications that require structural rigidity.
3. **Density:** Polymer composites are generally lightweight compared to metals, which is one of their most significant advantages. The low density contributes to their high strength-to-weight ratio, making them attractive for weight-sensitive applications.
4. **Fatigue Resistance:** Some polymer composites possess good fatigue resistance, meaning they can withstand repeated loading cycles without failure. This property is essential in applications where components are subject to dynamic and cyclic loading.

5. **Impact Resistance:** Polymer composites can exhibit excellent impact resistance, especially when reinforced with tough fibers. This property makes them suitable for applications where the material needs to absorb and dissipate energy during impact events.
6. **Thermal Properties:** The thermal behavior of polymer composites depends on the choice of the polymer matrix and reinforcing material. Some composites have low thermal conductivity, making them useful for thermal insulation applications.
7. **Electrical Properties:** Depending on the type of polymer and reinforcement, polymer composites can have electrical insulating properties, making them suitable for electrical and electronic applications.
8. **Chemical Resistance:** The chemical resistance of polymer composites depends on the choice of the polymer matrix and reinforcement. Some composites can resist corrosion and chemical attack, making them suitable for harsh environments.
9. **Dimensional Stability:** Polymer composites can exhibit good dimensional stability, meaning they are less prone to deformation under varying temperature and humidity conditions.
10. **Design Flexibility:** The manufacturing process of polymer composites allows for complex shapes and customization, giving engineers greater design flexibility to tailor the material for specific applications.
11. **Creep Resistance:** Creep is the time-dependent deformation that occurs under constant load or stress. Some polymer composites can have good creep resistance, making them suitable for applications that require long-term structural integrity. Water Absorption: The water absorption characteristics of polymer composites depend on the nature of the polymer matrix and reinforcing materials. Some composites can be engineered to have low water absorption, which is crucial for certain applications.

25.4.2 APPLICATIONS OF POLYMER COMPOSITES

1. **Aerospace:** Polymer composites are widely used in the aerospace industry for aircraft components like wings, fuselage sections, and tail structures due to their high strength-to-weight ratio.

2. **Automotive:** In the automotive sector, polymer composites are utilized to manufacture lightweight body panels, interior components, and suspension parts, contributing to fuel efficiency and reduced emissions.
3. **Marine:** Polymer composites find applications in boat hulls, decks, and other marine structures because of their corrosion resistance and lightweight nature.
4. **Wind Energy:** Wind turbine blades are often made from polymer composites due to their high strength and fatigue resistance.
5. **Construction:** Polymer composites are employed in construction for bridges, reinforcements, and architectural components due to their design flexibility and corrosion resistance.
6. **Sports and Recreation:** In sporting goods, such as tennis rackets, golf clubs, and bicycle frames, polymer composites provide a balance of strength, stiffness, and lightweight properties.
7. **Electrical and Electronics:** Polymer composites are used in electrical insulators, circuit boards, and electronic housings to take advantage of their electrical insulation properties.
8. **Medical:** Some medical devices, prosthetics, and dental materials are made from polymer composites due to their biocompatibility and customizable properties.

   The engineering properties of polymer composites, combined with their versatility and tailorable nature, make them a valuable class of materials for a wide range of applications in various industries.

25.5 EXPERIMENTAL RESULTS

25.5.1 POLYMER ENGINEERING PROPERTIES PREDICTION USING POLYMER GENOME WITH THE USE OF MULTI-TASK DEEP NEURAL NETWORKS (MT-NN) MODELS

Predicting polymer engineering properties using a polymer genome with multi-task deep neural networks (MT-NN) models is a cutting-edge approach that leverages the power of deep learning to efficiently analyze large-scale polymer datasets and make accurate predictions for multiple properties simultaneously. Here's how this process works:

1. **Polymer Genome Data:** The polymer genome refers to a comprehensive database that contains information about various polymer materials, including their chemical structure, composition, processing conditions, and a wide range of engineering properties. This dataset serves as the foundation for training and validating the MT-NN model.
2. **Data Preprocessing:** The polymer genome data may undergo preprocessing steps to clean the data, handle missing values, and extract relevant features. Feature engineering techniques may also be employed to represent the polymer structures effectively for input into the deep neural network.
3. **Multi-Task Deep Neural Network Architecture:** The MT-NN model is designed to handle multiple prediction tasks simultaneously. It typically consists of multiple interconnected layers of neurons, such as convolutional layers, recurrent layers, or fully connected layers. Each task corresponds to predicting a specific engineering property, such as tensile strength, melting temperature, or impact resistance.
4. **Training with Multi-Task Learning:** During training, the MT-NN model is fed with the preprocessed polymer genome data, and it learns to predict multiple engineering properties simultaneously. The model optimizes its internal parameters using backpropagation and gradient descent to minimize the combined loss across all the tasks.
5. **Task-Specific Layers:** In the MT-NN architecture, there may be layers specifically dedicated to each task to capture task-specific features and relationships. These task-specific layers help the model learn the unique aspects of each engineering property.
6. **Regularization and Optimization:** To prevent overfitting and improve generalization, regularization techniques, such as dropout or L2 regularization, may be applied. Additionally, various optimization algorithms are used to fine-tune the model's performance.
7. **Validation and Hyperparameter Tuning:** The MT-NN model is validated on a separate dataset to assess its performance on unseen data. Hyperparameter tuning may be performed to find the optimal configuration for the model's architecture and learning parameters.

8. **Predicting New Polymer Properties:** Once trained and validated, the MT-NN model can be used to predict engineering properties for new polymer materials. By inputting the relevant information about a particular polymer into the model, it can output predictions for multiple properties simultaneously.

This approach offers several advantages, including the ability to leverage shared information among different engineering properties, efficient handling of large-scale polymer datasets, and the potential for improved accuracy compared to single-task models. MT-NN models are an exciting development in polymer materials research, as they hold the promise of accelerating materials discovery and optimization for various applications.

The SMILES (Simplified Molecular-Input Line-Entry System) employs concise ASCII strings to depict the structure of chemical entities. The specific SMILES format detailed here is our custom version tailored for polymers, making it distinct from other standard SMILES formats. It's essential to adhere to the guidelines provided below to ensure accurate outcomes.

25.5.2 PREDICTING NEW POLYMER THROUGH SMILE

Using multi-task deep neural networks (MT-NN) models in polymer genome, we predicted polymer (Figure 25.1).

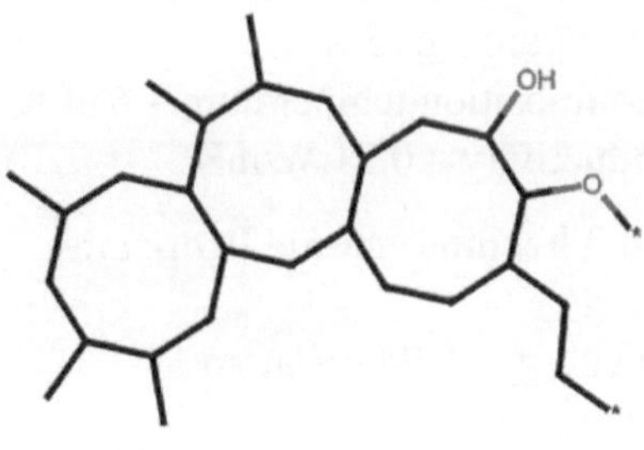

FIGURE 25.1 Predicted polymer.

- **SMILE Formula for Predicted Polymers**
- [*]CCC3CCC2CC1CC(C)C(C)CC(C)CC1C(C)C(C)CC2CC(O)C3O[*].
- **Electronic Properties**
  Bandgap (chain) – 4.5 eV
  Bandgap (bulk) – 4.3 eV

Electron affinity – 2.3 eV
Ionization energy – 6.2 eV

- **Dielectric and Optical Properties**

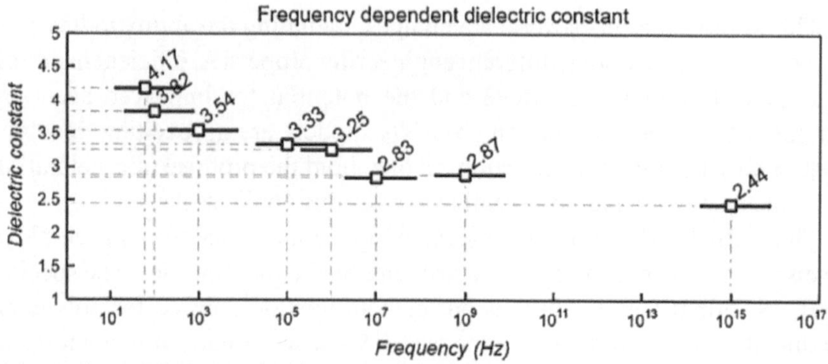

where:

Dielectric constant (crystal) – 4.6
Refractive index (bulk resin) – 1.6
Refractive index (crystal) – 1.9

- **Thermal Properties**
  Glass transition temperature – 400 K
  Melting temperature – 475 K
  Thermal decomposition temperature – 684 K
  Thermal conductivity – 0.21 W/mK
- **Physical and Thermodynamic Properties**
  Density – 1.2 g/cc
  Atomization energy –5.94 eV/atom
  Heat capacity – 1 J/gK
  Tendency to crystallize – 36%
  Limiting oxygen index – 34%
  Fractional free volume – 0.18
- **Mechanical Properties**
  Tensile strength – 80 MPa
  Young's modulus – 2 MPa
- **Permeability Properties**
  He gas permeability – 1.8 Barrer

H_2 gas permeability – 2.0 Barrer
CO_2 gas permeability – 1.7 Barrer
N_2 gas permeability – 0.8 Barrer
O_2 gas permeability – 1.1 Barrer
CH_4 gas permeability – 0.8 Barrer

25.6 CONCLUSION

In the given chapter, we have explored deep learning techniques that not only facilitate accurate predictions of the properties of polymer composites but also accelerate the iterative process of material design, optimization, and application. This ensures that the next generation of materials is not only superior in performance but also tailor-made for specific applications, catering to the evolving demands of industries ranging from aerospace to consumer electronics. In essence, the marriage of polymer composites and deep learning exemplifies the epitome of technological evolution, where empirical knowledge meets computational brilliance, setting the stage for a future where the possibilities in materials science seem boundless. The intersection of deep learning and materials science, specifically in the domain of polymer composites, represents a dynamic frontier in modern engineering and research. The intricate nature of polymer composites, characterized by their varied constituents and resultant properties, is aptly addressed by the robust capabilities of deep learning. This synergy promises to redefine the pace and precision of material discovery and design.

KEYWORDS

- **convolutional neural networks**
- **deep learning**
- **Monte Carlo simulations**
- **polymer chains**
- **polymer composites**

REFERENCES

1. Mirtskhulava, L. (2023). Deep learning applications in predicting polymer properties. In *Advanced Polymer Structures: Chemistry for Engineering Applications*. Apple Academic Press.
2. Mirtskhulava, L. (2024). Polymer composites with AI in environmental monitoring: A review. In *Sustainability in Polymer Technology and Plastic Engineering: Concepts, Strategies, and Opportunities*. Apple Academic Press.
3. Mirtskhulava, L. (2022). Deriving polymer properties using machine learning techniques. In *Proceedings of the XI International Scientific-Technical Conference "Advance in Petroleum and Gas Industry and Petrochemistry."* Lviv, Ukraine.
4. Chen, C., Zhang, Y., & Guo, Z. (2020). Prediction of mechanical properties of polymer composites using deep learning. *Composites Science and Technology, 192*, Article 108087.
5. Wu, Y., Zhang, Y., Liu, Y., & Gao, Y. (2020). Prediction of thermal conductivity of polymer composites using deep learning. *Composites Part B: Engineering, 187*, Article 107836.
6. Liu, Z., Lu, Y., Gu, J., Chen, S., & Cui, H. (2021). Deep learning-based prediction of electrical conductivity of polymer composites. *Materials & Design, 199*, Article 109350.
7. Li, H., Wang, J., Li, J., Li, Y., & Jia, Y. (2021). Analysis of mechanical properties of polymer composites using deep learning and finite element method. *Journal of Materials Science, 56*(7), 4659–4669.
8. Mannodi-Kanakkithodi, A., Pilania, G., Huan, T. D., Lookman, T., & Ramprasad, R. (2016). Machine learning strategy for accelerated design of polymer dielectrics. *Scientific Reports, 6*, Article 20952.
9. Huan, T. D., Mannodi-Kanakkithodi, A., Kim, C., Sharma, V., Pilania, G., & Ramprasad, R. (2016). A polymer dataset for accelerated property prediction and design. *Scientific Data, 3*, Article 160012.
10. Mannodi-Kanakkithodi, A., Chandrasekaran, A., Kim, C., Huan, T. D., Pilania, G., Botu, V., & Ramprasad, R. (2018). Scoping the polymer genome: A roadmap for rational polymer dielectrics design and beyond. *Materials Today, 21*, 785–796.
11. Kim, C., Chandrasekaran, A., Huan, T. D., Das, D., & Ramprasad, R. (2018). Polymer Genome: A data-powered polymer informatics platform for property predictions. *The Journal of Physical Chemistry C, 122*(31), 17575–17585.
12. Jha, A., Chandrasekaran, A., Kim, C., & Ramprasad, R. (2019). Impact of dataset uncertainties on machine learning model predictions: The example of polymer glass transition temperatures. *Modelling and Simulation in Materials Science and Engineering, 27*, Article 024002.
13. Venkatram, S., Kim, C., Chandrasekaran, A., & Ramprasad, R. (2019). A critical assessment of the Hildebrand and Hansen solubility parameters for polymers. *Journal of Chemical Information and Modeling, 59*(10), 4188–4194.
14. Lihua, C., Kern, J., Lightstone, J. P., & Ramprasad, R. (2021). Data-assisted polymer retrosynthesis planning. *Applied Physics Reviews, 8*(15).
15. Lightstone, J. P., Chen, L., Kim, C., Batra, R., & Ramprasad, R. (2020). Refractive index prediction models for polymers using machine learning. *Journal of Applied Physics, 127*(21), Article 215105.

16. Zhu, G., Kim, C., Chandrasekaran, A., Everett, J. D., Ramprasad, R., & Lively, R. P. (2020). Polymer genome–based prediction of gas permeabilities in polymers. *Journal of Polymer Engineering*, Article 20190329.
17. Tran, H. D., Kim, C., Chen, L., Chandrasekaran, A., Batra, R., Venkatram, S., Kamal, D., Lightstone, J. P., Gurnani, R., Shetty, P., Ramprasad, M., Laws, J., Shelton, M., & Ramprasad, R. (2020). Machine-learning predictions of polymer properties with Polymer Genome. *Journal of Applied Physics, 128*, Article 171104.
18. Chandrasekaran, A., Kim, C., Venkatram, S., & Ramprasad, R. (2020). A deep learning solvent-selection paradigm powered by a massive solvent/nonsolvent database for polymers. *Macromolecules*.
19. Kuenneth, C., Schertzer, W., & Ramprasad, R. (2021). Copolymer informatics with multitask deep neural networks. *Macromolecules, 54*, 5957–5965.
20. Kuenneth, C., Rajan, A. C., Tran, H., Chen, L., Kim, C., & Ramprasad, R. (2021). Polymer informatics with multi-task learning. *Patterns, 2*(4), Article 100238.
21. Huan, T. D., Mannodi-Kanakkithodi, A., & Ramprasad, R. (2015). Accelerated materials property predictions and design using motif-based fingerprints. *Physical Review B, 92*, Article 014106.
22. Mannodi-Kanakkithodi, A., Treich, G. M., Huan, T. D., Ma, R., Tefferi, M., Cao, Y., Sotzing, G. A., & Ramprasad, R. (2016). Rational co-design of polymer dielectrics for energy storage. *Advanced Materials, 28*, 6277–6283.

CHAPTER 26

Calculation Methodology of Stability Constants of Fulvate Complexes

TAMAR MAKHARADZE,[1,2,3] TAMARA TATRISHVILI,[3,4] GIA PETRIASHVILI,[2,3] GIORGI MAKHARADZE,[4] and KETEVAN CHUBINIDZE[2,3]

[1]*R. Agladze Institute of Inorganic Chemistry and Electrochemistry, Ivane Javakhishvili Tbilisi State University, Tbilisi, Georgia*

[2]*Institute of Cybernetics of Georgian Technical University, Tbilisi, Georgia*

[3]*Institute of Macromolecular Chemistry and Polymeric Materials, Ivane Javakhishvili Tbilisi State University, Tbilisi, Georgia*

[4]*Ivane Javakhishvili Tbilisi State University, Tbilisi, Georgia*

ABSTRACT

In the article, the sources of errors in the calculation of stability constants of fulvate complexes with heavy metals are discussed. It is shown that in the weak acid area (pH 3–4), when non-dissociated forms of fulvic acids participate in complex formation processes, to correctly display the stability constants of fulvate complexes, the equilibrium constant of the reaction should be multiplied by the dissociation constant.

When determining the values of stability constants of fulvate complexes, if hydroxo complexes are involved in the complex formation process with fulvic acids, the calculation of ionic forms using hydroxo complexes or hydrolysis constants leads to only one result: it artificially increases the value of the stability constant. This is because the smaller the denominator is in the expression of the stability constant, the greater its value.

Advanced Topics in Polymer Chemistry and Materials Science: Tamara Tatrishvili and Marc Jean Médard Abadie (Eds)

The article shows the optimal variants for calculating the composition and stability constants of fulvate complexes when using solubility and gel chromatographic methods.

26.1 INTRODUCTION

Fulvic acids are high molecular weight natural compounds. They take an active part in complex formation reactions with heavy metals and radionuclides through functional groups that have ionization ability, which stipulates their migration forms in natural waters, as well as in bottom sediments and soils [1–9].

26.2 EXPERIMENTAL METHODS AND MATERIALS

The spectra of methods used for investigating the complex formation process of fulvic acids with heavy metals are very diverse: ionometry, electron paramagnetic resonance (EPR), gel chromatography, spectrophotometry, ion exchange, potentiometric titration, etc. All of these methods have their advantages and limitations. For example, when using the ion exchange method, fulvic acids are adsorbed on the ionites. When using the gel chromatographic method, the free metal ions are adsorbed on the stationary phase. When using potentiometric titration, it is necessary to work with high concentrations of metals ($>>10^{-4}$ M), which in turn is a prerequisite for the formation of polynuclear hydroxocomplexes and hydrolysis. The main problem when using ionometry is the adsorption of fulvic acids on the electrodes, and in the solubility method, the choice of solid phase is very important. It should be characterized by stability over time and low solubility.

Thus, the choice of experimental method for the investigation of fulvic acids with heavy metals is the result of a compromise between the goal of the work and the limitations of the chosen method. From the discussed methods for the investigation of fulvate complexes, particularly, the determination of their composition and stability constants using solubility and gel chromatographic methods are the most optimal [1, 28–36].

26.3 RESULTS AND DISCUSSION

In fulvic acids, the main complex formation centers are the fragments of salicylic and phthalic acids. It is not entirely definitive which groups (carboxyl groups or phenolic groups) are more actively involved in the complex formation process. This issue is not resolved even in the simplest analogs of fulvic acids, such as salicylic acid. The participation of carboxyl or phenolic hydroxyl groups in the complex formation process, besides the value of pH, depends on the nature of the metal. For example, the coupling of fulvic acids to copper(II) is accomplished predominantly via a carboxyl group, and to zinc(II) via a phenolic hydroxyl group [10].

If we assume that fulvic acids are bidentate acids, in water solutions according to pH, theoretically the following types of reactions could be between heavy metals and fulvic acids:

$$Me^{2+}+H_2FA \leftrightarrows MeFA^0+2H^+ \quad (26.1)$$

$$Me^{2+}+2H_2FA \leftrightarrows Me(FA)_2^{2-}+4H^+ \quad (26.2)$$

$$Me^{2+}+H_2FA+H_2O \leftrightarrows MeOHFA^- +3H^+ \quad (26.3)$$

$$Me^{2+}+HFA^- \leftrightarrows MeFA^0 +H^+ \quad (26.4)$$

$$Me^{2+}+FA^{2-} \leftrightarrows MeFA^0 \quad (26.5)$$

$$MeOH^++FA^{2-} \leftrightarrows MeOHFA^- \quad (26.6)$$

$$Me(OH)_2+FA^{2-} \leftrightarrows Me(OH)_2FA^2 \quad (26.7)$$

The first four reactions, based on the dissociation constants of fulvic acids, characterize the acids' area (pH 2–4), where non-dissociated forms of fulvic acids dominate. Generally, according to pH stability constants, fulvate complexes differ from each other in several ranges [1–27]. When analyzing these data, the authors noted an important fact: in the weak acid area (pH 2–4), when studying the complex formation process, fulvic acids are typically presented as protonated ligands (HL^-). However, it is unclear and, moreover, incorrect to refer to the stability constants of fulvates as the equilibrium constants calculated under given conditions. To illustrate what was mentioned above, a simple scheme should be considered.

$$Me+HL^- \leftrightarrows MeL+H^+ \quad (26.8)$$

K is the Equilibrium constant of the reaction (26.8)

$$K=[MeL][H^+]/[Me][HL^-] \quad (26.9)$$

Numerator and denominator of (Eq. 26.8) reaction should be multiplied by [L].[L] is the concentration of ligand.

$$K=[MeL][H^+][L]/[Me][HL^-][L]= \beta_{MeL}/K_{HL} \quad (26.10)$$

$$\beta_{MeL} = K_{HL}K \tag{26.11}$$

where K_{HL} – dissociation constant of acid, β_{MeL} – stability constant of complex. $\beta_{ML} \neq K$. Thus, in such cases when non-dissociated forms of fulvic acids participate in the complex formation process, to obtain stability constants of fulvate complexes, the equilibrium constant of the given reaction should be multiplied by the dissociation constant of fulvic acids. Additionally, when calculating the stability constant according to the dissociation constant of fulvic acids, a correction of the concentration of [L] should be made. Another source of error is to calculate the concentration of free metal by using of hydroxo complexes constants(or hydrolysis constants) when calculation of stability constants of fulvate complexes.

$$\beta = [MeL] / \{[Me_{free}] [L]\} \tag{26.12}$$

Let's assume in given environment, heavy metal forms two hydroxo complexes $MeOH^+$ and $Me(OH)_2^0$. By simple transformations is obtained that the concentration of free metal

$$[Me_{free}] = [Me\ (total)]/(1+\beta_{1,1}[OH^-]+\beta_{1,2}[OH^-]^2) \tag{26.13}$$

Taking into consideration equations 26.13, 26.12 equation will take such form:

$$\beta = [MeL] / \{[Me\ (total)] /(1+?_{1,1}[OH^-]+\beta_{1,2}[OH^-]^2)]\} [L]\} \tag{26.14}$$

Equation (26.14) presents such universal formula, by means of is calculated stability constants of fulvate complexes in wide range of pH. At first glance, the formula is flawless; even the processes of hydrolysis are taken into account in it. However, if it is examined carefully, it is not difficult to see that Eq. (26.14) (as well as its analogues) fixes the ability of participation of just one form of heavy metal in the complex formation process. Meanwhile, fulvic acids can react with ionic and hydroxo complexes of heavy metals Eqs. (26.15–26.17) reactions. Besides, when using Eq. (26.14), in the case of two metals that have the same ability for complex formation, the stability constants of fulvate complexes could differ significantly from each other. The more hydrolyzable the metal ion, the greater the difference will be.

To put it simply, if hydroxo complexes participate in the complex formation reaction with fulvic acids, then calculating the ionic form by hydroxo complexes or hydrolysis constants leads to only one result: it artificially increases the value of the stability constant, because the smaller the denominator is in the expression of the stability constant, the larger its value is.

When using the solubility method, choosing the correct solid phase is practically impossible for polynuclear hydrolysis. The gel chromatographic method can determine those fractions where the possible existence of fulvate complexes may occur. When using these methods, the composition of fulvate complexes is determined in different ways [1, 35, 36]. In a heterogeneous system, for reaction (Eq. 26.15) (for simplicity, the charges of ions are not written in the reaction), (m) represents the numerical meaning of the stoichiometric coefficient or the amount of ligands in the inner coordination sphere, which equals the tangent of the tilt angle built in coordinates: lg[MFAm] and lg[FA]. The concentration of fulvate complexes in solution [MFAm] equals to the difference between the total amount of metal $[M(II)_{total}]$, received after formation of fulvate complex and initial concentration of metal $[M(II)_{free}]$ existed before adding ligand in solution.

$$M(II)_{free} + mFA \rightleftarrows [MFAm] \quad (26.15)$$

$$\beta = [MFAm]/ ([M(II)_{free}]x[FA]^{m} \quad (26.16)$$

$$[MFAm]= [M(II)_{total}] - [M(II)_{free}] \quad (26.17)$$

In homogeneous systems, where $[M(II)_{free}]$ is not constant value, numeral meaning of stoichiometric coefficient (m), equals to tangent of tilt angle built in coordinates: $lg([MFAm]/[M(II)_{free}])$ and lg[FA]. When using of gel chromatographic method, [MFAm] equals to the concentration of metal, determined in high weight molecular fraction and

$$[M(II)_{free}] = [M(II)_{total}] - [MFAm] \quad (26.18)$$

By the opinion of authors, using of Leden's method is the most optimal for the calculation of stability constants of fulvate complexes [37].

Leden's method is based on the using of Leden's F(L) function, which, in turn, is based on the using of the complex formation function:

$$F_1=(\Phi-1)/[FA]= \beta 1,1+ \beta 1.2 [L] \quad (26.19)$$

where β is the function of complex formation

$$\Phi=[Me_{total}]/[Me_{free}] \quad (26.20)$$

where $[Me_{total}]$ is the total quantity of metal in solution, $[Me_{free}]$ is the quantity of metal which is not included in the complex. Inserting Eq. (26.20) into Eq. (26.19), Leden's function is obtained, which in this case will look like this

$$F(L)=F(FA)=[MeFA]/([M(II)_{free}]x[FA_{free}])=([M(II)_{total}]-[M(II)_{free}])]/ ([M(II)_{free}]x[FA_{free}])= =\beta_1+\beta_2x[FA_{free}] \quad (26.21)$$

where $[FA_{free}]$ is free ligand, $[FA_{free}]$=[FA total] – [MFA]. By using the graphical variant of Leden's function, it is possible to find the stability constant of fulvate complexes. When the concentration of the ligand approaches zero, the section cut on the ordinate by the line built in the coordinates F(FA) and $[FA_{free}]$ equals the stability constant. In order to minimize systematic errors, it is advisable to use the least squares method to determine the values of both the composition and the stability constant:

$$tg\alpha = (n \Sigma xiyi - \Sigma xi \Sigma yi) / (n \Sigma xi^2 - (\Sigma xi)^2) \tag{26.22}$$

where

$$xi = lg[FA] \tag{26.23}$$

In heterogeneous systems

$$yi = lg([M(II)_{total}] - [M(II)_{free}]) = lg\ [MFAm] \tag{26.24}$$

and in homogeneous systems

$$yi = lg([MFAm]/[M(II)_{free}]) \tag{26.25}$$

$$\beta = (\Sigma yi - a\Sigma xi)/n \tag{26.6}$$

where $a=(n\Sigma xiyi - \Sigma xi\Sigma yi)/(n\Sigma xi^2-(\Sigma xi)^2)$, $xi=[FA_{free}]$ and yi=F(FA).

26.4 CONCLUSION

It is shown that in the weak acid area (pH 3–4), when non-dissociated forms of fulvic acids participate in the complex formation process, to correctly display the stability constants of fulvates $Me+HL^- \rightarrow MeL+H^+$, the equilibrium constant of the reaction should be multiplied by the dissociation constant of fulvic acids $\beta_{MeL} = K_{HL}K$. Alternatively, when calculating the stability constant of fulvate complexes according to the dissociation constant of fulvic acids, a correction of the concentration of [L] should be made.

It is determined that if hydroxo complexes participate in the complex formation process with fulvic acids, then when determining the values of stability constants of fulvate complexes, calculating the concentration of the ionic form using the constants of hydroxo complexes or hydrolysis leads to one result: it artificially increases the value of the stability constant, because the smaller the denominator is in the expression for the stability constant, the greater its value is.

It is shown that the use of solubility and gel chromatographic methods is the most optimal for investigating the complex formation process of fulvic acids with heavy metals.

Optimal variants for determining the composition of fulvate complexes and calculating the stability constants of fulvate complexes are given in the article.

ACKNOWLEDGMENTS

This research was supported by Shota Rustaveli National Science Foundation of Georgia (SRNSFG)[grant number №YS-21–3728]

KEYWORDS

- **electron paramagnetic resonance**
- **fulvic acids**
- **gel chromatographic method**
- **heavy metals**
- **ionometry**
- **phenolic hydroxyl groups**
- **solubility method**

REFERENCES

1. Varshal, G. M. (1994). *Migration Forms of Fulvic Acids and Metals in Natural Waters* [Dissertation, Vernadsky Institute of Geochemistry and Analytical Chemistry of Russian Academy of Sciences].
2. Varshal, G. M., Intskirveli, L. N., Sirotkina, I. C., Kolosov, I. V., & KoShcheeva, I. Y. (1975). On the association of fulvic acids in aqueous solutions. *Geokhimia*, 1581–1585.
3. Osadchyy, V., Nabyvanets, B., Linnik, P., Osadcha, N., & Nabyvanets, Y. (2016). *Processes Determining Surface Water Chemistry*. Springer International Publishing.
4. Dulaquas, G., Waeles, M., Gerringa, L. A., Midag, R., Rijkenberg, M., & Riso, R. G. (2018). The biogeochemistry of electroactive humic substances and its connection to

iron chemistry in the North East Atlantic and the Western Mediterranean Sea. *Journal of Geophysical Research: Oceans*, *123*(8), 5481–5499.
5. Makharadze, G. A., Supatashvili, G. D., & Varshal, G. M. (1988). The research of the forms of copper in surface waters. *Hidrochemical Materials*, *103*, 3–16.
6. Makharadze, G., Goliadze, N., Khaiauri, A., Makharadze, T., & Supatashvili, G. (2016). Fulvic and humin acids in surface waters of Georgia. In *High-Performance Polymers for Engineering-Based Composites* (pp. 167–179). Apple Academic Press.
7. Pisarek, I., & Glowacki, M. (2015). Quality of groundwater and aquatic humic substances from the main reservoir of Ground Water No. 333. *Journal of Ecology and Engineering*, *16*(4), 46–53.
8. Slaveykova, V. I., Wilkinson, K. J., Ceresa, A., & Pretsch, E. (2003). Role of fulvic acid on lead bioaccumulation by *Chlorella kesslerii*. *Environmental Science and Technology*, *37*(5), 1114–1121.
9. Moiseenko, T. I., Diny, M. I., Gashkina, N. A., & Kremleva, T. A. (2013). Occurrence forms of metals in natural waters depending on water chemistry. *Water Resources*, *40*(4), 407–416.
10. Kulik, F., Wieber, J., Pethica, B., & Zuman, P. (1986). Binding of copper (II) and zinc (II) ions on various lignins. *Journal of Electroanalytical Chemistry*, *214*(1, 2), 331–342.
11. Orsetti, S., Marco-Brown, J. L., Andrade, E. M., & Molina, F. V. (2013). Pb (II) binding to humic substances: An equilibrium and spectroscopic study. *Environmental Science and Technology*, *47*(14), 8325–8333.
12. Saar, R. A., & Weber, J. H. (1980). Lead (II)-fulvic acid complexes: Conditional stability constants, solubility, and implications for lead (II) mobility. *Environmental Science and Technology*, *14*(7), 877–880.
13. Sahu, S., & Banerjee, D. K. (1996). Complexation of copper (II), cadmium (II), and lead (II) with humic and fulvic acids of Yamuna River sediments. In *Chemistry for the Protection of the Environment* (pp. 375–388). New York, USA.
14. Turner, D. R., Varney, M. S., Whitfield, M., Mantoura, R. F. C., & Riley, J. P. (1986). Electrochemical studies of copper and lead complexation by fulvic acid. *Geochimica et Cosmochimica Acta*, *50*(3), 289–297.
15. Chakraborty, P., & Chakraborty, Ch. L. (2008). Competition from Cu(II), Zn(II), and Cd(II) in Pb(II) binding to Suwannee River fulvic acid. *Water, Air, and Soil Pollution*, *195*(1–4), 63–71.
16. Christl, I., Metzger, A., Heidmann, I., & Kretzschmar, R. (2005). Effect of humic and fulvic acid concentrations and ionic strength on copper and lead binding. *Environmental Science and Technology*, *39*(14), 5319–5326.
17. Gondar, D., López, R., Fiol, S., Antelo, J. M., & Arce, F. (2006). Cadmium, lead, and copper binding to humic acid and fulvic acid extracted from an ombrotrophic peat bog. *Geoderma*, *135*(1, 2), 196–203.
18. Quan, G., & Yan, J. (2010). Binding constants of lead by humic and fulvic acids studied by anodic stripping square wave voltammetry. *Electrochemistry*, *46*(1), 90–94.
19. Makharadze, G., Supatashvili, G., & Makharadze, T. (2018). New version of calculation of stability constant of metal-fulvate complexes on the example of zinc fulvate. *International Journal of Environmental Science and Technology*, *15*(10), 2165–2168.
20. Bertoli, A. C., Garcia, J. S., Trevisan, M. G., Ramalho, T. C., Matheus, P., & Freitas, M. P. (2016). Interactions fulvate-metal (Zn^{2+}, Cu^{2+}, and Fe^{2+}): Theoretical investigation of thermodynamic, structural, and spectroscopic properties. *Biometals*, *29*(2), 275–285.

21. Rey-Castro, C., Mongin, S., Huidobro, C., David, C., Salvador, J., Garces, J., Galceran, J., Mas, F., & Puy, J. (2009). Effective affinity distribution for the binding of metal ions to a generic fulvic acid in natural waters. *Environmental Science and Technology*, *43*(18), 7184–7191.
22. Schnitzer, M., & Kerndorff, H. (1981). Reactions of fulvic acid with metal ions. *Water, Air, and Soil Pollution*, *15*(1), 97–108.
23. Schnitzer, M., & Skinner, S. I. M. (1967). Stability constants of Pb, Ni, Mn, Co, Ca, and Mg fulvic acid complexes. *Soil Science*, *103*(4), 247–252.
24. Schnitzer, M., & Hansen, E. H. (1970). An evaluation of methods for the determination of stability constants of metal-fulvic acid complexes. *Soil Science*, *109*(4), 333–340.
25. Cheam, V., & Gamble, D. S. (1974). Metal-fulvic acid chelation equilibrium in aqueous $NaNO_3$ solution Hg (II), Cd (II), and Cu (II) fulvate complexes. *Canadian Journal of Soil Science*, *54*(4), 413–417.
26. Esteves da Silva, J. C. G., & Machado, A. S. C. (1996). Characterization of the binding sites for Al (III) and Be (II) in a sample of marine fulvic acids. *Marine Chemistry*, *54*(3, 4), 293–302.
27. Ephraim, J. H., Marinsky, A. J., & Cramer, J. S. (1989). Fulvic acid complexes with cobalt, zinc, and europium. *Talanta*, *36*(4), 437–443.
28. Mantoura, R. F. C., Dixon, A., & Rilly, J. P. (1978). The speciation of trace metals with humic compounds in natural waters. *Thalassia Jugoslavia*, *14*(1, 2), 127–145.
29. Makharadze, G., Supatashvili, G., & Makharadze, T. (2018). New version of calculation of stability constant of metal-fulvate complexes on the example of zinc fulvate. *International Journal of Environmental Science and Technology*, *15*(10), 2165–2168.
30. Makharadze, T. (2022). Measurement of complex formation process of lead (II) with fulvic acids at pH=8. *Current Topics on Chemistry and Biochemistry*, *1*, 148–154.
31. Makharadze, T., & Makharadze, G. (2021). Measurement of complex formation process of nickel (II) with freshwater fulvic acids using the solubility method. *Fine Chemical Engineering*, *2*, 54–61.
32. Makharadze, T., & Makharadze, G. (2021). The calculation of the stability constant of nickel fulvate, using an average molecular weight of active associate of fulvic acids at pH=8. *Challenges and Advances in Chemical Science*, *4*, 27–33.
33. Makharadze, T., & Makharadze, G. (2020). Investigation of complex formation process of copper with macromolecular organic substances, isolated from natural waters. *Organic Chemistry Plus*, *1*(1), 1–5.
34. Makharadze, T. (2019). The determination of average stability constant of zinc-fulvate complex by the gel filtration method. In *Science and Technology of Polymers and Advanced Materials: Applied Research Methods* (pp. 263–267). Apple Academic Press.
35. Makharadze, G., & Makharadze, T. (2014). Method of calculation of stability constants of fulvic complexes on the example of copper. *Journal of Chemistry and Chemical Engineering*, *8*(3), 108–111.
36. Makharadze, T., & Makharadze, G. (2021). Investigation of complex formation process of manganese (II) with fulvic acids at pH=5.0 by gel filtration method. In *Scientific Collection "interconf" N74: Chemistry and Materials Science* (pp. 339–347).
37. Beck, M. T., & Nagypal, I. (1990). *Chemistry of Complex Equilibria*. Chichester: Horwood.

CHAPTER 27

Quantum-Chemical Modeling and Physico-Chemical Research of the Hydride Addition Reaction of α,ω-bis(Trimethylsiloxy) Methylhydridsiloxane with Trimethylacryloxysilane

ZURAB PACHULIA[1] and NANA PIRTSKHELIANI[1,2]

[1]*Sokhumi State University, Faculty of Natural Sciences, Mathematics, Technologies, and Pharmacy, Tbilisi, Georgia*

[2]*Institute of Macromolecular Chemistry and Polymeric Materials, Ivane Javakhishvili Tbilisi State University, Tbilisi, Georgia*

ABSTRACT

Two reaction schemes were considered using quantum-chemical calculations. Based on the comparison of the activation energies and heat effects of the reaction, the experiment involving α,ω-bis(trimethylsiloxy)methylhydridosiloxane and trimethylacryloxysilane is energetically more favorable with the first scheme. The reaction of hydride addition of trimethylsiloxy end-blocked methylhydridesiloxane to trimethylacryloxysilane has been investigated, and comb-type oligomers with trimethylpropioniloxysilane groups in the side chain have been obtained. By ^{1}H NMR spectra, it was shown that the hydrosilylation proceeds according to Farmer's rule. The reaction order, rate constants, and activation energy of the hydrosilylation reaction were determined. Differential scanning calorimetric and X-ray analyses of the synthesized oligomers were carried out.

Advanced Topics in Polymer Chemistry and Materials Science: Tamara Tatrishvili and Marc Jean Médard Abadie (Eds)

27.1 INTRODUCTION

In the literature, methylsiloxane copolymers and oligomers with comb-type structures and various surrounding groups in the side chain are known [1, 2]. Such copolymers are characterized by liquid-crystalline properties. For example, methylpropylsiloxane copolymers are also characterized by liquid-crystalline properties. The present paper deals with the synthesis and physicochemical studies of the properties of comb-type oligomers containing trimethylpropioniloxysilane groups in the side chain [3].

27.2 EXPERIMENTAL METHODS

For the full characterization of the interaction reaction of α,ω-bis(trimethylsiloxy)methyl-hydridosiloxane with trimethylacryloxysilane, quantum-chemical calculations were performed using the non-empirical density functional theory (DFT) method. The program "Priroda-04" [3] was used.

27.3 RESULT AND DISCUSSION

According to Farmer's and Markovnikov's rules, two possible variants of the bimolecular reaction were considered. For the first time, we discussed a version of Farmer's rule.

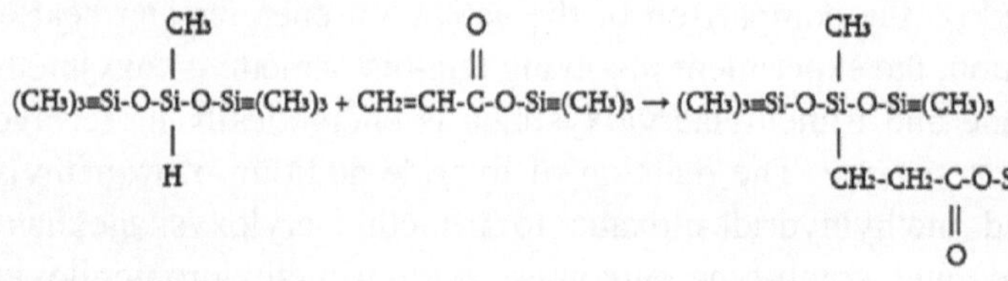

The initial state of the system is shown in Figure 27.1, and the final state is shown in Figure 27.2.

The distance between silicon and carbon atoms (Si_3- C_{35}) and between hydrogen and carbon atoms (H_{22}- C_{22}) was taken to be 1.0 Å longer than the bond distance in the final product. The distance between the atoms was changed at intervals of 0.05 Å.

The dependence of the energy change (ΔE) of the system on the distance between atoms is shown in Figure 27.3.

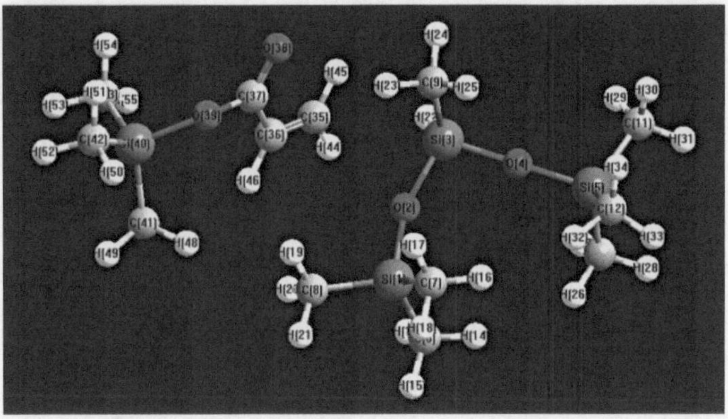

FIGURE 27.1 The initial state of the system.

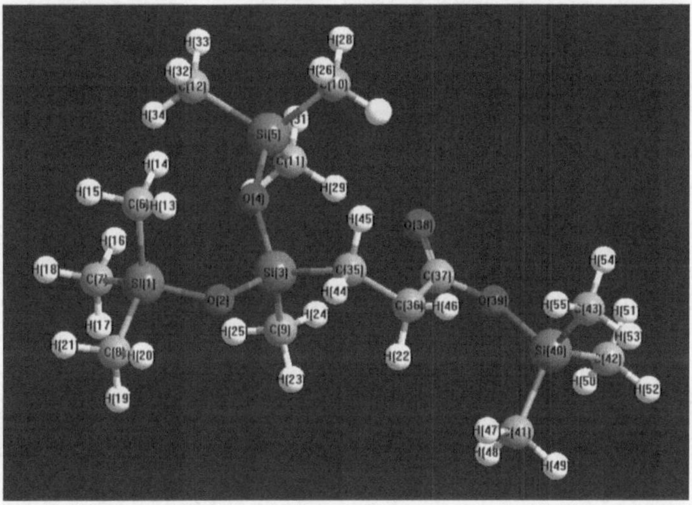

FIGURE 27.2 The final state of the system.

As can be seen from Figure 27.3, when the carbon C_{35} atom approaches the silicon Si_3 atom to the distance (R_{Si3}- c_{35}) = 2.31 Å, the energy of the system increases. In the trimethylacryloxysilane molecule, the bond order between C_{35} and C_{36} atoms decreases from 1.81 to 1.45. The order of the bond between the silicon Si_3 atom and the hydrogen H_{22} atom also decreases (0.86–0.66) and the formation of new bonds is observed (P_{Si3-} c_{35}= 0.24 and (Pc_{36}-H_{22}= 0.18). The double bond is converted to a single C¬C bond. The hydrogen atom is completely separated from the silicon Si_3 atom (P_{Si3}-H_{22} = 0.00) and joins with the carbon C_{36} atom (Pc_{36}-H_{22} = 0.92).

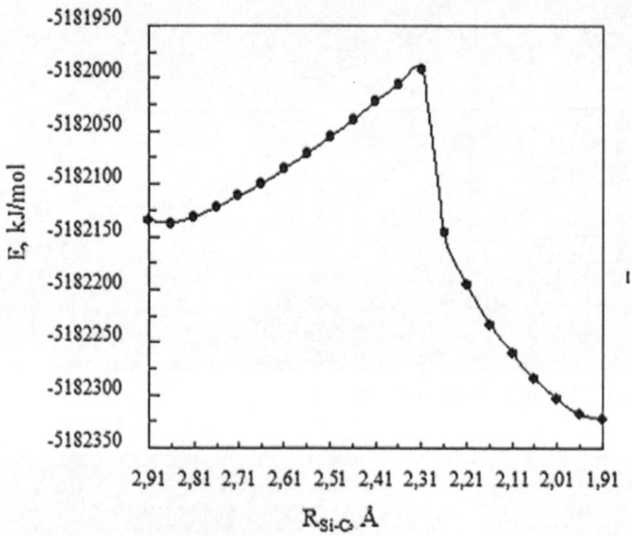

FIGURE 27.3 Dependence of system energy change (ΔE) on distance between silicon and carbon atoms ($R_{Si3-C35}$)

Activation energy ΔE* = 148.07 kJ/mol, and reaction heat effect ΔE = –183.74 kJ/mol. As we can see, the reaction is exothermic.

Taking into account that the system has 4 reaction centers [broken bonds (2): Si_3-H_{22}, C_{35}-C_{36}; bonds formed (2): Si_3-H_{35}, C_{36}-H_{22}], the said activation energy must be divided by 4. In this case, the activation energy calculated for one reaction center is ΔE* = 148.07 kJ/mol/4 == 37.02 kJ/mol, which means that the obtained value corresponds to the energy characteristic of chemical reactions.

The second time we discussed the variant of Markovnikov's rule.

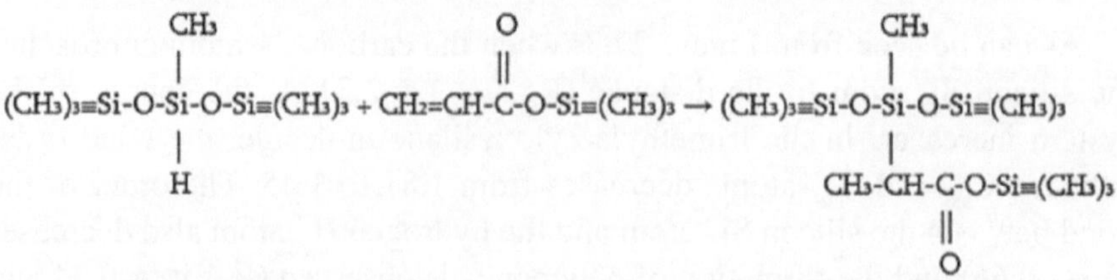

The initial state of the system is shown in Figure 27.4, and the final state is shown in Figure 27.5.

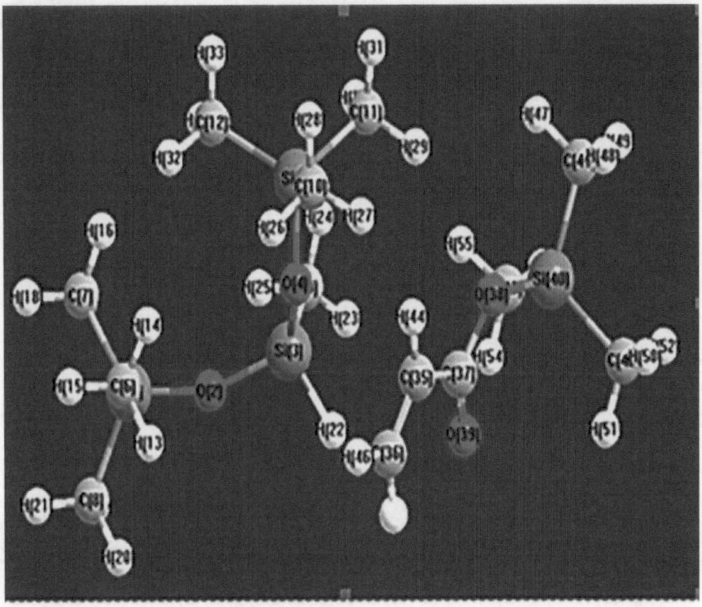

FIGURE 27.4 The initial state of the system.

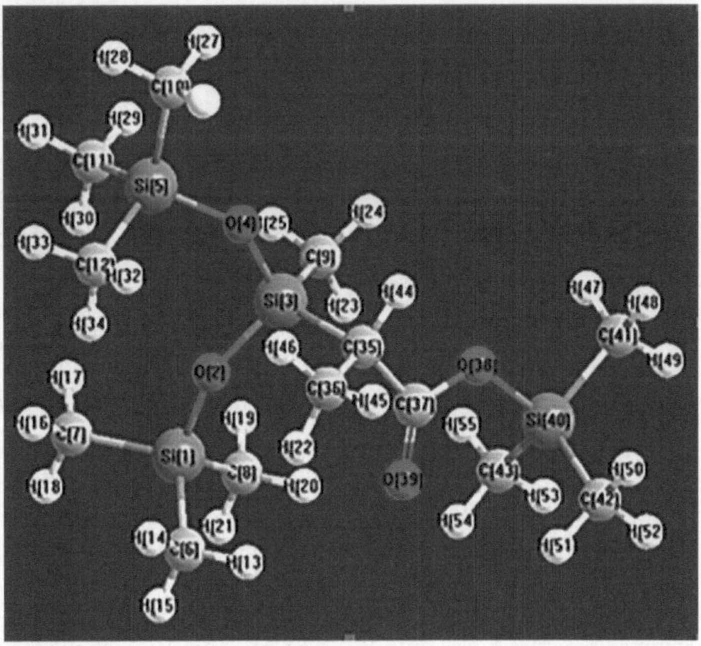

FIGURE 27.5 The final state of the system.

The distance between silicon and carbon atoms Si_3-C_{35} and between hydrogen and carbon atoms (H_{22}-C_{22}) was taken to be 1.0 Å longer than the bond distance in the final product. The distance between atoms was changed with an interval of 0.05 Å.

The dependence of the energy change (ΔE) of the system on the distance between atoms is shown in Figure 27.6.

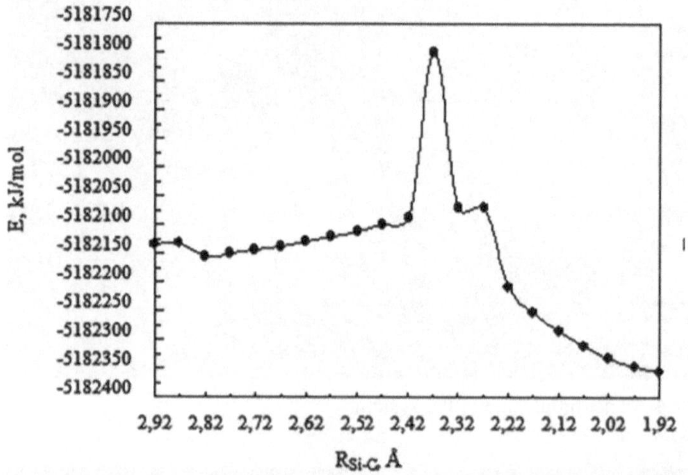

FIGURE 27.6 Dependence of system energy change (ΔE) on distance between silicon and carbon atoms (Rsi_3-c_{35}).

As can be seen from Figure 27.6, when the carbon C_{35} atom approaches the silicon Si_3 atom to the distance Rsi_3-c_{35} = 2.37 Å, the energy of the system increases. In the trimethylacryloxysilane molecule, the bond order between the C_{35} and C_{36} atoms decreases from 1.78 to 1.36. The order of the bond between silicon Si_3 atom and hydrogen H_{22} atom also decreases (0.92–0.53) and the formation of new bonds is observed (Psi_3-c_{35} = 0.25 and (Pc_{36}-H_{22} = 0.24). The double bond turns into a single C¬C bond. The hydrogen atom is completely separated from the silicon Si_3 atom (Psi_3-H_{22} =0.00) and joins with the carbon C_{36} atom (Pc_{36}-H_{22} = 0.96).

Activation energy ΔE* = 354.70 kJ/mol, and reaction heat effect ΔE = –200.67 kJ/mol. As we can see, the reaction is exothermic.

Taking into account that the system has 4 reaction centers [broken bonds (2): Si_3-H_{22}, C_{35}-C_{36}; bonds formed (2): Si_3-C_{35}, C_{36}-H_{22}], the said activation energy must be divided by 4. In this case, the activation energy calculated for one reaction center is ΔE* = 354.70 kJ/mol/4 = = 88.68 kJ/mol, which means that the obtained value corresponds to the energy characteristic of chemical reactions.

From the comparison of the activation energies and heat effects of the reaction, it seems that the experiment involving α,ω-bis(trimethylsiloxy) methylhydridsiloxane and trimethylacryloxysilane is energetically more favorable with the first scheme.

In the literature, methylsiloxane copolymers and oligomers with a comb-type structure and various surrounding groups in the side chain are known [1, 2]. Such copolymers are characterized by liquid-crystalline properties. For example, methylpropylsiloxane copolymers are also characterized by liquid-crystalline properties. The present paper deals with the synthesis and study of the properties of comb-type oligomers containing trimethylpropioniloxysilane groups in the side chain.

For the purpose of synthesizing comb-type methylsiloxane oligomers with liquid-crystalline properties, containing trimethylpropioniloxysilane groups in the side chain, the hydrosilylation reaction of a,ω-trimethylsiloxydimethylsiloxane (n " 35) with trimethylacryloxysilane, at a 1:35 ratio of initial compounds, in the presence of a Pt/C catalyst (0.1% by weight) was investigated. The starting compound, trimethylacryloxysilane, was synthesized by heterofunctional condensation of trimethylchlorosilane with acrylic acid, in the presence of pyridine as an acceptor of hydrogen chloride [4].

The reaction of hydrosylilation was examined at different temperatures 40, 50 and 60°C in delute solutions of anhydrous toluene ($C \approx 5.3 \times 10^{-2}$ mol/l). Preliminary heating of the initial compounds for 5–6 hours in the presence of catalyst platinum on carbon in the temperature range of 40–600°C showed that polymerization and polycondensation of the initial compounds, breakage of the siloxane backbone, or elimination of methane and hydrogen in methylhydrogensiloxane, or other changes do not take place. It was shown that the conversion of all active ≡Si-H groups does not proceed completely, and various linked oligomers are obtained. With the rise in temperature, the conversion of active ≡Si-H groups (the depth of hydrosilylation) and the rate of the hydrosilylation reaction increase from 70% (40°C) to 85% (60°C).

Hydrosilylation reaction proceeds according to the following scheme:

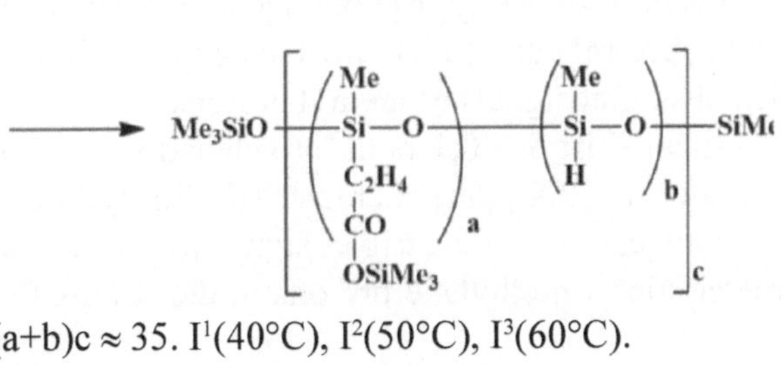

where: $m = (a+b)c \approx 35$. I^1(40°C), I^2(50°C), I^3(60°C).

As a result, transparent, colorless oils soluble in ordinary organic solvents with $\eta_{sp} \approx 0.03$–0.05 were obtained. The structure and composition of the synthesized oligomers were determined by means of elemental analysis, determination of molecular masses, and by IR and ^{1}H NMR spectral data. Some physicochemical properties of the synthesized oligomers are presented in Table 27.1.

TABLE 27.1 Yields and Some Physical-Chemical Properties of Comb-Type Methylsiloxane Oligomers

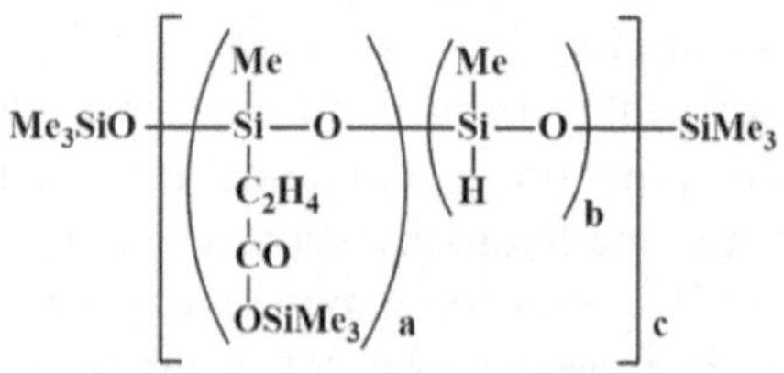

#Oligomer	Yield, %	η_{sp}*	d_1, Å	$\bar{M}\omega$x10–3**	Tg, °C	Elementary*** Composition,%		
						C	H	Si
I^1	77	0,03	8,65	_	–80	38,77	7,80	29,92
						38,72	7,50	29,53
I^2	84	0,04	_	_	_	38,95	7,81	29,75
						39,16	7,46	29,02
I^3	89	0,05	8,65	4,42	–72	39,76	7,84	29,00
						40,03	7,43	28,15

*In 1% solution of dry toluene, at 25°C.

**Average molecular weights were determined by gel permeation chromatographic method.

***In numerator there are calculated values, in denominator experimental values.

The structure of the synthesized oligomers was confirmed by IR, ^{1}H and ^{13}C NMR spectral data, on the basis of which it was determined that the hydrosilylation reaction mainly proceeds according to Farmer's rule ≈ 83% and Markovnikov's rule ≈ 17%. [5, 6] obtained results and conclusions are in agreement with spectral data. In ^{1}H NMR spectra of oligomer I^3 one can observe broadened singlet signals of methyl protons in $-SiMe_3$ and ≡Si-Me groups with chemical shift $\delta \approx 0,1$ ppm, broadened signal with chemical shift and with center $\delta \approx 0,8$ ppm, characteristic for hydrogen protons in group ≡Si-CH_2-. One can observe a triplet signal with a chemical shift $\delta \approx$ 2.3 ppm, characteristic of methylene protons in the -CH_2-CO- group. The

^{1}H NMR spectrum also shows a low-intensity signal with a chemical shift of δ ≈ 4.2 ppm, characteristic of ≡Si-H bonds that have not entered into the hydrosilylation reaction, which indicates the multi-unit nature of the oligomeric chains. Thus, the ^{1}H NMR spectral data show that hydrosilylation proceeds by Farmer's rule.

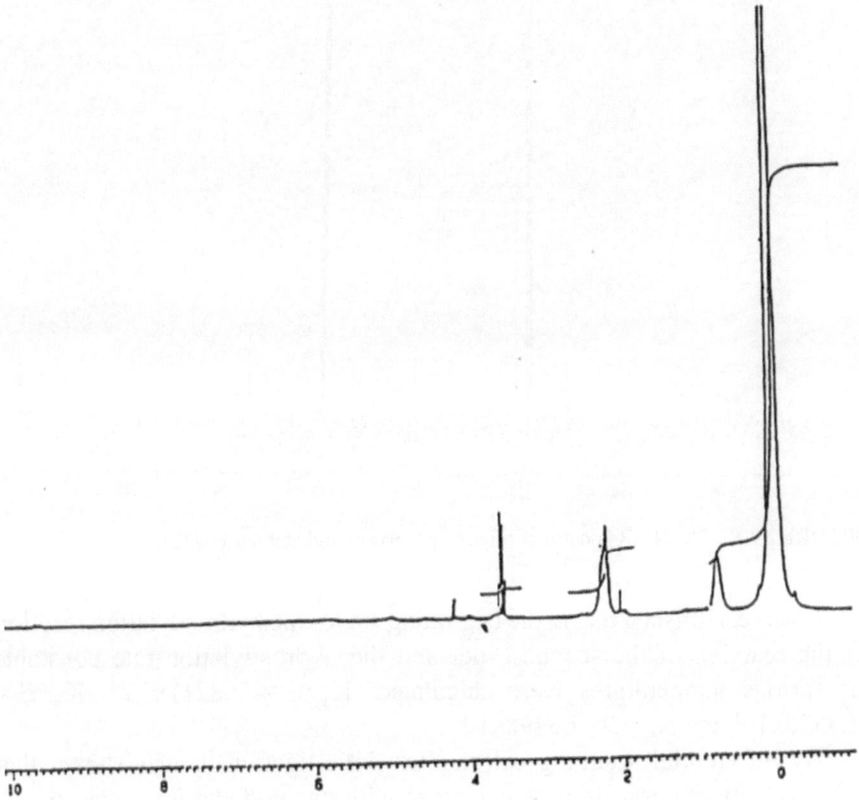

FIGURE 27.7 ^{1}H NMR spectra of oligomer (solvent and standard $CDCI_3$).

In the ^{13}C NMR spectrum of the oligomer, a resonance signal is observed, characteristic of the carbon of the carboxyl group in the region of δ ≈ 178.3 ppm, for ?SiMe- a signal with a chemical shift of δ ≈ 7.5 ppm. and for -$SiMe_3$ _ signal with a chemical shift δ ≈ 7.28 ppm; resonance signal with a chemical shift of δ ≈ 25.7 ppm, for the -CH_2- group on silicon, a signal with a chemical shift of δ ≈ 56.7 ppm for methylene protons –CH_2-CO- at carbonyl groups.

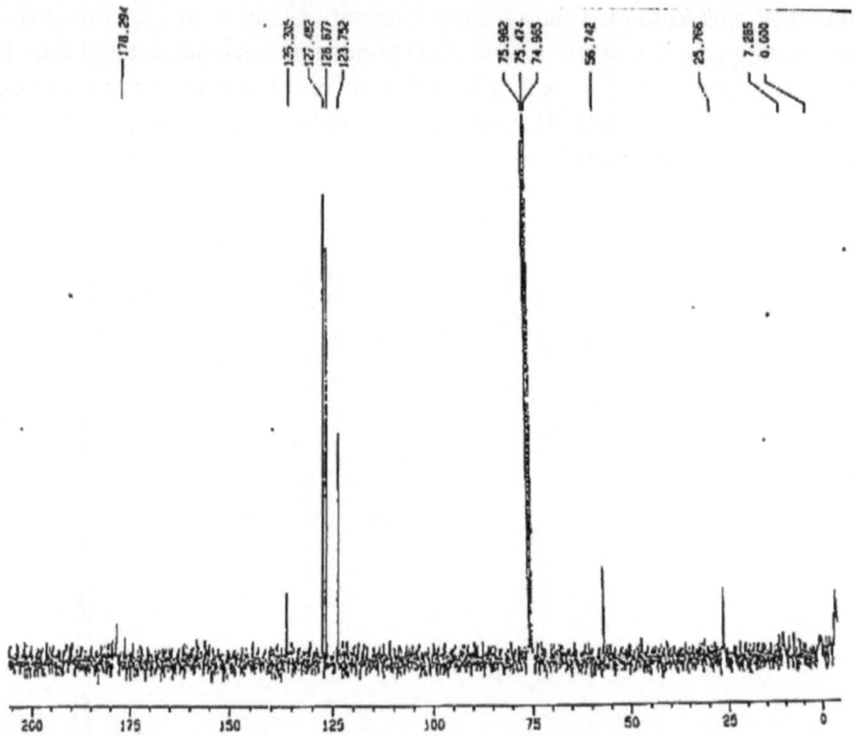

FIGURE 27.8 ^{13}C NMR spectra of oligomer (solvent and standard $CDCl_3$).

It was established that at the beginning stages the hydrosilylation reaction is the reaction of the second order and the hydrosilylation rate constants at various temperatures were calculated: $K_{40}°C \approx 2.2217x10^{-1}$, $K_{50}°C \approx 3.3528x10^{1}$ and $K_{60}°C \approx 5.5498x10^{-1}$.

By differential scanning calorimetric investigation, it was shown that there is only one transition temperature with one endothermic peak, which corresponds to the glass transition temperature of oligomers I^1 and I^3, changing in the temperature range of –80 to –90°C.

The X-ray investigation of oligomer I^3 showed that the oligomers are one-phase amorphous systems with an interchain distance of $d_1 \approx 8.65$.

The order of the hydrosilylation reaction, rate constants, and activation energy were determined. The activation energy of the hydrosilylation reaction was calculated to be $E_{activ} \approx 40.0$ kJ/mol. Synthesized oligomers were studied using DSC, HPLC, and wide-angle X-ray methods.

Thus, for the first time, we synthesized comb-type silicon-organic oligomers containing trimethylpropionoxysilane groups in the side chain.

These oligomers are interesting products for obtaining water-soluble silicon-organic compounds.

27.4 CONCLUSION

The reaction of hydrosilylation of α,ω-trimethylsiloxymethylhydridesilo xane oligomer (n ≈ 35) to trimethylacryloxysilane at a 1:35 ratio of initial compounds, in the presence of platinum on carbon catalyst (0.1% weight), was investigated, and comb-type methylsiloxane oligomers containing trimethylpropionylsiloxysilane groups in the side chain were obtained. Based on the comparison of the activation energies and heat effects of the reaction, the experiment between α,ω-bis(trimethylsiloxy)methylhydridesiloxane and methylacrylate is energetically more favorable with the first scheme. The hydrosilylation reaction order, rate constants, and activation energies were calculated. The structure of the synthesized oligomer was confirmed by with functional and elemental analysis IR, ^{1}H and ^{13}C NMR spectral data. By NMR spectra data it was shown that the reaction proceeds by Farmer rule. The synthesized oligomers were studied by differential scanning calorimetric and wide-angle roentgenographic methods. The obtained oligomers are transparent products that are well soluble in common organic solvents (η_{sp} “ 0.05). To establish the truth, it is necessary to conduct a simple experiment and use more complete non-empirical methods of quantum-chemical calculations.

KEYWORDS

- **activation energy**
- **hydrosilylation**
- **methylhydridesiloxane**
- **oligomer**
- **silicon and carbon atoms**
- **trimethylacryloxysilane**
- **trimethylchlorosilane**

REFERENCES

1. Mukbaniani, O. V., Gurgenidze, G. N., Meladze, S. M., & Khananashvili, L. M. (2001). *Russian Polymer News*, *6*(4), 18.
2. Mukbaniani, O. V., Scherf, U., Gurgenidze, G. N., Karchkhadze, M. G., Meladze, S. M., & Khananashvili, L. M. (2001). *International Journal of Polymeric Materials*, *48*(3), 267.
3. Laikov, D. N., & Ustynyuk, Yu. A. (2005). PRIRODA-04: A quantum-chemical program suite. New possibilities in the study of molecular systems with the application of parallel computing. *Russian Chemical Bulletin, International Edition*, *54*, 820–826.
4. Andreev, D. N., & Kukharskaya, E. V. (1960). Zhurnal Obshchei Khimii, *30*, 2782.
5. Mukbaniani, O., Pirtskheliani, N., & Tatrishvili, T. (2007). Hydrosilylation reactions of methylhydridesiloxane to trimethylsilylated and triethoxysilylated: Modification reactions of oligomethylhydridesiloxanes. In *Modification Reactions of Oligomethylhydridesiloxanes* (pp. 37–56). Nova Science Publishers, Inc.
6. Pachulia, Z., & Pirtskheliani, N. (2014). Quantum-chemical modeling of hydride addition of α,ω-bis(trimethylsiloxy)methylhydridesiloxane to trimethylmethacryloxysilane. *Bulletin of the National Academy of Sciences of Georgia–Chemistry Series*, *40*(1), 62–64.

CHAPTER 28

Kinetics on Thermal/UV Curing and Mechanical Properties of Epoxy-Based Composites

JAMES LEE WAI LEONG[1] and MARC JEAN MÉDARD ABADIE[1,2]

[1]*Department School of Materials Science and Engineering, Nanyang Techological University, Singapore*

[2]*ICGM Institute Charles Gerhardt, Montpellier Université de Montpellier, Département Chimie des Matériaux, Nanostructures, UMR 5253 CNRS-UM ENSCM, Pôle Chimie Balard Recherche, Campus CNRS, Route de Mende, Montpellier, France*

ABSTRACT

Polymer composite materials are becoming more widely-used on the fabrication of various components in aircrafts and spacecrafts. However, the main factor that is still of the greatest concern to its widespread use in constructing primary components in aircrafts and spacecrafts is high cost. Recent advances in polymer curing technology, such as UV curing and electron-beam curing technology, have received much attention. This allows manufacturers to find cheaper alternatives to the currently available materials in the market, with minimal compromise in mechanical properties. Epoxy is one of the materials that has a lot of potential, especially regarding adhesion. The paper aims to perform a kinetics study on EPICLON HP4700, a high-performance epoxy available in the market. EPICLON HP4700 has been cured thermally by modulated differential scanning calorimetry (MDSC) and with ultraviolet (UV) radiation by differential photocalorimetry (DPC). Kinetics analysis was performed on three formulations for thermal and UV curing. Comparisons of the kinetics

Advanced Topics in Polymer Chemistry and Materials Science: Tamara Tatrishvili and Marc Jean Médard Abadie (Eds)

of curing will be made between the two methods of curing and subsequently mechanical characterization. Generally, it was shown that UV curing required much less activation energy for curing compared to thermal curing.

28.1 INTRODUCTION

Polymer composite materials are fast gaining ground as preferred materials for the construction of aircraft and spacecraft [1]. The emergence of strong and stiff reinforcements like carbon fiber or aluminum, along with advances in polymer research to produce high-performance resins as matrix materials, has helped meet the challenges posed by the complex designs of modern aircraft [2].

The current generation of military aircraft features approximately two-thirds of its structural weight composed of advanced composites, including critical and primary structures such as the wings, vertical stabilizer, wing box, control surfaces, and radome. In contrast, the proportion of composites in civil aircraft was much lower in the 2000s (up to 20%), primarily limited to secondary structures. However, this proportion has now increased to around 50%, as seen in Boeing's Dreamliner 767 (50%) and even 52% in the Airbus A350. This significant increase in the use of composites, which provides appreciable weight savings without compromising the mechanical and physical properties of structures, can be attributed to advancements in composite materials. These include the development of new woven fabrics, such as stitched (non-crimp) fabrics, and materials incorporating aluminum foils. New aero-composites, such as Fiber Metal Laminate (FML) and Hybrid Metal Laminate (HML)—e.g., glass fiber-reinforced GLARE laminate, aramid fiber-reinforced ARALL laminate, and carbon fiber-reinforced CARALL laminate—are increasingly used in aircraft fuselages, such as the Airbus A380 and A350 [3].

Current research favors bonding over riveting for the reinforcement of structural parts, such as the stiffened FML skin panel in the fuselage (stringer, stringer coupling, end strap, frame, clip, and castellation). Adhesive bonding distributes disruptive forces across the interface between the two bonded parts, thereby significantly reducing riveting tension levels. This paper focuses on both thermal and UV curing of the epoxy resin used as an adhesive in new aircraft manufacturing. Kinetics analysis was first performed on thermally cured epoxy using Modulated Differential Scanning Calorimetry (MDSC) and on UV-cured epoxy using Differential Photocalorimetry (DPC). Subsequently, the mechanical properties of the epoxy were studied using a Dynamic Mechanical Analyzer (DMA).

28.2 EXPERIMENTAL METHODS AND MATERIALS

The cross-linking reaction of the epoxy resin will be followed and compared between thermal-cured reactions and UV-cured reactions.

28.2.1 MATERIALS

The epoxy resin used, EPICLON HP-4700 (556 g/mol), is a tetra-functional epoxy resin containing a rigid naphthalene skeleton with high symmetry. Such an ultra-high Tg is able to satisfy recent severe needs in the aerospace and electronic materials fields. It is obtained from Dainippon Ink and Chemicals (DIC). Figure 28.1 illustrates the chemical structure of EPICLON HP-4700.

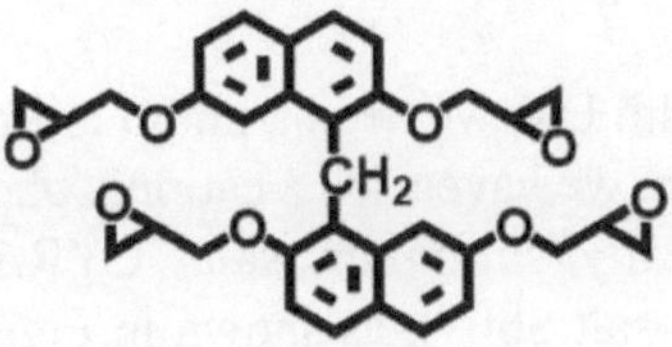

FIGURE 28.1 Chemical structure of EPICLON HP-4700.

It comes in solid orange-brown beads and has to be stored in a cool and dry place with a shelf life of at least 6 months if stored in that conditions.

28.2.1.1 FORMULATION FOR THERMAL KINETICS

EPOLAM 5015 hardener (170.25 g/mol) is used as the curing agent for EPICLON HP-4700 for experimentation with the MDSC thermal analysis. The ingredient mainly responsible for the curing process of the epoxy is isophorondiamine (3-aminomethyl-3,5,5 trimethylcyclohexylamine). Figure 28.2 shows the chemical structure of EPOLAM 5015 hardener.

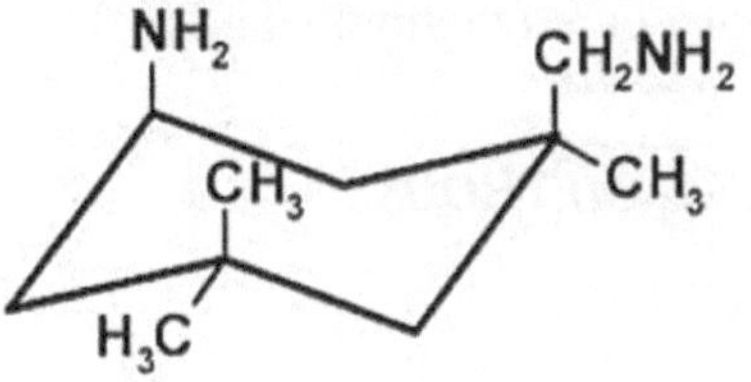

FIGURE 28.2 Chemical structure of isophorondiamine EPOLAM 5015.

28.2.1.2 FORMULATION FOR UV KINETICS

For curing EPICLON HP-4700, we have used a co-solvent, Syna-Epoxy 08 [Bis (3,4-epoxycyclohexylmethyl) adipate], to dissolve the EPICLON HP-4700 solid resin for experimentation with DPC and the DMA. Figure 28.3 shows the chemical structure of adipate.

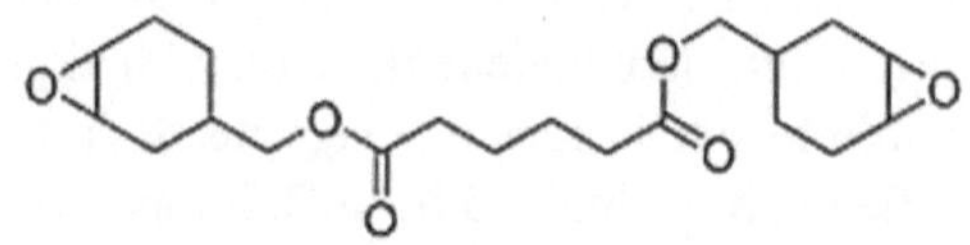

FIGURE 28.3 Chemical structure of Syna-Epoxy 08 or Bis (3,4-epoxycyclohexylmethyl) adipate.

To cure the epoxy with UV, we need a photo initiator [4]. As the mechanism involved is cationic, we have used a cationic photo initiator at different concentrations, specifically sulfonium salts CYRACURE UVI-6974 (a mixture of mono and di-salt SbF_6^-), as shown in Figure 28.4. Upon irradiation, these salts decompose and liberate a proton (H^+), which initiates the photo-crosslinking reaction.

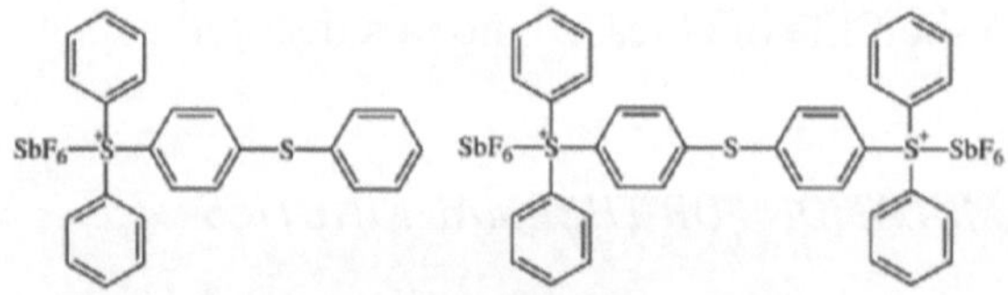

FIGURE 28.4 Chemical structure of CYRACURE UVI-6974.

28.2.1.3 MECHANICAL ANALYSIS ON CURED EPOXY

The DMA will be used to determine the glass transition temperature of the cured resin. A comparison will be made between the glass transitions temperatures of the epoxy cured thermally and by UV [5].

28.2.2 EXPERIMENTAL METHODS

28.2.2.1 THERMAL KINETICS

The curing mechanism of epoxy resins by amine hardeners is illustrated in Figure 28.5, where each NH group reacts with one epoxide group.

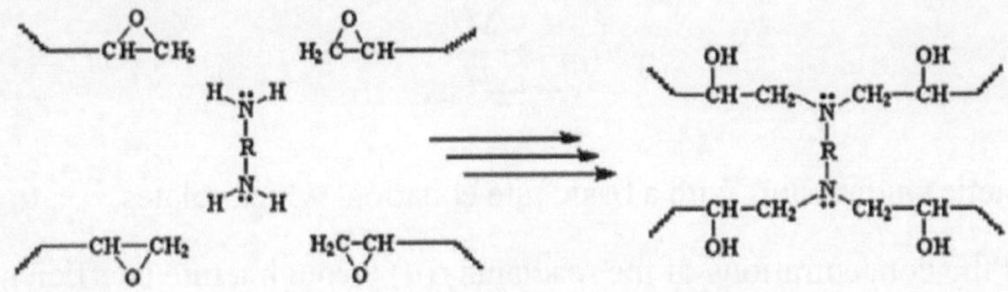

FIGURE 28.5 A chemical reaction of a diamine molecule (EPOLAM 5015) with four epoxide groups.

As the curing process of epoxy resin with an amine hardener is highly exothermic, the kinetics were studied using differential scanning calorimetry (DSC). After mixing the epoxy resin and the amine hardener, the mixture was subjected to a constant temperature ramp in the DSC, starting from room temperature and increasing up to 200°C at different heating rates. The signals detected by the DSC were plotted, and the kinetics of the curing process were analyzed.

Three formulations were prepared for the kinetic analysis, as summarized in Table 28.1.

TABLE 28.1 Formulations Prepared for Thermal Curing Kinetics Analysis

System	EPICLON HP -4700 (wt.%)	EPOLAM 5015 (wt.%)
1	100	40
2	100	80
3	100	100

28.2.2.1.1 *Kinetics of polymerization of epoxy*

The kinetic study of a curing process [6, 7] is based on the assumption that the measured heat flow ($\frac{dH}{dt}$) is proportional to the rate of conversion at a constant temperature ($\frac{d\alpha}{dt}$). The constant temperature $\frac{d\alpha}{dt}$ can be defined as follows:

$$\frac{da}{dt} = \frac{dH/dt}{\Delta H_{Rxn}} \tag{28.1}$$

where: ΔH_{Rxn} is the total heat flow of the reaction. The conversion at any time $t(\alpha_t)$ can be defined as follows:

$$\alpha_t = \frac{\Delta H_t}{\Delta H_{Rxn}} \tag{28.2}$$

All kinetic studies start with a basic rate equation, which relates $\frac{d\alpha}{dt}$ to a function of the concentrations of the reactants $f(\alpha)$ through a rate coefficient (k):

$$\frac{d\alpha}{dt} = kf(\alpha) \tag{28.3}$$

where: α is chemical conversion or extent of reaction and $f(\alpha)$ is assumed to be independent of temperature. k is assumed to follow an Arrhenius equation [8]:

$$k = A\exp\left(-\frac{E}{RT}\right) \tag{28.4}$$

where: E is the activation energy, R is the gas constant (8.314 J/mol. K), T is the absolute temperature, and A is the pre-exponential or frequency factor. Taking *ln* on both sides, E can be obtained with the following equation:

$$\ln k = \ln A - \frac{E}{RT} \tag{28.5}$$

The function f(α) is expressed in the form of two equations, an nth-order equation (follows the Borchandt and Daniels BDs approach) [8] and an simplified autocatalytic equation (Sesbak-Berggren equation) [9], respectively:

$$\frac{d\alpha}{dt} = k(1-\alpha)^n \tag{28.6}$$

and

$$\frac{d\alpha}{dt} = k\alpha^m (1-\alpha)^n \tag{28.7}$$

The autocatalytic equation assumes that we only consider the beginning of the reaction and neglect termination reactions. The objective of the kinetic study of the curing process is to determine the reaction equation, the reaction orders m (order of initiation) and n (order of propagation), E (activation energy), and A (collision factor).

28.2.2.1.2 Determining the activation energy for the epoxy system

Multi-heating-rate non-isothermal methods are used to determine the activation energies of the various epoxy systems under study. The multi-heating-rate method is an iso-conversional method, where it is assumed that the activation energy (*E*) is temperature-independent but may vary with conversion. This method is particularly suitable for systems with multiple reactions (including the overlapping of degradation and curing) and for which the thermal analysis baselines are often not resolvable. Two widely used kinetic analysis methods for multi-heating-rate were implemented: the Ozawa method and the Kissinger method.

The Ozawa method [10] relates the activation energy (*E*) to the heating rate (φ) and the iso-conversional temperature (T_i) by the following equation:

$$E = \frac{-R}{1.052}\frac{\Delta \ln \varphi}{\Delta(1/T_i)} \tag{28.8}$$

The advantage of the Ozawa method is that the activation energy (*E*) can be measured over the entire course of the polymerization reaction.

The Kissinger method [11] is based on the assumption that the conversions at the peak curing exotherms are constant and independent of the heating rate. Activation energy (*E*) can be calculated using the peak exotherm temperature (T_P) and the heating rate (φ) as follows:

$$\frac{d\left[\ln\left(\varphi/T_P^2\right)\right]}{d(1/T_P)} = \frac{E}{R} \tag{28.9}$$

The Kissinger method [11] assumes that the DSC peak exotherms for a certain formulation are iso-conversional. Thus, their value is not dependent on the heating rate. It also helps to detect changes in the reaction mechanism via changes in *E* with conversion.

For both nth-order and autocatalytic reactions, the obtained values of *E*, ln A, and $\frac{d\alpha}{dt}$ are comparable to those of isothermal methods.

28.2.2.2 UV KINETICS

The curing mechanism and kinetics of UV polymerization [12] were investigated by differential photocalorimetry, known as DPC, which is used to observe the exothermic peaks of UV polymerization [13, 14]. In other words, it is used for real-time monitoring of UV curing reactions with highly exothermic processes. Figure 28.6 shows a schematic diagram of a DPC machine [15]. It consists of a differential scanning calorimetry (DSC) unit coupled with a

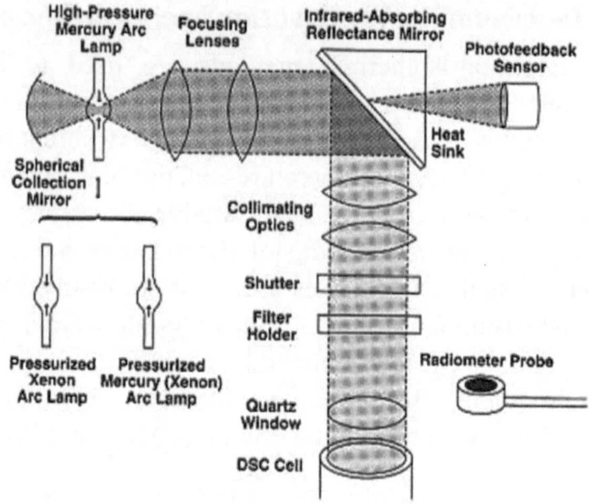

FIGURE 28.6 Schematic diagram of DPC instrumentation.

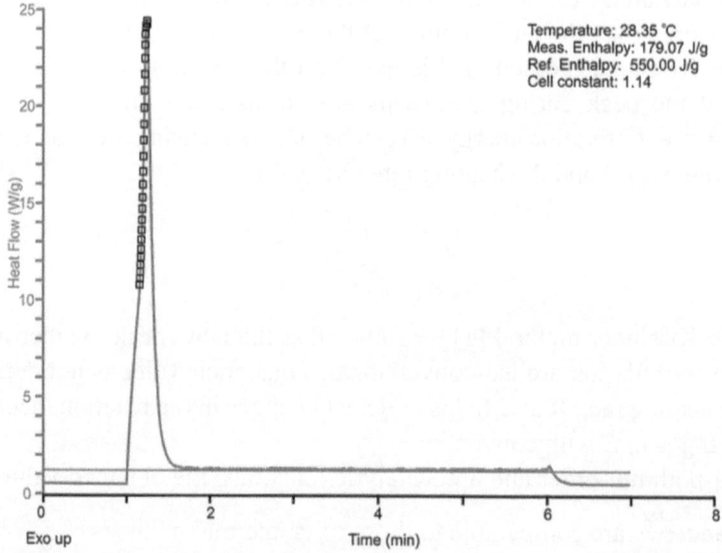

FIGURE 28.7 Typical DPC plot of autocatalytic systems.

high-pressure mercury lamp. The intensity of UV light can be controlled and measured using a radiometer. After isothermal curing of a sample, the result of a DPC plot was analyzed using TA Instruments software, and the kinetics of UV curing can be calculated. An example of a general DPC plot is shown in Figure 28.7 [13]. The presence of an exotherm means that curing has taken place, and the area of the peak equals the enthalpy heat of reaction.

TABLE 28.2 Formulations Prepared for UV Curing Kinetics Analysis

System	EPICLON HP -4700 (wt.%)	Adipate (wt.%)	Photo-Initiator (wt.%)
1	20	80	3
2	30	70	3
3	40	60	3

From the DPC plot, we obtain the induction time (the time required for the consumption of 1% of the monomer), the peak maximum (the time needed for curing to take place), allowing the calculation of k (rate coefficient). The "reacted at peak" refers to the amount reacted at the time when the exotherm peaks. Thus, the induction time and the peak maximum give us an idea of how fast the curing is taking place. Solid EPICLON HP-4700 resin is dissolved in adipate, with as much as 40% weight of resin powder and 60% weight of adipate co-solvent. Thus, three different formulations were prepared for kinetics analysis by the DPC, as shown in Table 28.2.

28.2.2.2.1 *Kinetics of photopolymerization of epoxy*

Calculations of the kinetics of photopolymerization of the epoxy are based on the Sestak and Berggren equation [8]:

$$\frac{d\alpha}{dt} = k(T)\alpha^m (1-\alpha)^n (-\ln[1-a])^P \tag{28.10}$$

where: α is the degree of conversion, *k* is the rate coefficient, m is the order of initiation reaction, n is the reaction order and *p* is the order of termination reaction.

In order to simplify (Equation 28.11), we consider only the outset of the polymerization process. In so doing, the value of *p* can be taken as 0. A simplified autocatalytic kinetic equation can thus be obtained which gives us the following rate equation:

$$\frac{d\alpha}{dt} = k\alpha^m (1-\alpha)^n \tag{28.11}$$

Finally, if the reaction follows *n*th order kinetics, the general equation of rate will be as follows:

$$\frac{d\alpha}{dt} = k(1-\alpha)^n \tag{28.12}$$

Just like the kinetics analysis for polymerization of epoxy with hardeners, the values of *k* and *n* can be determined from a *ln* curve – *ln* plot of $\frac{d\alpha}{dt}$ vs $[\alpha^{m/n}(1-\alpha)]$:

$$\ln\left(\frac{d\alpha}{dt}\right) = \ln k + n \ln \alpha^{m/n} (1-\alpha) \tag{28.13}$$

Both the autocatalytic and *n*th order models were applied, and it was found that the reaction appeared to be better modeled by the autocatalytic representation. While applying the autocatalytic equation, the value of n was kept fixed at 1.5 and the values of $m + n$ at 2. Maintaining fixed values of n and $m + n$ gave more consistent results [13].

$$k = A \exp\left(-\frac{-E_a}{RT}\right) \tag{28.14}$$

where: A is the frequency factor or collision factor, E_a is the activation energy, R is the ideal gas constant (8.314 J/mol. K) and T is the temperature measured in Kelvin.

28.2.3 MECHANICAL PROPERTIES OF CURED EPOXY

The DMA will be used to determine the glass transition temperature, storage modulus E,' loss modulus E',' and loss factor tan δ of the cured resin. A comparison will be made between the glass transition temperatures of the epoxy cured thermally and by UV. Dual cantilever bending analysis is used (Figure 28.8). The sample is clamped firmly at both ends and at the midpoint to the push-rod.

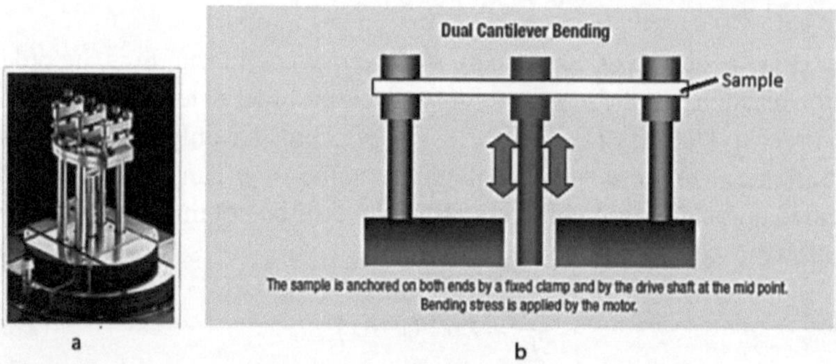

FIGURE 28.8 (a) Dual Cantilever Clamp used in DMA tests, (b) Schematic of Dual Cantilever test

Source: (a) EAG Laboratories. https://www.eag.com/app-note/characterization-of-polymers-using-dynamic-mechanical-analysis-dma/ (b) Original.

A Teflon mold was used to fabricate the sample for DMA testing. The dimensions of the samples made from the mold are to be 17.50 mm by length, 10.50 mm by width and 2 mm by thickness.

28.3 RESULTS AND DISCUSSION

28.3.1 THERMAL CURING KINETICS

All formulations prepared for thermal curing were analyzed in the MDSC Figure 28.9.

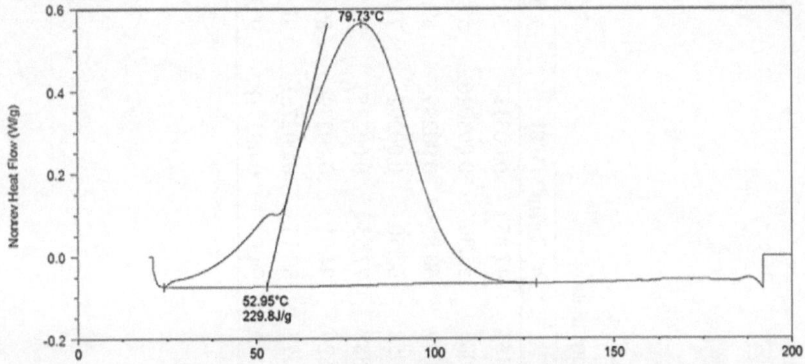

FIGURE 28.9 Calculating area under exotherm with universal analysis software. Measured enthalpy of curing is 229.8 J/g.

For each formulation, five samples were thermally cured using different ramp rates (3°C, 4°C, 5°C, 6°C, and 7°C/min). These five plots were compiled, and kinetics analysis was performed using the Ozawa [10] and Kissinger [11] methods to determine the activation energy for that particular formulation.

28.3.1.1 OZAWA METHOD OF KINETICS ANALYSIS [10]

After the enthalpy of curing is known, the total area under the exotherm was taken to be the 100% conversion point. Subsequently, temperatures that correspond to the conversions of 5%, 10%, 15%, 30%, 50%, 70%, 80%, and 90% were measured, as shown in Figure 28.10. For example, the temperature at which the area under the exotherm corresponds to 5% of the total area of the exotherm is taken to be the temperature where conversion was at 5%.

These two steps were repeated for plots of different ramp rates. Table 28.3 shows the tabulated results for one formulation.

These results were then plotted on a graph. Figure 28.11 shows an example of the plots for one formulation. Linear trendlines were drawn to connect the iso-conversional points, and the slopes of these trendlines represent the activation energy of the epoxy system at a specific conversion percentage. The activation energy values for all three formulations at multiple conversion percentages were plotted in Figure 28.12, according to Eq. (28.8).

TABLE 28.3 Tabulated Results for Ozawa Method Of Kinetics Analysis for One Formulation

Conversion	3°C/min	(1/T)	4°C/min	(1/T)	5°C/min	(1/T)	6°C/min	(1/T)	7 °C/min	(1/T)
5%	38.53	0.00321	40.65	0.003188	44.35	0.003151	47.85	0.003117	50.75	0.003089
10%	44.26	0.003152	4655	0.003129	5120	0.003085	55.30	0.003046	57.93	0.003022
15%	48.25	0.003113	50.63	0.00309	56.04	0.003039	60.85	0,002995	62.87	0.0002977
30%	57.39	0.003027	59.55	0.003007	65.24	0.002956	69.50	0.00292	7153	0.002903
50%	64.62	0.002962	67.26	0.002939	73.28	0.002888	77.55	0.002853	79.87	0.002834
70%	71.14	0.002906	74.56	0.002877	80.49	0.002829	84.91	0.002794	87.55	0.002774
80%	74.84	0.002875	78.99	0.002841	84.56	0.002797	89.13	0.002761	91.96	0.00274
90%	79.76	0.002835	85.72	0.002788	89.99	0.002755	94.85	0.002718	97.95	0.002696

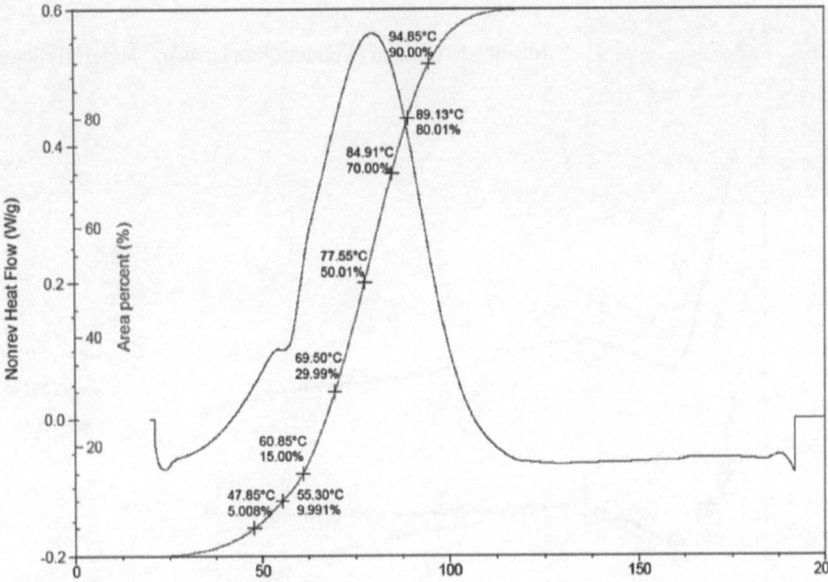

FIGURE 28.10 Measuring temperature at fixed conversion points for the Ozawa method of kinetics analysis for one formulation.

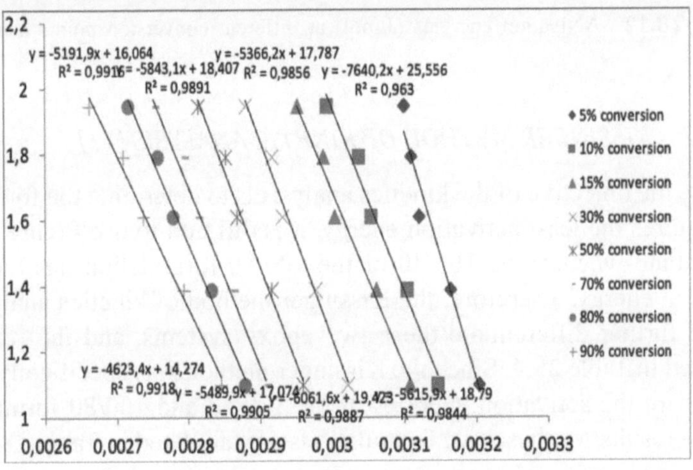

FIGURE 28.11 Plotting the tabulated results for the Ozawa method of kinetics analysis.

We do observe, in Figure 28.12, that for all three formulations, the activation energy always experiences a huge initial drop. This may be due to the amine groups formed during the reaction that facilitate ring-opening at the very beginning of the cure, where a non-autocatalytic reaction occurs. Once the conversion reaches 30%, the activation energy for the three systems remains fairly constant throughout the rest of the curing process. Thus, the relatively constant value of activation energies at conversions above 30% is associated with the autocatalytic reaction.

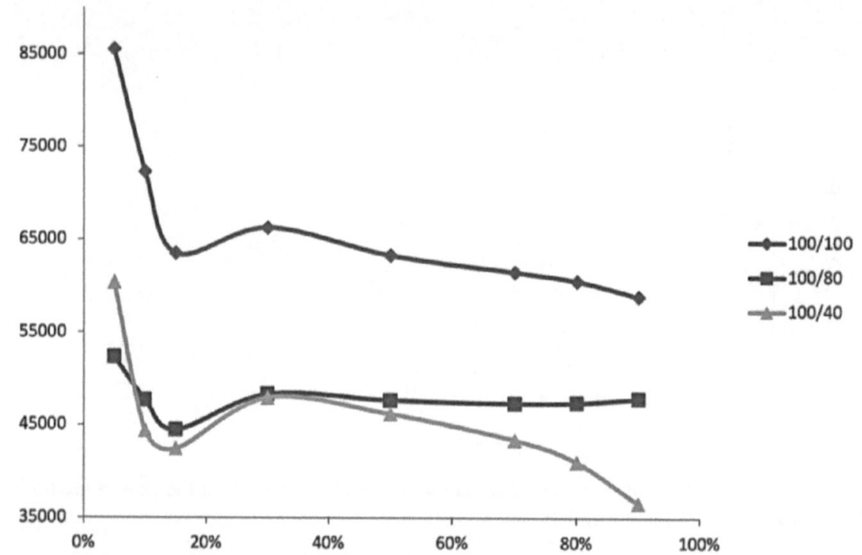

FIGURE 28.12 Activation energies (J/mol) at different conversion points using Ozawa method.

28.3.1.2 KISSINGER METHOD OF KINETIC ANALYSIS [11]

Because the objective of the kinetics analysis is to determine the formulation that requires the least activation energy, it is difficult to use Figure 28.12 to differentiate whether the 100/40 or the 100/80 formulation has the lowest activation energy. Therefore, the Kissinger method of kinetics analysis was used to further differentiate these two epoxy systems, and the results are presented in Table 28.4. Since the Kissinger method was specifically applied to compare the activation energies of the 100/40 and 100/80 formulations, only the results for these two formulations are tabulated in Table 28.4.

TABLE 28.4 Tabulated Results for the Kissinger Method of Kinetics Analysis

Heat. Rate	100/40				100/80			
	Conversion	Tp	(1/Tp)	in(ϕ/ Tp2)	Conversion	Tp	(1/Tp)	in(ϕ/ Tp2)
3°C/min	57.23%	67.10	0.002940	-10.55987	57.37%	67.00	0.002941	-10.5993
4°C/min	55.66%	71.80	0.002900	-10.29963	52.65%	68.25	0.002930	-10.2789
5°C/min	54.94%	77.35	0.002854	-10.10843	56.98%	75.80	0.002867	-10.0996
6°C/min	53.04%	78.69	0.002843	-9.93374	56.27%	79.82	0.002834	-9.94016
7°C/min	51.33%	82.90	0.002810	-9.80339	55.51%	81.79	0.002819	-9.79714

According to Eq. (28.9), the values of the activation energy and collision factor for the two formulations, as determined using the Kissinger method, are given in Table 28.5.

TABLE 28.5 Comparison of Activation Energies and Collision Factors Between 2 Formulations

Formulation	Activation Energy (kJ/mol)	Collision Factors (× 10^3 min^{-1})
100/40	48.27	3926.0
100/80	42.81	556.5

Although the difference is small, we can conclude that the formulation of 100/80 EPICLON HP-4700 and EPOLAM 5015 hardener, respectively, provides the system with the lowest activation energy compared to the other two formulations.

28.3.2 UV CURING KINETICS

Using the simplified Sestak and Berggren equation (28.7), considering that we are at the beginning of the process, the termination reaction can be ignored. Therefore, values of k (rate coefficient) and *n* can be derived from the logarithmic plot of Equation (28.13) (*see* Figure 28.13).

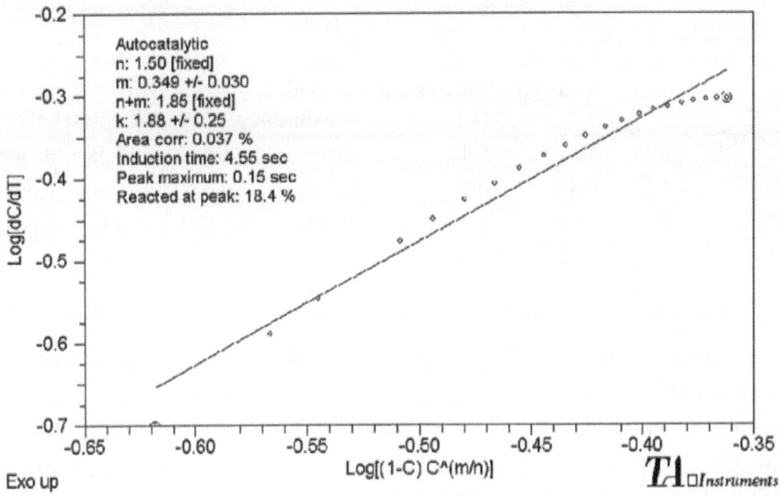

FIGURE 28.13 Logarithmic plot of Eq. (28.11) using TA software for 20/80 formulation at 30°C.

Table 28.6 compiles the kinetics data for the UV photopolymerization of all three formulations prepared and analyzed at different temperatures.

TABLE 28.6 Compilation of Results From DPC Kinetics Analysis

Formulation				**20/80**			
Temperature	**k(s^{-1})**	**Induction Time (s)**	**Peak Maximium (s)**	**Reacted @ Peak Maximum**	**k (min^{-1})**	**In k**	**(1/T)**
30°C	1.88	4.55	0.15	18.4%	112.80	4.73	0.003300
40°C	2.25	4.29	0.15	19.7%	135.00	4.91	0.003195
50°C	2.32	4.39	0.16	20.7%	139.20	4.94	0.003096
60°C	1.99	4.76	0.17	18.3%	119.40	4.78	0.003003
70°C	3.08	3.55	0.12	13.0%	185.40	5.22	0.002915
Formulation				**20/80**			
Temperature	**k(s^{-1})**	**Induction Time (s)**	**Peak Maximium (s)**	**Reacted @ Peak Maximum**	**k (min^{-1})**	**In k**	**(1/T)**
30°C	1.31	6.36	0.24	21.7%	78.60	4.36	0.003300
40°C	1.84	6.05	0.22	20.5%	110.40	4.70	0.003195
50°C	2.44	4.47	0.17	21.4%	146.40	4.99	0.003096
60°C	2.12	3.77	0.12	12.7%	127.20	4.85	0.003003
70°C	3.95	5.41	0.15	8.8%	237.00	5.47	0.002915
Formulation				**20/80**			
Temperature	**k(s^{-1})**	**Induction Time (s)**	**Peak Maximium (s)**	**Reacted @ Peak Maximum**	**k (min^{-1})**	**In k**	**(1/T)**
30°C	0.49	10.70	0.41	22.8%	29.52	3.39	0.003300
40°C	1.07	7.83	0.29	22.0%	64.20	4.16	0.003195
50°C	1.73	6.08	0.20	20.3%	103.80	4.64	0.003096
60°C	1.52	6.05	0.23	22.9%	91.20	4.51	0.003003
70°C	2.24	4.69	0.16	16.2%	134.4	4.90	0.002915

As expected, we do observe in Table 28.6 that the rate coefficient k increases with temperature. The induction time (the time at which 1% of monomer is consumed) decreases when the temperature increases, indicating that the photo reactivity of the formulation becomes more and more reactive. Moreover, it appears that the percentage of resin that reacted at the peak maximum is more or less constant with temperature but depends on the composition of the formulation, i.e., approximately 18% for 20/80, approximately 20% for 30/70, and approximately 22% for 40/60.

Using the Arrhenius equation (Equation 28.4), the activation energy and collision factor can be calculated from ln k against (1/T) (the slope of the plot gives us the value of $-\frac{E_a}{R}$, (*see* Table 28.7).

TABLE 28.7 Tabulates the Activation Energies and Collision Factors for All Three UV Curing Formulations Being Studied

Formulation	Activation Energy (Kl/mol)	Collision Factors (x 10^3 min^{-1})
20/80	10.26	6.75
30/70	21.49	426.4
40/60	29.39	4000

The formulation 20/80 (resin/co-solvent) gives us the lowest activation energy as expected, compared to the two other formulations. Cycloepoxy co-solvent (EPOLAM 5015) is more photoreactive than epoxy resin, so the less EPICLON HP-4700 there is, the more reactive the system will be. To be able to compare the activation energy values for a thermal system (Table 28.5) and a UV system (Table 28.7), we need to consider pure EPICLON HP-4700 without co-solvent. This is not possible because the DPC technique requires working with a solution. However, from the values given in table 28.7, we can extrapolate to obtain the activation energy of the solid EPICLON HP-4700, such as Ea ≈ 43 kJ/mol, a value very close to that of Ea = 42.81 kJ/mol given for formulation 100/80 (resin/hardener) in the thermal system (Table 28.5).

28.3.3 MECHANICAL PROPERTIES

28.3.3.1 DYNAMICAL MECHANICAL ANALYSIS

Both thermally cured and UV cured epoxy samples were subjected to dual-cantilever testing. As shown in Figure 28.8, the sample is clamped at both ends and subjected to minor oscillations under an increasing temperature of 5°C/min up till 180°C. The applied stress elicits a corresponding strain whose amplitude and phase shift can be determined and thus properties such as the storage modulus E’, loss modulus E”, and tan δ (or loss factor) can be determined. Figure 28.14 shows a typical DMA plot obtained from the analysis of our samples.

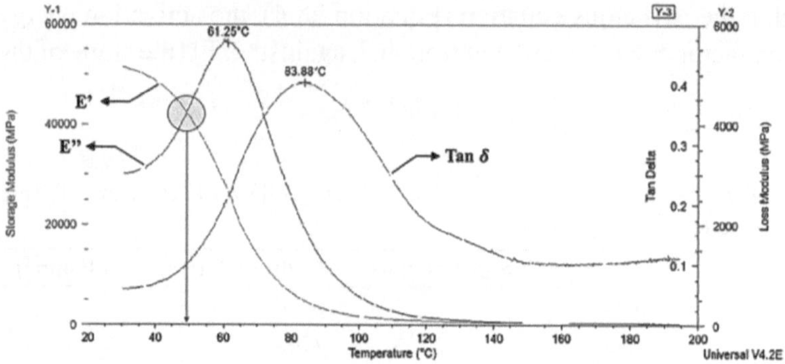

FIGURE 28.14 A typical DMA analysis plot obtained from DMA tests.

As we can see from Figure 28.14, the storage modulus *E*' decreases with increasing temperature, whereas the loss modulus *E*" increases with increasing temperature. This means that the sample is becoming less elastic and more viscous, implying that the material is already undergoing its glass transition phase. It is transitioning from a glassy state to a rubbery state, becoming less stiff. The maximum loss modulus gives the commonly used glass transition temperature T_g = 61.25°C, but a better indicator of glass transition temperature is the tan delta peak because the maximum loss modulus is the point where energy is dissipated the most, meaning that it is the point where the polymer is the most viscous. Note that the crossover of *E*' and *E*" gives the gel temperature T_{gel} = 50°C.

Table 28.8 shows the effects of thermal and UV regimes, comparing *E*" and tan *δ* in both cases. We have added the effect of annealing, submitting samples at 100°C for 2 hours in the case of UV curing. Considering tan *δ*, the results are very conclusive that UV curing of annealed EPICLON HP-4700 gives a material that has a higher glass transition temperature. This indicates that for the thermal curing, the crosslinking was not complete and the network was not achieved. These results are supported by the values observed for the maximum loss modulus (*E*") that increase with tan *δ*.

TABLE 28.8 Tabulated Results from DMA Tests

	Thermal Curing		UV curing		
	Maximum Loss Modulus (E")	Tan δ	Tan δ	Tan δ (Annealing)	MLM (E") (Annealing)
Sample 1	60.15°C	81.32°C	70.25°C	112.63°C	77.72°C
Sample 2	61.62°C	83.71°C	80.15°C	114.78°C	83.71°C
Sample 3	69.22°C	94.63°C	91.74°C	127.27°C	89.49°C

28.3.3.2 FLEXURAL TESTING OF COMPOSITE PANELS

Flexural testing was used to characterize the composite panels according to ASTM D 790 (*see* Figure 28.15).

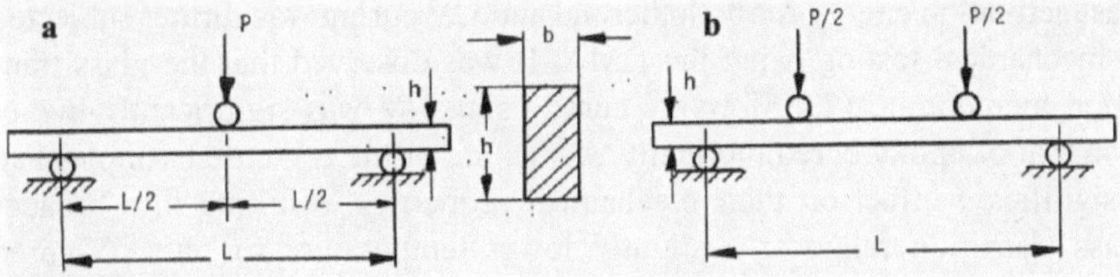

FIGURE 28.15 Schematic diagrams of three-point bending test (a) and four-point bending test (b).

We compared the differences in mechanical properties between the composite panels fabricated by thermal curing and those fabricated by UV curing (*see* Table 28.9).

TABLE 28.9 Displays the Tabulated Results of the Flexural Tests of the Composite Panels

Sample	Flexural Strength (MPa)	Flexural Modulus (GPa)
EPICLON +EPOLAM (Thermal)	327.67 ± 15.44	8.56 ± 0.58
EPICLON+UVI 6990 (UV)	221.54 ± 23.45	8.39 ± 0.93

From Table 28.9, we can conclude that composite panels that were fabricated via thermal curing exhibit much higher flexural strength than the panels that are fabricated via UV curing. In this case, the flexural strength of composite panels fabricated via thermal curing is 48% higher due to a higher density of crosslinking. On the other hand, it seems that neither method of curing provides a distinct advantage over the other in terms of the flexural modulus of the composite panels, with a mere difference of 2% in values of flexural modulus.

28.4 CONCLUSION

The use of MDSC and DPC to perform kinetics analysis of the high-performance epoxy, EPICLON HP-4700, proved to be successful. Kinetics analysis was performed on three formulations for thermal curing and three formulations for UV curing. For thermal curing, the optimum activation energy was determined for a formulation of approximately 100 parts resin (EPICLON HP-4700) to 80 parts hardener (EPOLAM 5015) by weight. For UV curing, the activation energy increased with the concentration of resin dissolved in adipate. Generally, it was shown that UV curing required much less activation energy compared to thermal curing. The formulation with the least activation energy for both thermal and UV curing was further subjected to mechanical testing using the DMA. It was observed that the glass transition temperature (T_g) of epoxy cured using UV was significantly higher than that of epoxy cured thermally. Annealing of the UV-cured samples had a significant effect on their mechanical properties, resulting in a broader glass transition range at a slightly lower temperature for the UV-cured samples. Finally, it was demonstrated that both thermal and photochemical cross-linking are possible, although annealing is always necessary to ensure complete cross-linking, leading to maximum Tg and improved mechanical properties.

KEYWORDS

- **differential photo calorimetry**
- **dynamic mechanical analyzer**
- **epoxy resin**
- **kinetics**
- **photo curing**
- **thermal curing**

REFERENCES

1. Rana, S., & Fangueiro, R. (2016). *Advanced composite materials for aerospace engineering: Processing, properties and applications*. In S. Rana & R. Fangueiro (Eds.), *Woodhead Publishing Series in Composites Science and Engineering*. Elsevier. https://doi.org/10.1016/C2014-0-00846-5.
2. Jawaid, M., & Thariq, M. (2018). Sustainable composites for aerospace application. In M. Jawaid & M. Thariq (Eds.), *Woodhead Publishing Series in Composites Science and Engineering*. Elsevier. https://doi.org/10.1016/C2016-0-01691-1.
3. Ding, Z., Wang, H., Luo, J., & Li, N. (2021). A review on forming technologies of fiber metal laminates. *International Journal of Lightweight Materials and Manufacture, 4*(1), 110–126. https://doi.org/10.1016/j.ijlmm.2020.06.006.

4. Abadie, M. J. M., & Voytekunas, V. Yu. (2004). New trends in UV curing. *Eurasian ChemTech Journal, 6*, 67–77.
5. Chen, C., Li, B., Wang, C., Iwasaki, S., Kanari, M., & Lu, D. (2018). UV and thermal cure epoxy adhesives. In F. Yilmaz (Ed.), *Paint and Coatings Industry*. IntechOpen. https://doi.org/10.5772/intechopen.82168.
6. Um, M. K., Daniel, I. M., & Hwang, B. S. (2002). A study of cure kinetics by the use of dynamic differential scanning calorimetry. *Composites Science and Technology, 62*(1), 29–40. https://doi.org/10.1016/S0266-3538(01)00188-9.
7. Chang, T., Hsueh, K. H., Liu, C. C., Cao, C. R., & Shu, C. R. (2022). A method to derive the characteristic and kinetic parameters of 1,1-bis(tert-butylperoxy)cyclohexane from DSC measurements. *Processes, 10*(5), 1026. https://doi.org/10.3390/pr10051026.
8. Kohout, J. (2021). Modified Arrhenius equation in materials science, chemistry and biology. *Molecules, 26*(23), 7162. https://doi.org/10.3390/molecules26237162.
9. Sestak, J., & Berggren, G. (1971). Study of the kinetics of the mechanism of solid-state reactions at increasing temperatures. *Thermochimica Acta, 3*, 1–11.
10. Cui, H. W., Suganuma, K., & Uchida, H. (2015). Using the Ozawa method to study the thermally initiated curing kinetics of vinyl ester resin. *Royal Society of Chemistry Advances, 5*, 2677–2683.
11. Schulz, H. (2018). From the Kissinger equation to model-free kinetics: Reaction kinetics of thermally initiated solid-state reactions. *ChemTexts, 4*, 9. https://doi.org/10.1007/s40828-018-0062-3.
12. Saenz-Dominguez, I., Tena, I., Sarrionandia, M., Torre, J., & Aurrekoetxea, J. (2020). An analytical model of through-thickness photopolymerization of composites: Ultraviolet light transmission and curing kinetics. *Composites Part B: Engineering, 191*, 15. https://doi.org/10.1016/j.compositesb.2020.107963.
13. Voytekunas, V. Yu., Ng, F. L., & Abadie, M. J. M. (2008). Kinetics study of the UV initiated cationic polymerization of cycloaliphatic diepoxide resins. *European Polymer Journal, 44*(11), 3640–3649.
14. Boey, F. Y. C., Rath, S. K., Ng, A. K., & Abadie, M. J. M. (2002). Cationic UV cure kinetics for multifunctional epoxies. *Journal of Applied Polymer Science, 86*(2), 518–525.
15. Appelt, B. K., & Abadie, M. J. M. (1988). Calorimetric characterization of photosensitive materials. *Polymer Engineering and Science, 28*, 367–371.

Index

D

N

O

P

T

Z